D1145790

Plants and Wildlife of the Mediter...

Biology and Wildlife of the Mediterranean Region

Jacques Blondel

and

James Aronson

CNRS, Montpellier
France

Original Drawings by René Ferris

OXFORD
UNIVERSITY PRESS

This book has been printed digitally and produced in a standard specification
in order to ensure its continuing availability

OXFORD
UNIVERSITY PRESS

Great Clarendon Street, Oxford OX2 6DP

Oxford University Press is a department of the University of Oxford.
It furthers the University's objective of excellence in research, scholarship,
and education by publishing worldwide in

Oxford New York

Auckland Cape Town Dar es Salaam Hong Kong Karachi
Kuala Lumpur Madrid Melbourne Mexico City Nairobi
New Delhi Shanghai Taipei Toronto
With offices in
Argentina Austria Brazil Chile Czech Republic France Greece
Guatemala Hungary Italy Japan South Korea Poland Portugal
Singapore Switzerland Thailand Turkey Ukraine Vietnam

Oxford is a registered trade mark of Oxford University Press
in the UK and in certain other countries

Published in the United States
by Oxford University Press Inc., New York

ISBN 0-19-850035-1

Antony Rowe Ltd., Eastbourne

Foreword

Peter H. Raven, Missouri Botanical Garden, St. Louis, Missouri, USA

A lovely book, and one to ponder. A geological pivot between Africa and Eurasia, the crossroads of the great human migrations over the past two million years, the stage on which so many civilizations have come and gone, the Mediterranean has been loved, exploited, and inhabited by human beings for longer than any other part of the world outside of Africa. It is humans who have very largely made it what it is today, and the marks of this domination are visible everywhere.

With deteriorating global climates over the last 15 million years or so, and with the cold offshore currents along the western sides of the continents drying out the prevailing westerlies in the course of their journeys onto land, summer rains have largely disappeared, and habitats have become increasingly varied. Where once an almost unbroken continuum of evergreen vegetation existed, there is now a panorama of habitats, rain shadows, outcrops, dunes, deserts and semi-deserts, maquis, wooded patches, moisture-concentrating cliffs – all in a region that affords a richness of biodiversity with few parallels anywhere on Earth. When one superimposes on these habitats the tree-cutting, land-clearing, mining, building, and recreational activities of humanity, one arrives at a rich, beautiful, and bewildering complexity that is explored well in the pages of this outstanding volume.

The modern flora and fauna of the Mediterranean have been altered greatly. Increasing drought and seasonality have affected the vegetation, as have climatic shifts tied to the waxing and waning of polar ice caps over the past several million years; and human activities have been profoundly influential, especially in the last few tens of thousands of years. Flocks of animals were domesticated, which led to the grazing, clearing, and eroding of wide stretches of vegetation; and, ultimately, plants were cultivated, in one of the earliest centres of domestication. Flooding and desertification followed in extreme cases of overgrazing or land overuse, and the production of so much food so efficiently in such small areas created the wealth that allowed the formation and growth of villages, towns, and finally, cities, and led to war and the ultimate consolidation of the modern political state, a trend in which events around the Mediterranean played such a key role.

But the modern vegetation of the Mediterranean also contains relict animals and plants, those that have by reason of their adaptations been able to survive in the forest patches, on the cliffs, at higher elevations, or in habitats for which they are so exquisitely suited. It is even more extensively characterized, though, by the extensive formations of weedy, often annual vegetation, and degraded shrublands that have replaced the once more extensive woods of the region. Those weeds, linked to the extensive cultivation of, and other human pressures on, the Mediterranean vegetation, have travelled around the world and are now the predominant weeds of wild and cultivated lands in California, South Africa, Chile, and areas of Australia – the other parts of the world that have 'mediterranean' climates.

The authors give clear and illuminating descriptions of the physical features of the Mediterranean and the ways in which they have evolved; of the climates that characterize this regions, and the considerable degree to which they vary from place to place; of the biological and human sense of the Basin, where the cultivation of olives, pomegranates, grapes, and the occurrence of evergreen shrubs such as the holm-oak present such a characteristic appearance; its subdivision into bioclimatic subregions; the complexity of its flora and of the origins of the plant genera that comprise it, and the richness of the biotas that occupy the region at present – all these are parts of the story unfolded here. As many as 25 000 species of plants are estimated to occur in the Mediterranean region, half of them found nowhere else; and similar degrees of endemism are characteristic of other groups of organisms well enough known from these statistics to be reliable. The region, then, may be home to something like 1/12 of the world's biodiversity, some 400 000 to 600 000 species, most very poorly known and many unknown. These organisms are not particularly richly represented at any one location in the Mediterranean, but change rapidly from place to place, thus presenting an ever-changing kaleidoscope of biological richness unparalleled elsewhere.

The dynamics of these communities are explored in a most informative way in this book, the discussion making it clear that it is not a set of particular habitats that exist, and will continue to exist, in the Mediterranean, but rather a panorama that is never quite the same, whether at the genetic level of individual populations or in terms of the mixtures of plants, animals, fungi, and micro-organisms that make up particular communities. The ways in which this volume deals with the dynamics of the situation are unique and especially helpful to anyone wishing to make sense out of the Basin's incredibly rich biodiversity.

For the future – for this beautiful, rich area that so many poets and philosophers have loved, and where so many people have lived in such relative harmony with the developing communities that constitute its nature at present – there are many difficult questions. Humanistic values that have proved inspirational in the past must be re-emphasized and rejuvenated, but they will change, as will the Mediterranean Basin itself. A kind of Edenic Mediterranean, densely clothed with forests, will never be regained, least of all in the face of globally changing climates; but the Mediterranean of the future, like the entire world, will be what humans make of it. The conservation of individual species has an important role to play, but the implicit realization of what the Mediterranean is today, which involves a deep knowledge of how it came to appear as it does now, will be the essential ingredient. By understanding and respecting the landscapes of the Mediterranean we will be able to serve them best, and to shape the world that we would like our children and grandchildren to be able to enjoy in the future. The ecological complexity of the situation is summarized well by the authors in their concluding chapter, but it is basically the human spirit, the attitude, the willingness to visualize the future and to take the steps to realize it, that will surely prevail. To the end of understanding what the Mediterranean is now, in physical, biological, and human terms, and thus to lay the foundation for the future, this volume makes an invaluable contribution, and one that will be deeply appreciated by all those who take the time to study its contents.

Preface

In 1995, J.B. was invited by Oxford University Press to prepare a book on biological diversity in the Mediterranean area. After much hesitation, given the immensity of the task, he decided to accept provided he could find a co-author with complementary interests. J.A. became that co-author. J.B.'s fields are primarily biogeography, community ecology, and the population biology of animals. J.A. is mostly interested in vegetation dynamics and the interactions between man and living systems, including the ecological restoration of Mediterranean and dryland ecosystems.

Wading through the immense amount of published material on biological diversity in the Mediterranean is in itself an enormous undertaking, which has been a time-consuming and sometimes discouraging enterprise. It soon became apparent that we had to limit ourselves to selected aspects of biodiversity and make a number of somewhat arbitrary choices. Reviewing and analysing in detail the biological diversity of the Mediterranean is a daunting task, and it would be unrealistic to attempt to summarize the state of the art in all aspects of biodiversity. We agreed at once that two points of paramount importance had to be considered in order to understand the patterns observed today, namely the historical components contributing to the evolution and establishment of living systems, and the preponderant role of humans in shaping and endlessly redesigning habitats and landscapes. Although these two points are more or less touched upon in all chapters, we devote the first two chapters to the historical context in order to 'set the scene', and a full chapter to human influences on biological diversity, given the number of peoples and societies that have succeeded one another over the centuries, and have left their mark on different parts of the Basin.

To avoid misunderstandings and prevent disappointments we should point out that this book is not intended to be an encyclopaedia or a field manual. Nor is it in any way exhaustively representative of the vast literature on the subject. It should be considered as an introductory text for ecologists, naturalists, students, scholars and, more generally, for people with a natural sciences background who come to the Mediterranean for the first time. Our goal was to introduce the reader to the kaleidoscopic aspects of all living beings in this fascinating area, where there is often more biodiversity in a single square kilometre than in any area 100 times larger in the northern parts of Europe. Topics include biological diversity at many spatial scales of space, from the entire Basin to minute surface areas, and from entire floras and faunas to population variations. Rather than trying to report as much as possible on all that is known in the field of Mediterranean biodiversity, we selected some relevant examples to illustrate salient aspects of diversity at different scales of space and biological integration.

Many times we have been frustrated because we had to restrain ourselves from developing certain aspects, giving more data, or searching deeper in the literature. We wanted to tell a story of the Mediterranean area today, from two biologists' points of view. Inevitably, the book suffers from our biases. In particular we apologize for not having covered equally all the dominant groups of organisms. We gave disproportionate attention to birds and vascular plants, the groups of organisms with which we are most familiar. Our sole excuse for the many flaws, omissions, and mistakes that specialists will inevitably find in this book, and which are ours alone, is that we *dared* to tackle an enterprise which has not been attempted before from a modern biologist's viewpoint. Many pure scientists would have declined the invitation to write such a book.

In the course of writing, we tried to adopt a used-friendly style, informative but informal, and to avoid technical jargon insofar as possible. Terms that may be unfamiliar to some readers have been starred and defined in the Glossary. Some sections will nevertheless appear too technical for some readers, while other parts will irritate specialists by their brevity and paucity of detailed references. Throughout the book we use 'Mediterranean' to refer to the Basin itself, and 'mediterranean' when dealing with mediterranean-type features, keeping in mind that such features also occur in other regions of the world.

We would like to particularly acknowledge the skill and generous help of many friends and colleagues who provided us with material, encouragement, and comments on drafts of various parts of the book: M.-C. Anstett, J.-C. Auffray, M. Barbéro, A. Charrier, E. Crégut-Bonnoure, F. Catzeflis, E. Garnier, B. Girerd, L. Hoffmann, F. Kjellberg, E. Le Floc'h, R. Lumaret, P. Quézel, J.-Y. Rasplus, F. Romane, J. Roy, and T. Shulkina. Of particular value has been the help of M. Cheylan, M. Debussche, J. Thompson, and F. Vuilleumier who read several chapters, and gave most valuable criticism. We gratefully acknowledge publishers for permission to reproduce illustrations, especially Backhuys, Birkhauser, Blackwell Science, Cambridge University Press, Delachaux et Niestlé, Elsevier Science, the Israel Academy of Humanities and Sciences, Masson, and Springer-Verlag. We are also most grateful to M. Arbieu, A. Carrière, C. Lacombe, D. Lacombe, and especially to R. Ferris, who undertook the heavy task of preparing so many figures and drawings with much care and cheerfulness.

Montpellier   J.B. and J.A.
July 1998

To Chantal and Joëlle

Contents

Glossary

Terms defined here are marked in the text with an asterisk on first appearance.

Aestivation	dormancy during summer or dry seasons.
Allogamous	mating system in which plants must be cross-pollinated.
Allopatric	species or subspecies not occurring together, that is, having different areas of distribution.
Amphihaline	species living in both fresh and sea water.
Anemochoric	plant species whose seeds are dispersed by wind.
Anthracology	the study of fossilized charcoal.
Aphyllous	leafless plants in which green stems are photosynthetically active.
Autogamous	mating system in which plants are self-pollinated.
Autopolyploidy	polyploidy in which all the chromosomes in the nucleus come from the same species.
Autotetraploid	form of polyploidy in which the nucleus includes four times the haploid number of chromosomes coming from the same species.
Bath'a	a type of low matorral common in the north-east Mediterranean quadrant.
Boreal	major life zone covering the northern part of the continents of the Northern Hemisphere.
Calciphobe	a plant that does not support active calcium carbonate, and thus is rarely or never found on limestone-derived soil types.
Cenozoic	the most recent geological era (from 65 Myr BP to present); includes the Tertiary period, and the Pleistocene and Holocene epochs.
Chamaephytes	woody shrubs with perennial buds borne more than 25 cm above soil level (a class of Raunkiaer's life forms).
Chasmophytes	plants growing in cracks and crevices of cliffs or walls, independent of surface soil.
Commensal	species living in close association, for example in same burrow, shell, or house without mutual influence; that is, not symbiotic.
Congener	species belonging to the same genus.
Daya	temporary marsh in arid regions (mostly North Africa) unpredictably filled with water.
Diadromous	species (usually fish) that move from fresh to salt water according to the stages of their life cycle.
Diaspore	dispersal organ of plants or animals, for example seeds, fruits, eggs, or spores.
Dioecious	unisexual, the male and female reproductive organs being

	borne on different individuals.
Diploid	having chromosomes in pairs in a nucleus. Chromosomes in a pair are homologous so that twice the haploid number is present.
Disharmony	refers to changes in the relative proportions of different taxa or trophic levels among communities on islands compared with those on the nearby mainland.
Disturbance regime	all sorts of spontaneous disturbance events that characterize a region and contribute to the spatial and temporal dynamics of species and communities.
Djebel	the Arabic word for mountain.
Ecomorphology	the form and shape of organisms in relation to their ecology.
Ecosystem trajectories (see trajectories)	alternative concept to the 1930s–1970s successional models of ecosystem development or dynamics.
Ecotone	mixed habitat formed by the overlapping areas or transition zone between two habitats. Transitional strip separating two communities.
Ecotypic variation	variation of the subunits of a species in relation to variation of environmental conditions. Ecotypic variation is usually genetically determined.
Edaphic	refers to the physical and chemical conditions of a soil and their influence on the growth of plants.
Endemic	indigenous or native in a restricted locality, area, or region.
Endoreic marsh	body of water, often temporary, that does not empty into any river or larger body of water.
Entomogamous	plants whose mating systems require pollination by insects.
Ephemerals	very short-lived organisms, especially plants. Usually found in desert or arid Mediterranean regions, germinating especially after rain events.
Eremean (Eremic)	desert-dwelling, of desert origins. The Eremean region is the wide arid belt stretching from Mauritania to Arabia and from north-western India into central Asia.
Euryhaline	a species that is resistant to great changes in salinity over a season or life cycle.
Evapotranspiration	the process of water loss through evaporation and plant transpiration.
Evergreen	perennial plants that bear living leaves all year round.
Evergreenness	the habit of plants having photosynthetically active leaves all year round.
Feedback systems	the modification or control of a process or system by its results and effects.
Feralization	the process whereby domesticated animals escaping human control create wild populations, eventually returning to ancestral forms.
Ferruginous	soils containing iron.
Forb	any herbaceous plant which is not a grass nor a sedge.
Fractals	a curve or geometrical figure, each part of which has the same statistical character as the whole. Useful in describing partly

random or chaotic natural phenomena.

Functional group   group of species having similar functions within a given ecosystem.

Garrigue   a type of vegetation composed of perennial readily sprouting shrubs and trees. Sometimes restricted to calcareous soils and usually seen as low scrubland with patches of bare ground.

Geophyte   herb with perennial buds below soil surface (a class of Raunkiaer's life forms).

Germplasm   term for plant material of any kind useful for propagation or long-term storage.

Gondwana   historical super-continent that included all the present-day continents of the Southern Hemisphere before their splitting apart as a result of plate tectonics.

Guelta   small permanent body of water in temporary rivers of the arid zones (mostly North Africa).

Guild   a group of organisms, generally closely related, that share similar resources.

Gynodioecious   plants having female and hermaphroditic flowers on separate plants.

Halophyte   a plant that tolerates very salty soils and water, at least for certain periods of its life cycle; a condition typical of plants growing near the seashore, and in coastal salt flats and river estuaries.

Hemicryptophytes   herbs with perennial buds at or near soil level, protected during dry season by the soil itself or by dry, dead portions of the plant (a class of Raunkiaer's life forms).

Herbivory   the process whereby animals eat living parts of plants.

Heterozygosity   having two or more alleles at a same genetic locus.

Holocene   the geological epoch following the Pleistocene and consisting of recent times since the end of the last ice-age, about 11 000 years ago.

Hysteranthous   a plant flowering in autumn in a leafless state. Usually found in geophytes.

Inertia   the property of matter, or a system, by which it continues in its existing state unless changed by an external force.

Introgression   infiltration of genes of one species into the genotype of another.

Isohyet   the line of equal amount of average rainfall over a wide region.

Karst   limestone regions that are highly fissured as a result of calcareous dissolution.

Laurasia   historical super-continent that included all the present-day continents of the Northern Hemisphere before their splitting apart as a result of plate tectonics.

Lauriphyllous   plants bearing leaves that resemble the glossy evergreen leaves of the laurel tree.

Laurisylva   Mediterranean forests dominated by trees of the laurel family or else with leaves similar to those of the laurel tree. In the Mediterranean area today, mostly occur in Macaronesia.

Life-form spectrum   the range of vegetative growth forms that plant species can

	present, including trees, shrubs, herbs, forbs, grasses, bulbs, etc. Life forms are often found to be closely linked to life-history traits that are adaptive to conditions of a given environment.
Lithology	refers to the science of rock substrates.
Local specialization	the process whereby a population evolves life-history traits that adapt it tightly to its local environment.
Maghreb	region formed by the three countries of north-western Africa: Morocco, Algeria, and Tunisia. Belongs to the Palaearctic biogeographical realm.
Mesic	characterized or pertaining to conditions of medium moisture supply.
Matorral	any kind of shrubby, predominantly evergreen Mediterranean vegetation.
Mesozoic	the geological era dating from 225 to 65 Myr BP; it includes the Triassic, Jurassic, and Cretaceous periods.
Metapopulation	a series of local populations interconnected through processes of extinction and recolonization.
Mimicry	protective similarity in appearance of one species of animal to another (generally insects). In Batesian mimicry, the imitated species is poisonous and often conspicuously marked. In Müllerian mimicry both species are protected from predators, gaining mutually from having the same warning coloration.
Molecular clock	the theory that molecules evolve at an approximately constant rate through time. The difference between the form of a molecule in two species is therefore assumed to be proportional to the time elapsed since the species diverged from a common ancestor.
Monotypic	a genus with only one species.
Neogene	the second part of the Tertiary period, from the Miocene epoch onwards.
Neotenic	an animal that retains larval characteristics throughout its life cycle. Neoteny is common in amphibians.
Oogon	female sex organ of certain algae and fungi containing one or more oospheres.
Outbreeding depression	the reduction of viability due to the mating of individuals from populations that largely differ genetically.
Palaeogene	the first part of the Tertiary period, from the early Palaeocene epoch (65 Myr BP) to the Miocene which ended *c.* 5 Myr BP.
Palynology	the study of ancient floras and vegetations from fossil pollens that accumulate in soil profiles.
Pedology	the science of soils.
Phenotype	the way in which the genotype of an individual is expressed in its morphology, physiology, and behaviour.
Phenotypic plasticity	phenotypic variation expressed by a single genotype in different environments.
Phenology	study of periodical events, for example the timing of spring leaf appearance, flowering, or the timing of migration in birds.
Phrygana	low matorral mostly composed of prickly and stiff undershrubs.

Phyllode	a flattened leaf-like petiole.
Phylogeny	the genealogy of a group of taxa. Study of the branching relationships among species deriving from a same ancestor. Phylogeny aims to reconstruct the evolutionary history of a lineage.
Phylogeography	the reconstruction from molecular markers (e.g. mitochondrial DNA in animals and chloroplastic DNA in plants) of the history of the spatial distribution of species and populations.
Phytophagy	(see herbivory).
Ploidy	the haploid number of chromosomes in a nucleus.
Polyploid	having three or more times the haploid number of chromosomes in a nucleus.
Propagule	any part of an organism or stage in the life cycle (seeds, individuals, at least one pregnant female) that can reproduce the species and thus establish a new population (see diaspore).
Refugia	in historical biogeography, regions where species have persisted during harsh climatic periods (e.g. Pleistocene glaciations) while becoming extinct elsewhere.
Resilience	the ability of an ecological system to recover after disturbance.
Retamoid	from the genus *Retama*; refers to broom-like plants with photosynthesizing stems and highly reduced leaves.
Riparian	plants and vegetation types growing along rivers or streams and taking water mostly from ground water.
Ruderals	plant species commensal to man usually frequent in and around man-made habitats such as roadsides, ditches, and other frequently disturbed habitats; usually annuals (cf. segetals).
Sarmatic	organisms belonging to the coastal fauna that in upper Tertiary times inhabited the shallow, brackish, or salt Sarmatic inland sea which formed an eastern continuation of the Mediterranean Sea.
Sclerophylly	a leaf form that is generally evergreen, leathery, and spiny.
Sebkha	depressions which contain brackish water after rainfall, but are dry and covered by salt incrustations in the summer (common in North Africa and the Near East).
Segetals	plant species, mostly annuals, common in cereal crop fields and tied to this habitat.
Semi-species	two closely related species which hybridize where they come into contact.
Spectra	(see life form spectrum)
Spinescence	the habit of leaves or stems to be spiny.
Stochastic	random, unpredictable.
Successional	refers to the progressive replacement of one community by another in space or in time, e.g. during the development of vegetation, from initial colonization to old mature forest (a stage of a succession is often called a sere).
Summergreenness	the habit of plants having leaves that are shed in winter-time (deciduous).
Supersaturated	communities with more co-existing species than expected.

	Refers to island communities which are not at equilibrium.
Sympatric	two or more species or populations occupying the same geographical area such that the possibility of interbreeding occurs.
Synanthous	flowering during the leafing phase. Usually refers to geophytes.
Synanthropic	closely associated with man (plants, animals, environments).
Therophytes	herbs that complete their life cycle from germination to seed maturation within a single vegetative period and which survive unfavourable seasons as seeds (a class of Raunkiaer's life forms).
Threshold	a term used in ecology for more or less irreversible (except over very long periods) changes or perturbations induced in a community or ecosystem. Often provoked by human disturbances.
Trajectory	the path followed by an ecosystem whose dynamics is driven by given forces.
Vicariance	the geographical separation of two or more closely related taxa.
Wadi	temporary rivers in arid regions. The term is mostly used in North Africa and the Near East.
Xeromorphic	plant possessing morphological characters adapted to prolonged drought.
Xerophytic	adapted to extremely arid conditions.
Zoochory	the process whereby plant species are disseminated by animals. Plants may be epi-zoochoric when seeds or fruits are transported in the plumage of birds or the fur of mammals, or endo-zoochoric when fruits are swallowed by the animal and then the seeds are regurgitated or defecated.

1 Setting the scene

The birth of the Mediterranean

The Mediterranean Basin, with its fantastic, endlessly evolving variety of land-scapes, peoples, plants, and animals defies imagination for anyone who has never been there. It is one of the most complex regions on Earth in terms of geological history, geography, morphology, and natural history. Imagine a keen naturalist spending long holidays in this region in May–June, moving from high mountains to seashores, coastal wetlands to small islets, dry shrublands to moist fir forests. In each of these habitats, he or she will be amazed by the diversity of living things to be discovered while walking, turning over stones, peering into ponds and brooklets, or scanning the sky for soaring raptors. Beautiful orchids, strange lizards, tropical-like birds, flamboyant butterflies, and secretive mammals will fill the eye and mind with wonder. Should this naturalist be interested in the natural history, evolutionary ecology, and biogeography of all these organisms, they will naturally want to learn something about their history: when did they originate, and why then, and what caused their evolution to unfold in the way it has.

Analysing and understanding biodiversity in space and time, which is the aim of this book, first requires a brief explanation of how and when the features we observe today have developed. Our keen naturalist must also remember, when beginning their journey, that the influence of humanity is incommensurably heavier in the Mediterranean than in any other part of the world, and at some times and in some places may override all other factors. Some events that have marked the history of the Basin are indicated in Fig. 1.1.

From the Tethys Sea to the Mediterranean

From the beginning of the Mesozoic* era, some 200 million years Before Present (hereafter, Myr BP), the Mediterranean Sea's giant ancestor, the Tethys Ocean, formed a vast wedge-shaped unbroken seaway which subdivided the super-continent Pangaea into the Laurasian* proto-continent to the north and the Gondwanian* proto-continent to the south. The physical geography of the future Mediterranean area changed continuously during this epoch as a result of several periods of continental convergence and collisions of tectonic plates, while seafloor spreading repeatedly rearranged the relative positions of the continental plates and oceans in the ancient Tethys. Between the Middle Jurassic and the Upper Cretaceous (165–80 Myr BP), an eastward motion of Africa relative to

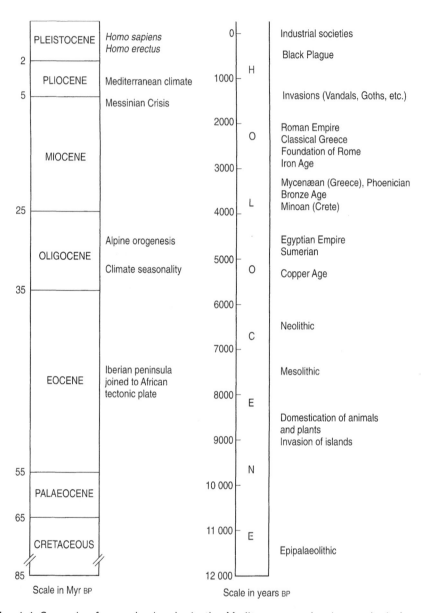

PLEISTOCENE	*Homo sapiens* *Homo erectus*	0		Industrial societies	
2			H	Black Plague	
PLIOCENE	Mediterranean climate	1000			
5	Messinian Crisis			Invasions (Vandals, Goths, etc.)	
		2000	O	Roman Empire Classical Greece Foundation of Rome Iron Age	
MIOCENE		3000			
				Mycenæan (Greece), Phoenician Bronze Age	
25		4000	L	Minoan (Crete)	
	Alpine orogenesis			Egyptian Empire Sumerian	
OLIGOCENE		5000	O		
	Climate seasonality			Copper Age	
35		6000			
			C	Neolithic	
		7000			
EOCENE	Iberian peninsula joined to African tectonic plate			Mesolithic	
		8000	E		
		9000		Domestication of animals and plants Invasion of islands	
55		10 000	N		
PALAEOCENE					
65		11 000	E		
CRETACEOUS				Epipalaeolithic	
85		12 000			
Scale in Myr BP			Scale in years BP		

Fig. 1.1 Synopsis of some landmarks in the Mediterranean, showing geological and climatic events during the Tertiary period (left) and human events in the Holocene* epoch (right).

Europe, which was then still joined to North America, permitted the formation of the Atlantic Ocean (Fig. 1.2). Later, the movement of Africa became westward as Europe began to separate from North America in the lower Tertiary. Again, Africa moved eastward relative to Europe from the upper Eocene (40 Myr BP) onwards, enlarging the Atlantic. This led to the closing of the gap between Europe and

Middle Jurassic (180 Myr BP)

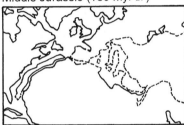

Lower Cretaceous (120 Myr BP)

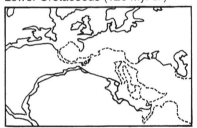

Upper Cretaceous (80 Myr BP)

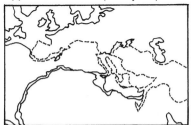

Lower Miocene (20 Myr BP)

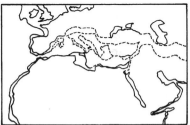

Fig. 1.2 Relative position of the African and European tectonic plates at various epochs from the Middle Jurassic to the lower Miocene (After Smith and Woodcock 1982).

Africa and the elimination of much of the original Tethys. At the same time, a northward movement of Africa initiated the creation of the mountain ranges that encircle the Basin (Hsü 1971).

All these complicated movements of continental drift contributed to the intricate puzzle-like geography of the Mediterranean area. Mountain configurations, in particular, reflect the tectonic interactions and collisions between the blocks of North Africa and those of south-west Eurasia. These movements generated the rotation of micro-continents or 'micro-tectonic plates' that separated from the main continental blocks. Dynamic reconstruction models show an ever-changing mosaic of smaller and larger tectonic plates that produced ridges, trenches, and island arcs (Cohen 1980). During the Eocene (40 Myr BP), after more than 100 million years of transcontinental shoving and reshuffling, the Tethys seafloor began to buckle between the colliding tectonic blocks. The Greco-Italian micro-continent, which was joined to the African plate during the Lower Cretaceous (120 Myr BP), moved with it until their joint collision with the Eurasian plate in the lower Eocene. Then Italy collided with the Balkan block, giving rise to the Apennines. The Iberian peninsula was joined to the African plate and shared its westward movement, which prompted the deformation of the Pyrenees in the lower Eocene. Later, in the upper Oligocene (28 Myr BP), the south-eastward motion of Africa relative to Europe caused the rotation of the Iberian plate, which included all the large islands of the western Mediterranean and several crystalline blocks that were subsequently connected either to Africa or to Europe. At that time, Italy, Corsica, Sardinia, the Balearic Islands, and Spain were all still crowded in south-west Europe. The Corsico-Sardinian block and Italy drifted counterclockwise and left behind the Balearic and Tyrrhenian basins, while various chunks of the Iberian microplate became detached to form the Balearic Islands. In the lower Miocene (23 Myr BP), these microplates shifted westward and progressively took their present positions, sometimes with large changes of orientation. Although still uncertain, the kinematics of the Corsico-Sardinian block suggests that this micro-continent was partly connected to southern France during the lower Oligocene, but then became detached and rotated 90° from its original east–west axis to its present north–south orientation some time around 20 Myr BP.

The squeezing of Iberia and the buckling of the Tethys seafloor between the African and Eurasian plates caused the uplift of the early Pyrenees, the Baetic Cordillera of southern Spain (Sierra Nevada), and the Moroccan Rif mountain, facing Gibraltar (see Fig. 1.5). In the upper Oligocene (25 Myr BP), the African plate, including Arabia, first made contact with south-west Asia, thus dividing the Tethys Sea into two parts: the southern part, ancestor of the modern Mediterranean Sea, and the second, further north and east, which was a shallow, brackish sea that geologists call the Paratethys.

Thus, during the Tertiary, the Mediterranean Sea was progressively reduced into a network of epicontinental seas as the gap between the African and European tectonic plates was progressively reduced. As a result of these tectonic movements, mountain building produced the Alps in the north, and the Atlas and Anti-Atlas ranges in the south. Subsequent mountain building in south-west Asia yielded the Carpathian, Pontic, Taurus, and Zagros ranges (see Fig. 1.5), while diminishing the Paratethys Sea and dividing it into smaller parts, whose sole surviving descendants today are the Black, Caspian, and Aral Seas. All connections between the Mediterranean Sea and the Indian Ocean were lost when the isthmus of Suez closed the sea at the lower Miocene, 20 Myr BP. It was only in 1869, when the Suez Canal was dug, under the supervision of Ferdinand de Lesseps, that a direct passage from the Mediterranean to the Red Sea was artificially restored. This great event marked 'the end of the Mediterranean lake, and the transformation of the inner Sea into a sea-way essentially turned towards the Indian Ocean' (Braudel 1985).

The northward drift of Africa continued during the Miocene with a general westerly movement up to the present. About 6 Myr BP, Africa collided with south-western Europe. The newly wedded blocks drifted along together from that point on, raising the Pyrenees to great heights and, for the first time in history, closing the Mediterranean Sea at its western end. Thus came into being the *Mare Medi-Terraneum*, or 'sea among-the-lands', sometimes also called the 'inland Sea'.

One dominant feature that helps bring into focus the Mediterranean area as a whole is the striking contrast between its northern half, with many large peninsulas—Iberian, Italian, Aegean, and Anatolian—which are strips of continental crust caught in the Jurassic collisions of Eurasia and Africa, and the southern half, with its more or less rectilinear shorelines.

The Messinian Salinity Crisis

This short but crucial period, which followed the Mediterranean's enclosure, occurred in the last phase of the Miocene era, some 5.6 to 5.3 Myr BP, and was one of the most spectacular geological events on Earth during the entire Cenozoic*. Huge seabed salt deposits ('evaporites') at offshore sites near Sicily, Calabria, and North Africa had long intrigued researchers, but it was not until the early 1970s that an international team, aboard the American research vessel *Glomar Challenger*, scrutinized the Mediterranean seafloor with innovative deep-sea drilling methods, and determined the contents of these thick deposits, in places over 1500 metres deep. Researchers found salt domes that contained not only sodium chloride but also many other evaporites, like those found today at the southern end of the Dead Sea. These deposits included fossilized remains of light-demanding algae (cyanobacteria) such as occur only in shallow water. This proved that the Mediterranean Sea dried up more or less completely during this brief but critical period in the making of the Mediterranean world. Only some sebkhas* and alkaline lakes were left, in depressions (Fig. 1.3).

Annual water loss by evaporation from the Mediterranean Sea is currently estimated at 1600 m^3 per year, of which only 10% is replaced by rainfall and the influx of rivers. The remaining 90%, therefore, must come from the Atlantic Ocean through the Gibraltar Strait. Thus, it is not surprising that, when the Gibraltar strait was closed and prevailing climatic conditions in the lower Pliocene were much warmer than at present, the Mediterranean seafloor could have dried up in less than a thousand years (Suc 1984) (Fig. 1.3).

The Messinian Crisis also had consequences for the earth's crust at both the northern and southern shores of the Mediterranean Sea. Enormous fissures opened up, earthquakes shook the ground, and ancient volcanoes were reactivated while several new ones were born. Many coastal areas were rising, transformed into isolated 'mesas' or islands towering thousands of metres above the arid, increasingly saline flats below. Concurrently, great rivers such as the Rhône and the Nile continued to feed the nearly dry Mediterranean, shooting over high cliffs and gradually digging out deep gorges in the thick granitic crust and limestone blocks at the Sea's edges. One such gorge lies 900 m below sea level, at the mouth of the Rhône river, near Marseilles, while another is more than 2000 m beneath Cairo, which itself lies more than 100 km inland and upstream from the Nile River delta near Alexandria.

As the current ecological and biogeographical complexity of the Mediterranean began to take shape during the Messinian Crisis, yet another geological surprise was in store. Some 5.3 Myr BP, a new series of westerly oriented tectonic shudders shook the region, breaking open the land bridge between Morocco and Spain and allowing the waters of the Atlantic Ocean to once again surge through the Strait of Gibraltar and into the dried up Basin. The water cascaded down drops fifty times the height of Niagara Falls; gigantic waterfalls poured onto the floor of the Basin 3000–4000 m below. It has been calculated that as much as 65 km^3 of seawater rushed in each day, enough to refill the entire Basin in less than a century.

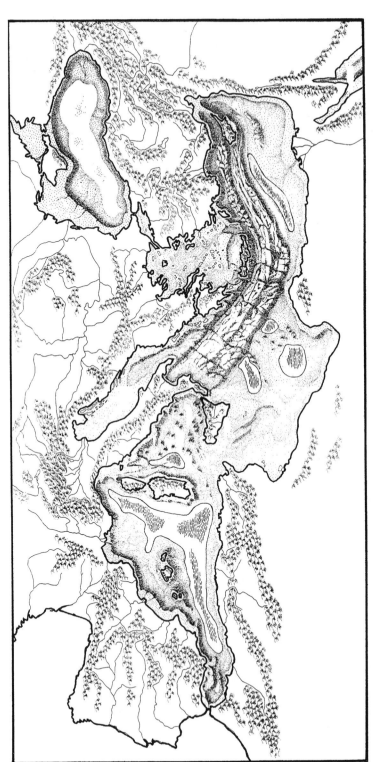

Fig 1.3 Probable aspect of the Mediterranean Sea when it almost completely dried up during the Messinian Crisis, 5.6–5.3 Myr BP. (After Hsü 1972 and Attenborough 1987.)

Following this awe-inspiring event, the present-day size and shape of the Mediterranean, as well as its main physiographic and geomorphological features, were finally established roughly 5 Myr BP.

The physical background

The Mediterranean Basin stretches over *c*. 3800 km west–east and 1000 km north–south, between 30°N and 45°N. A marked geographic boundary runs north–south through the Sicily–Tunisia 'sill', which is only 140 km wide and 400 m deep, between the southern tip of Sicily and Cap Bon (Tunisia). This feature results in, or corresponds to, a marked biogeographical contrast between the western and eastern halves of the Mediterranean, the former being shifted somewhat north with respect to the latter. The boundary separating the two north–south ranges in each half of the Basin runs approximately along the 36th parallel in the western half and the 33rd in the eastern half. In the western half, west of the Sicily–Cap Bon line, biota are more boreal in character and overlap to a large degree with those of central Europe. To the east, they have greater affinities with central Asia. The clear geographic and biological distinction between the two halves of the Basin prompted De Lattin (1967) to recognize a western 'Atlanto-Mediterranean' subregion and an eastern 'Ponto-Mediterranean' one. In fact, the Basin can be divided further, into four quadrants (see Fig. 1.4), taking into account the latitudinal shift between its two halves: the north-west, north-east, south-west, and south-east. We will use this abbreviated terminology throughout this book.

Volcanic activity

Tectonic remodelling, stimulated by the Alpine and other uplifts, as well as by the seismic cataclysms provoked by colliding African and Eurasian plates (which continue to move towards each other at a rate of about 2 cm per year), all contribute to an exceedingly complex geomorphological situation. In many parts of the Basin, tectonic events have been accompanied by renewed bouts of volcanic activity, such as the eruption of Vesuvius in AD 62 that resulted in the vast, ash-covered mausoleum of Pompeii. Other well-known examples are Stromboli and Mt. Etna. In fact, the Mediterranean is a huge seismic 'hearth', especially in its eastern part. The eruption of Santorini, in the Greek Cyclades, some 3000 years ago, was one of the most significant recent 'catastrophes' in the Earth's history, along with the eruption of Krakatoa (Indonesia) in 1883. Among other things, the Santorini eruption may have helped precipitate the collapse of the renowned Minoan civilization, which flourished in Crete during the Bronze age between 5000 and 3000 yr BP.

Historical records also reveal the occurrence of devastating earthquakes: at least one hundred were recorded during the past thousand years in the Mediterranean

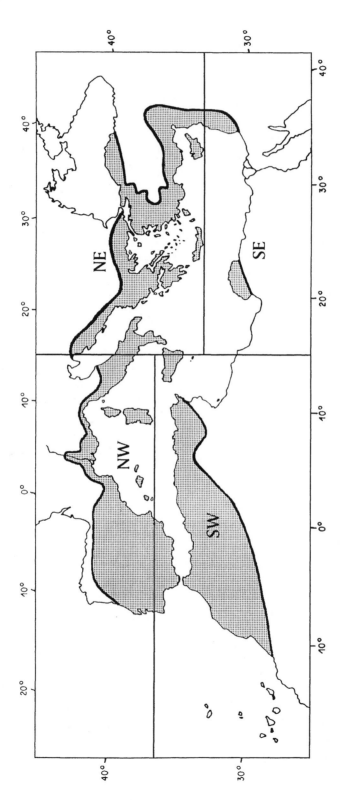

Fig. 1.4 The four quadrants of the Mediterranean area.

Basin. At the site of the ancient temple of Olympia, in south-western Greece, there are rows of nearly identical disks, laboriously carved from granite and marble in the first millennium BC to form the columns of a temple, which were all knocked over by an earthquake in AD 426. Still visible today, lying just where they fell, these columns give a vivid impression of the destructive strength of earthquakes. Devastating earthquakes in the Mediterranean in still more recent times include those of Lisbon in 1877, Al Asnam (formerly Orléansville), Algeria, in 1954, Agadir, Morocco, in 1960, Skopje (ex-Yugoslavia) in 1963, Ardebil, Iran in 1997, and the most recent one, which partly destroyed the historic frescos of the cathedral of Assisi, in August 1997. Very few places in the Mediterranean region are entirely free of seismic risk since the effects of large earthquakes may occur several hundreds of kilometres from the epicentre. In Turkey alone, no fewer than 60 000 people have been killed and 380 000 houses destroyed by earthquakes since 1930 (Kolars 1982).

The sea

The Mediterranean Sea, the largest inland sea in the world, extends over 2 969 000 km^2. With its satellites, the Black and Azov seas, it is really a miniature ocean. Stretching over 3800 km from Gibraltar to Lebanon, the Sea is 740 km wide between Marseilles and Algiers, and 400 km wide between southern Greece and Tripolitania (Libya). Its coastline is 46 000 km in length and several 'interior seas' are delimited by the shorelines of the larger archipelagos and northern peninsulas: the Balearic Sea between the Balearic Islands and continental Spain, the Tyrrhenian Sea between Corsica–Sardinia and the Italian peninsula, the Adriatic and Ionian seas between Italy and the Aegean peninsula, the Aegean Sea between Greece and the Anatolian peninsula, and still more. To simplify, in this book we will speak of three main 'banks' bordering the Sea: southern Europe, northern Africa, and the Near East.

A striking geomorphological feature of the Basin is the steepness of the coast, which almost everywhere sinks abruptly to the seafloor at 2500–4000 m. Average depth of the sea is 1460 m but deeper troughs occur in some places, for example at the southern tip of the Peloponnese (5100 m) and south-east of Sicily (4100 m). Average salinity is 3.7–3.9% (as compared to 3.5% in the Atlantic Ocean and 4.2% in the Persian Gulf). Since evaporation greatly exceeds incoming water supply, the Atlantic Ocean must replenish the Mediterranean Sea each year with an amount of water corresponding to a layer one metre thick. Tides in the Mediterranean are of very small amplitude except in some southern regions and, to a lesser degree, at the northern end of the Adriatic Sea. Around the island of Djerba, Tunisia, tides may reach 3 metres. The continental shelf is usually very narrow except in the Golfe du Lion (southern France), between Tunisia and Sicily, and in the large Aegean area between Greece and Turkey.

Table 1.1 Some of the highest mountain peaks, moving clockwise around the Mediterranean Basin.

Mountain(s)	Country	Altitude (m)	Mountain(s)	Country	Altitude (m)
Sierra Nevada	Spain	3481	Elburz	Iran	5600
Canigou	France	2300	Mt. Lebanon	Lebanon	3090
Mt. Bégo	France	2870	Troödos	Cyprus	1950
Etna	Italy	3260	Mt. Chambi	Tunisia	1540
Olympus	Greece	2920	Hodna-Aures	Algeria	2330
Taurus	Turkey	3920	High-Atlas	Morocco	4165

Topography

The topographic diversity of the Mediterranean results in large part from the numerous mountain chains (see Table 1.1 and Fig. 1.5) that define the Basin's various continental contours, except in the south-east (Egypt, Libya) where the lowland desert meets the sea. These mountain ranges, whose geography and history were vividly described by Braudel (1949), Houston (1964) and McNeil (1992), among others, are the Alps, Pyrenees, Apennines, Dinaric Alps in ex-Yugoslavia, the Caucasus, Pontic and Taurus mountains of Anatolia, the mountains of Lebanon, and the Rif and Kabylie, Atlas, and Anti-Atlas ranges of North Africa, not to mention the many cordilleras of the Iberian peninsula. Mountain ranges of high elevation are found in Spain's Sierra Nevada (the Baetic cordillera) and on some of the larger islands, including Majorca, Corsica, Crete, and Cyprus. Some of these peaks reach 4000 m or more, and completely isolate upland basins or plateaux, as well as create 'hidden' valleys opening only towards the sea. Several authors have suggested that a more appropriate name for the Mediterranean might be the 'Sea-among-the-Mountains'! Except for the long strip of lowland between Tunisia and the Sinai, a visitor to the Mediterranean is almost never out of sight of some mountain. All Mediterranean landscapes are within the main mountain ranges that mark the geological frontiers built over millions of years by the colliding of the African and Eurasian tectonic plates. Flat lands are almost absent, and slopes steeper than 20% occur in more than half of the territory.

The Mediterranean mountain systems are the source of most of the rivers that traverse the Mediterranean lands, first in torrential cascades, then spreading and meandering through vast alluvial plains to extensive deltas. Prominent examples are the Guadalquivir, Ebro, Rhône, Pô, Evros, and Nestos rivers and watersheds. Among the main rivers traversing Mediterranean lands, only the Nile is 'non-native' since its headwaters arise in tropical East Africa. Thus, the Mediterranean region, with its islands, coastal lands, rivers, and high mountains, provides a veritable cornucopia of habitats, all finely distinguished by local topographies and soil types, an intricate filigree of microclimates related to altitude, rainfall, and exposition of slopes.

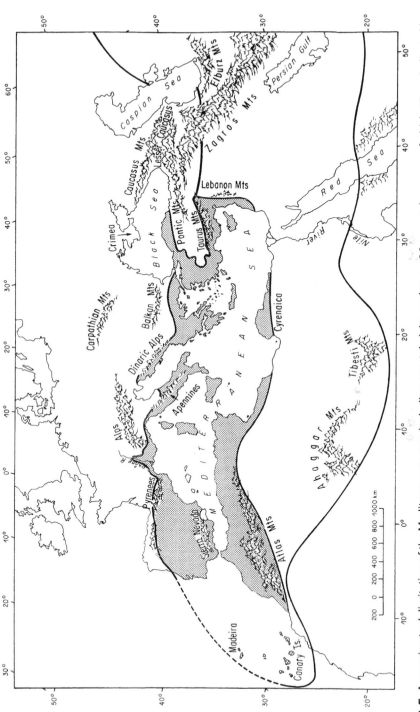

Fig. 1.5 Approximate delimitation of the Mediterranean area including coastal plains and some 20 mountain ranges, of which 12 are indicated here, as well as the Macaronesian region off the Atlantic coast of Morocco. The lower black line to the south indicates the isoclimatic region defined by Daget (1977).

Islands and archipelagos

With almost 5000, the Mediterranean is richer in islands and islets than anywhere else in the world apart from the Caribbean and the archipelagos of the Pacific Ocean. The total coastal length of Mediterranean islands is 18 000 km. In no other region of the world are land–sea interactions more intimately established in a kind of 'symbiotic' relation. Each of the large and smaller islands shows a unique array of physical, bioclimatic, and geobotanical features. Most of the larger islands have been entirely disconnected from any continent since at least the Messinian Crisis. Several, in fact, are ancient submarine volcanoes. For example, Mt. Troödos in eastern Cyprus emerged as an oceanic island at the end of the Cretaceous, 65 Myr BP (Gass 1968).

   To the list of true islands must be added a series of 'biological islands' such as the Peloponnesian peninsula, artificially isolated at the end of the last century by the opening of the Corinthian canal. The Maghreb* of North Africa (Morocco, Algeria, Tunisia) is by itself a large biological island of Palaearctic character on the African continent. Moreover, the Maghreb contains many biological islands, scattered from the Rif range in northern Morocco to upland areas of Tunisia and Algeria. Further east, in Cyrenaica, Libya, is another biological island, the Djebel* Akhdar, 700 m above sea level. Cyrenaica is the only portion of Libya with a truly mediterranean climate, with an annual average rainfall of 400 mm, as compared to less than 25 mm in the rest of the country. Its regional flora is much more similar to that of the Near East than to the rest of Mediterranean North Africa, and includes a remarkable number of endemics* for such a small area. Reaching much higher elevations, the isolated Hoggar (3003 m) and Tibesti (3415 m) ranges in the middle of the Sahara are usually considered, on the basis of their vegetation, as outlying Mediterranean enclaves, along with the long narrow chain of the Elburz Mountains just south of the Caspian Sea.

Soils

Most Mediterranean substrates are limestone of marine origin, and one can find fossil seashells well above the timberline in most Mediterranean mountains. However, unusual soil types and discontinuous geological substrates, including volcanic soils, also contribute to the local and regional diversity of habitats. Metamorphic granitic and siliceous (acidic) parent rocks occur locally as do also occasional ultrabasic rocks in the Balkans, ex-Yugoslavia, and other eastern territories. As lime content and degree of alkalinity have a great influence on plant growth, it is not surprising that different vegetation types occur on calcareous vs. non-calcareous substrates. High rates of endemism are often related to unusual rock types, particularly when they are imbedded in a geological matrix of a different nature. For example, outcrops of ultrabasic serpentine soils occurring in Cyprus, Greece, and other parts of the eastern Mediterranean nurture many endemic species. Recent studies from the Gibraltar region also show that many species of plants there are restricted to acid soil 'islands' in a surrounding geobotanical 'matrix' of limestone substrates.

Many soil types, especially in the northern part of the Basin, are ferruginous*
brown soils known as 'terra rossa' but dolomite (from degraded calcites), clayey
marls, rendzines, loess, regisols, lithosols, and alkaline and gypsum outcrops also
occur more or less sporadically in many regions. In some parts of the Basin,
especially in Spain, along the Adriatic coast of ex-Yugoslavia, and in Anatolia,
large karstic* outcroppings occur, where rainfall infiltrates rapidly and then
reappears far away as 'Vauclusian springs' at the foot of mountain ranges. These
springs are the outcome of complicated networks of underground water resulting
from the dissolution of thick calcareous deposits.

Biological delimitation of the Mediterranean area

Defining and mapping the Mediterranean realm from a biological point of view is
much more difficult than situating it geographically. This has been a subject of hot
debate among biogeographers for more than a century. There are no sharp borders
with neighbouring regions, and many factors must be considered, including
vegetation, climate, latitude, and altitude. Any approach we might adopt is
condemned to a certain amount of arbitrariness, but the one we have chosen
seems to be the most appropriate for the purposes of this book. We shall briefly
explain our reasoning.

In Europe, the main areas of discussion, insofar as limits are concerned, occur
in the higher zones of the mountain ranges in Spain, France, the Alps, the Apennines, and the Aegean peninsula. To the east and south, in addition to the
similarly controversial alpine zones of Turkey and North Africa, there are vast
areas covered by steppe vegetation which some authors include in the Mediterranean region while others adamantly exclude them. These occupy the interior
drylands of Turkey as well as vast areas in the Near East and North Africa.

Ecologists, historians, and geographers agree that what provides unity to the
region, and also its particularity, is its climatic pattern of hot, dry summers and
humid, cool or cold winters, a climate type that is also found in parts of California
and Chile, the Cape province of South Africa, and two unconnected regions of
southern and south-western Australia. In the Mediterranean Basin, this bimodal
weather pattern represents a sharp discontinuity, or anomaly, in the sequence of
climate types occurring from the Equator to the North Pole. Within the
Mediterranean Basin itself, climate also changes with rising altitude in mountain
ranges and when travelling from west to east. On the whole, a sharp gradient
exists between the colder, wetter north-western and north-eastern extremities of
the Basin and the hotter, more arid, south-eastern and south-western parts in
North Africa and the Near East. In association with these climatic patterns, an
evergreen* shrubland vegetation type occurs, which is considered as characteristic of the Mediterranean, but many 'grey' areas also occur. Does it make sense,
for instance, to talk about Mediterranean steppes, deserts, rivers, or mountains?
How should one consider areas where vegetation has been profoundly
transformed by humans? The sheer size of the area complicates matters of

definition further still. Depending on the definition used, 'the Mediterranean' comprises between 2.3 and 9.5 million square kilometres! At its outer limits, the area also shows great biogeographical and climatic complexity where it meets the boreal* forests in central and northern Eurasia, the vast steppe regions of central Asia and North Africa, and the hot subtropical deserts of north-east Africa and the Middle East.

Bioindicators

One historical approach to mapping and defining the region has been to rely on so-called 'bioindicator' plant species. Such species are thought to provide a reliable index to Mediterranean-type ecosystems, thriving despite long, hot, dry summers and cool wet winters, but unable to survive prolonged periods of frost. The main candidates as bioindicators are the olive tree (*Olea europaea*), the holm-oak (*Quercus ilex*), and various types of *Citrus*, especially the orange tree.

Pliny the Elder (AD 23–79) was probably the first to use the area of cultivation of the olive tree to define the limits of the Mediterranean. To a biologist, the olive tree is indeed characteristic of a certain biological type that seems particularly well adapted to the long dry summers of the Mediterranean. The thick, waxy leaves that remain on the trees for 2–3 years or more represent an outstanding water-saving system that is shared by many evergreen trees and shrubs in the Basin. Deep frost, however, acts like a sword of Damocles for olive groves in the northern parts of the Basin, devastating plantations over huge areas. Yet the olive tree's regenerative properties after severe stress are also typical of many Mediterranean trees and shrubs. For example, unusually cold weather in February 1956 apparently killed almost all the olive trees of southern France. However, it soon appeared that although above-ground parts had frozen, the root systems had survived. When warm weather returned in spring, the trees grew back vigorously from their root crowns.

However, many arguments can be advanced to rule out the olive tree as a bioindicator. First, it is controversial to use *any* cultivated plant to delimit a biogeographical area, even if, like the olive tree, it is considered native to that region. Incidentally, fossil pollen analyses do not yield reliable information on whether the olive tree is indeed native to the Mediterranean Basin, as it is not possible to distinguish between the pollen of wild and cultivated olive trees.

An alternative bioindicator for the Mediterranean realm was first proposed by the geographer Drude in 1884. His choice was the holm-oak whose stiff, evergreen leaves are somewhat similar to those of the olive tree. This tree appears rather suddenly at about 43°N latitude as one travels through south-western Europe, and at about 600–800 m above sea level as one comes down from the tops of Mediterranean mountains. Holm-oak and several congeners* dominate huge expanses of vegetation around the Mediterranean Basin. However, here again problems exist. The holm-oak is absent from large portions of the eastern half of the Basin, where it is restricted to warm plains and foothills near the coast. It also extends well outside the Mediterranean Basin in some areas, as for example along the Atlantic coast of France and along the Rhône River almost to Lyon. It also penetrates the high rainfall areas of north-west Spain and northern Portugal, and occurs sporadically along the southern shores of the Black Sea in northern Turkey, two areas that are often excluded from the Mediterranean region by ecologists. Furthermore, recent studies indicate that the holm-oak, like most oaks in fact, is a rather complex botanical entity, and is best seen as part of a hybrid swarm involving at least three other species (Michaud *et al.* 1995).

The next approach, historically, to define the Mediterranean area was to take certain plant 'associations', including the holm-oak or other similar evergreen oaks, as Mediterranean markers. Historically, the French botanists Emberger (1930a), Flahaut (1937), and their followers emphasized as diagnostic the presence of the evergreen formations dominated by the holm-oak, or closely related species. Like the olive tree, these oaks are long-lived, resprout readily after fire or cutting, and show numerous ecophysiological adaptations that recur among many dominant woody plant species in mediterranean-type ecosystems around the world. They are slightly more tolerant to extreme cold than the olive trees, as witnessed by the fact that they did not freeze in the 1956 winter. Holm-oaks also share the so-called sclerophyllous* leaf structure which occurs in a great many species in these same associations, such as the laurel (*Laurus nobilis*), strawberry tree (*Arbutus* spp.), and lentisk (*Pistacia lentiscus*). Together these sclerophyllous plants constitute a characteristic dense, evergreen woodland or, more often, a shrubland, the likes of which are not found elsewhere except in other mediterranean-climate regions.

However, a broad range of life forms* other than evergreen sclerophyllous shrubs and trees also occurs in the Mediterranean Basin, not only at higher altitudes, but also within the evergreen formations themselves. When climbing any of the Mediterranean mountains, on both dry and especially wetter slopes, the evergreen oaks are gradually replaced by deciduous oaks, chestnut (*Castanea sativa*), and beech (*Fagus sylvatica*), as well as by many other broad-leaved deciduous trees, and a wide range of conifers such as pines (*Pinus*), firs (*Abies*), and cedars (*Cedrus*). At lower altitudes as well, conifers are mixed with broad-leaved trees, such as junipers (*Juniperus*), cypress (*Cupressus*), and many species of pines. The conifers can also be considered as bearing sclerophyllous leaves but they are gymnosperms and share a very different palaeohistory and biogeography from angiosperms. Thus, the Basin's vegetation can not be defined solely on the basis of evergreen oak shrublands, even if these do indeed represent the most remarkable and most characteristic structures in the circum-Mediterranean region. Many other types of plants and associations in fact share the terrain, as is readily apparent both in autumn and winter, when leaves of the numerous deciduous species present change colours and fall, just as in temperate forests.

What is more, it now appears that only in the inter-glacial period which we are presently enjoying have these sclerophyllous oak forests come to dominate large areas of the Mediterranean. Pollen analyses and palaeoanthracology—the study of fossilized ancient charcoal—indicate that the coastal lands, as well as the lower mountain slopes now covered by evergreen shrublands, were once forested with a mixture of deciduous, sclerophyllous evergreen, and conifer species as recently as 4500–7500 years ago (Reille and Pons 1992). As we shall see, there are only three small areas where remnants of such forests remain: a subtropical form in the larger Canary Islands, and temperate ones in the Pontic Mountains of northern Turkey and the Elburz chain of northern Iran.

A bioclimatic approach

A climate is generally considered to be 'mediterranean' when summer is the driest season, during which there is a prolonged period of drought. Several authors have attempted to use climatic data to delimit the Mediterranean. At one extreme, some scientists have emphasized temperature and the range of mean annual rainfall. But this leads to very narrow lines being drawn around a littoral band characterized by mild winters, excluding high mountain areas. At the other extreme, in the so-called 'isoclimatic' Mediterranean definition given by Emberger (1930b), and mapped by Daget (1977), the only criteria are that summer should be the driest season, and that there should be a period of effective physiological drought during that season. This approach encompasses Mediterranean mountaintops and adjacent steppe regions where there is no appreciable rainfall in summer. Using this approach, about 8 million km^2 (and up to 9.5 million on some counts), are brought together under the term 'Mediterranean area', including the central Asian steppes as far east as the Aral Sea and the Hindus Valley, all the northern half of the Sahara desert, and most of the Arabian peninsula, where mean annual precipitations are often below 100 mm per year, and little or no rainfall occurs in summer.

Climatic criteria alone are certainly not the best tools to define the Mediterranean area. A more realistic approach combines both climatic (temperature and precipitation) and vegetation factors, as advocated by Gaussen (1954). In addition to climatic analysis, typical plant assemblages are identified in this 'bioclimatic approach' by indicating two or more dominant tree or shrub species whose combined presence invariably characterizes one of a series of altitudinal vegetation zones which replace each other according to altitude, latitude, and slope exposition; that is, wetter, north-facing vs. drier, south-facing slopes. This approach makes it possible to identify cases of typical Mediterranean vegetation hundreds or even thousands of kilometres from the Basin itself. Examples of such enclaves may be found in the mid-Saharan mountains, the Elburz Mountains of northern Iran, and highland regions on either side of the southern Red Sea. The bioclimatic approach defines a series of types or zones, ranging from hyper-arid, arid and semi-arid, to sub-humid, humid, and per-humid. In parallel, a series of altitudinal vegetation belts or life zones are used to compare altitudinal belts on slopes of Mediterranean mountains. These life zones will be described in detail in Chapter 4. Although it is tempting to seek direct correspondences between a given bioclimatic zone and a given altitudinal life zone, variation occurs as a result of latitude, exposure, and soil types. Furthermore, there is no satisfactory answer to the question of what is a

'Mediterranean mountain', as compared to a mountain range simply marking a regional boundary. What about ranges, or parts of ranges where the bioclimate on one side is clearly Mediterranean, but not on the other side?

A certain amount of ambiguity and controversy also exists in areas where vegetation has been drastically altered by people, either very early in the Holocene, or much more recently. In such areas, where very little remains of the original Mediterranean habitats, sufficient evidence exists to suggest that 4–5 millennia ago a typical Mediterranean vegetation did exist over large areas, only to be subsequently removed by human activities. However, following many authors, we would rather recognize habitats on the basis of their present-day composition, and not on an inferred hypothetical and controversial 'historical climax'.

Several field guides to the Mediterranean flora and fauna draw a limit at about 1000 m altitude, but this is not justified from a broad historical–ecological perspective. In our view, delimitation of the Mediterranean phytogeographical territory should include not only the 'basal' zone with its evergreen shrubland formations, but also the altitudinal zones above it, as shown in Fig. 1.5. According to this approach, the extreme eastern Pyrenees and most of the Apennine ranges, for example, are included in the Mediterranean area, even if they are not uniformly endowed with a Mediterranean climate. The western Pyrenees and the northernmost part of the Apennines, however, are excluded. We also include the Canary Islands and Madeira, the so-called Macaronesian region, in our delimitation of the Mediterranean realm. This is inevitable in view of the largely mediterranean-type bimodal climate prevailing in these islands, as well as the important insights their extant vegetation provides into ancient plant lineages and even forest types that were formerly widespread around the Mediterranean Basin in Miocene–Pliocene times (see Chapter 2).

Macaronesia includes the seven Canary Islands and the tiny Madeira archipelago. Especially in the wetter Canary Islands (La Palma, Tenerife, and Gran Canaria) a series of 'living fossils' occurs, sole survivors of a more tropical, arboreal Mediterranean flora and vegetation of Tertiary times. These include many tropical families now entirely or mostly absent in the Mediterranean, such as the palm family (Arecaceae), sapote family (Sapotaceae), and tea family (Theaceae), as well as a large number of taxa in the laurel and olive families (Lauraceae and Oleaceae). Many of these trees played important parts in the Tertiary floras of southern Europe, in the Miocene–Pliocene period. A remarkably large collection of endemics are also found in these islands, with at least 15 genera in 12 different families showing 10 or more endemic species (Quézel 1995).

Using bioclimatic criteria, the Mediterranean territory covered in this book extends over $c.\,2\,300\,000$ km^2 in 18 countries. The exclusion of some areas, for example those in Libya, and the total absence of certain countries such as Iraq, are questionable. In both these countries, small but important areas of mediterranean-type woodland exist. If the broad isoclimatic region first recognized by Emberger were adopted, then *all* of Libya and *all* of Iraq would be included in the Mediterranean. By contrast, the narrow approach cited above would lead us to

Table 1.2 Extension of mediterranean-type territories in the 18 countries encircling the Sea and the contribution of each to the total Mediterranean area.

Country	Extension (km² × 10³)	%	Country	Extension (km² × 10³)	%
Spain	400	17.3	Syria	50	2.1
Portugal	70	3.0	Lebanon	10	0.4
France	50	2.1	Israel	10	0.4
Italy	200	9.0	Jordan	10	0.4
Ex-Yugoslavia	40	1.7	Egypt	50	2.1
Albania	20	0.8	Libya	100	4.3
Greece	100	4.3	Tunisia	100	4.3
Turkey	480	20.8	Algeria	300	13.3
Cyprus	9	0.1	Morocco	300	13.3

Source: after Quézel 1976b

reduce to about 500 km² the portion of Libya included but also to include some of the highlands of northern Iraq. In sum, the bioclimatic approach we adopt constitutes a compromise which excludes the steppe areas outside the first ring of Mediterranean mountains. The approximate results are shown in Table 1.2 and in Fig. 1.5.

Adjacent and transitional provinces

The next step in 'setting the scene' is to identify important 'neighbours' of the Mediterranean area (Fig. 1.6).

The Pontic province

The 'Euro-Siberian' region corresponds to the Palaearctic deciduous and coniferous forests—in places interspersed with steppes that cover all of central and northern Eurasia. The ancient flora of this region is often called 'Arcto-Tertiary'. This vast region includes the so-called 'Pontic' province with the Pontic Mountains of northern Turkey that stretch all along the southern edge of the Black Sea. Here, the Euro-Siberian region comes into contact with the Mediterranean, and the vegetation types of the two intermingle in a variety of ways. Most remarkably, there occurs in some areas an assemblage of forest types which may resemble formations that covered cooler parts of southern Europe in the late Tertiary. As Davis (1965) wrote in the introduction to his monumental *Flora of Turkey*, the interrupted chain of what can be called 'Mediterranean enclaves' along the Black Sea coast occurs in a very narrow belt from sea level to 200 or 300 m,

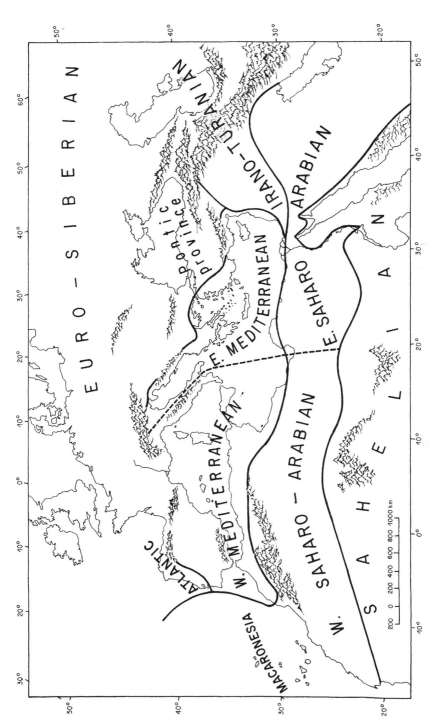

Fig. 1.6 Subdivisions of the Mediterranean area and delineation of the major adjacent biogeographical regions and provinces. (After Quézel 1985 and Zohary 1973.)

particularly on thin soils and southern exposures. Elsewhere in the Euxine province, Mediterranean elements occur as intrusive scrubby invaders where the indigenous forest has been destroyed. Conversely, there are numerous enclaves of Pontic vegetation further south in the Anti-Taurus, the Syrian Amanus range, and even parts of northern Lebanon.

About 500 kilometres to the south-east of the Pontic province, the so-called 'Hyrcanian' province occupies the flanks and summits of the coastal mountains at the southern extremity of the Caspian Sea. This province apparently occupied a considerably larger area during the late Tertiary and Pleistocene (Zohary 1973), and descended much further south towards Kurdistanian Iraq and northern Syria, as shown both by fossil records and extant relicts subsisting outside this territory today. The deciduous species of maple (*Acer*), birch (*Betula*), hazelnut (*Corylus*), beech (*Fagus*), ash (*Fraxinus*), lime (*Tilia*), and elm (*Ulmus*) that occur in the northern Mediterranean regions, as well as dominant parts of the fauna there, all show very clear Arcto-Tertiary origins. A certain number of the evergreen trees commonly found here also occur in wetter parts of southern Europe despite their strong Euro-Siberian affinities. Examples include boxwood (*Buxus sempervirens*), holly (*Ilex aquifolium*), and yew (*Taxus baccata*). Conversely, just as in the Pontic province, in the Hyrcanian there are many enclaves of Mediterranean elements which invade disturbed forest areas and may become locally dominant. The Pontic and Hyrcanian provinces are actually quite similar and are often treated as a single province of the Euro-Siberian region.

The Irano-Turanian province

To the east of the Mediterranean Basin lies the Irano-Turanian region, which includes the high steppes stretching eastward and north-eastward to Mesopotamia and central Asia. Winters here are extremely cold, and summers are extremely hot, but for all intents and purposes no rain falls in summer. For this reason, among others to be developed in later chapters, a great many faunistic and floristic elements of predominantly Irano-Turanian distribution are prominent in low rainfall areas of the Mediterranean Basin, not only in the eastern half, but as far west as Iberia and North Africa.

Examples of widespread plants in this region are the wormwood (*Artemisia* spp.), *Ephedra*, saltbushes (e.g. *Atriplex*, *Suaeda*, and *Salsola*), tamarisks (*Tamarix*), and perennial bunch grasses such as *Stipa* and *Piptatherum*. Furthermore, there are a very large number of vicariant* pairs or series involving taxa occurring in the Mediterranean and the Irano-Turanian region, and sometimes the Saharo-Arabian region as well.

The Saharo-Arabian province

To the south of the Mediterranean Basin lies the immense zone called the Saharo-Arabian biogeographical region. It extends about 4800 km, from Mauritania to the Red Sea, with a width of about 2000 km. With an area of 9.6 million km^2, the

Sahara, the world's largest desert, is more than twice the size of the entire Mediterranean Basin. Desert conditions in the Sahara appeared as early as the lower Pliocene. Today, the northern part of the Sahara is characterized by a mediterranean-type climate with rainfall occurring between autumn and spring, while its southern part has a tropical climate with rainfall occurring in summer when temperatures are the highest. Beyond the Red Sea this gigantic region extends much further eastward, across the Arabian peninsula and on to parts of southern India.

Climate

The Mediterranean climate is a transitional regime between cold temperate and dry tropical climates. The bimodal Mediterranean climate we know today only began to appear during the late Pliocene, about 3.2 Myr BP, as part of a global cooling trend (Suc 1984) and became firmly established throughout the region about 2.8 Myr BP. Since then, the contrast between the alternating hot-dry, and cold-wet seasons has continued to steadily intensify. However, this most recent period has itself been punctuated by alternating glacial and interglacial periods, with the former far outlasting the latter.

Given this unique combination of hot, dry summers and cool (or cold) humid winters, little or no surface water is available during the months when the sun is at its strongest. Accordingly, the short spring and autumn seasons are critical periods for plant growth. They can also be hazardous periods because unseasonable cold spells can strike hard and unexpectedly. Apart in the mountains, snow rarely falls in the Mediterranean, but periods of hard frost are not infrequent, and can cause much damage and mortality if they strike immediately after a period of balmy weather.

Regional and local variation

Despite its invariable bimodal pattern, the Mediterranean climate shows much wider variation in temperature and rainfall regimes than those encountered in areas further north or south. Mean annual rainfall ranges from less than 100 mm at the edge of the Sahara and Syrian deserts to more than 4 m on certain coastal massifs of southern Europe.

For at least two months each year in the western Mediterranean, and five to six months in the east, when there is no precipitation at all, most plants and animals experience a water deficit to which they must respond with ecophysiological or behavioural adaptations (see Chapter 7). For living organisms, the summer heat and drought are often more significant than the cold winter so that the long aestivation* periods may be more hazardous than hibernation for survival.

Although mean annual temperature by itself is not of great biological significance, its geographical variation gives a good idea of the range of climatic conditions in the Mediterranean Basin. Mean annual temperatures range from

2–3 °C in certain mountain ranges, such as the Atlas and the Taurus, to well over 20 °C at certain localities along the North African coast. At a local scale, the Mediterranean is well known for pronounced climatic differences over very short distances. Such variability is under the influence of factors that include slope, exposition, distance from the sea, steepness, and parent rock type. As the innovative colony of English and French gardeners on the Côte d'Azur noticed at the end of the last century, this area is typically marked by a 'thousand and one microclimates' crowded cheek by jowl.

Perhaps the most useful system for characterizing mediterranean bioclimates was proposed by Gaussen (1954). In this scheme climatic patterns are represented with curves expressing annual changes in temperature (T, in °C) and precipitation (P, in mm) on the same graph. The 'ombrothermic' (precipitation and temperature) graph is drawn with $P \times 2$ on one ordinate and T on the other. The rationale for multiplying P by a factor of 2 is that plants are assumed to suffer from drought when the rainfall curve drops below the temperature curve. The classification retained by UNESCO (1963), based on this system, yields a sum of monthly indices of dry months, where a 'dry' month is defined as one having precipitation less than or equal to twice the average temperature for the same month. Thus, by definition, in a dry month $P < 2T$. Such indices make it possible to construct the highly useful ombrothermic diagrams, a few examples of which are shown in Fig. 1.7. Not everyone accepts this system of classification because it is rather empirical. Some authors argue that the best method for defining the limits of the dry period in a mediterranean climate region is to use the ratio P/ETP (P = precipitation, ETP = potential evapotranspiration*). A ratio of 0.35 is defined as the threshold* behind which dryness does not allow the growth of plants. However, this ratio gives very similar results to those of Gaussen's empirical system.

Wind

The Mediterranean is a windy area. On a broad scale, wind regimes on the northern side of the Basin are mainly northerly, caused by seasonal differences in temperature between land masses and the sea. As a rule, northerly winds predominate in summer when the overheated African continent creates a southward in-draught whereas an opposite trend occurs in winter. But there are many local winds and variants. For example, in southern France the dominant winds are the *mistral*, descending from the cold Massif Central and whistling down the Rhône River valley, freezing the Côte d'Azur and Provence and then petering out at sea between Corsica and the Balearic Islands. Another important wind is the dry *tramontane*, which comes in from the north-west, and can be just as fierce as the mistral. Both these continental winds can provoke sudden springtime weather anomalies, including diurnal temperature swings of 10 °C or more. Somewhat less common, but locally important throughout the north-western quadrant of the Mediterranean, are humid winds from the east (*levant* and *grec*), from the south-west (*ponant* or *poniente*) and from the south

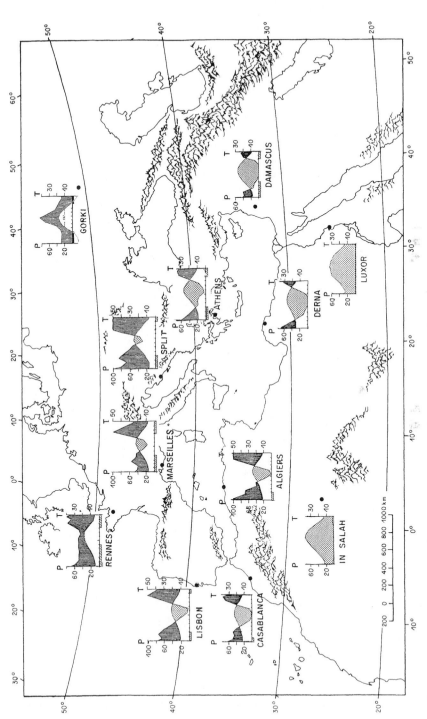

Fig. 1.7 The range of local climate patterns occurring around the Mediterranean Basin, and in a few adjacent regions, as illustrated by a series of ombrothermic diagrams drawn according to Gaussen (1954). A Mediterranean climate occurs wherever the rainfall curve crosses the temperature curve. (Data from Walter and Lieth 1960.)

(*marin* and *sirocco*). There are also occasional spells of the hot dry Saharan winds (*harmattan, foehn*), especially in summer, that at times can deposit red sand as far inland as the highest crests of the Alps. Wind strongly increases evaporation, hence aggravating the effects of drought and high temperatures. In areas with strong dominant winds, direct effects on vegetation include trees with a wind-oriented flag-like structure.

Unpredictability

To all these prominent features of the Mediterranean climate must be added the less obvious but very characteristic one of unpredictability, which emerges in analyses of long-term weather patterns. It now appears that a wide range of climatic variations, from long-term cycles of several centuries to extreme sudden unpredictable events, undoubtedly plays an important role in shaping life cycles in Mediterranean species. From one year to the next, between seasons of a given year, and even within the course of a single day, temperature extremes, pre-cipitation, winds, and other climatic factors can vary dramatically. The wide range of diurnal temperature fluctuations at certain seasons, the violence of certain winds, and short-lived but powerful deluges, make the Mediterranean climate notoriously capricious and unpredictable. The violent climatic events are life-threatening for all living things. During such violent storms, what are normally insignificant streams can be transformed in a matter of hours into devastating torrents, capable of carrying away houses and their terrified inhabitants. In 1876, 128 people were drowned at Saint Chinian, in southern France, by a stream so small that even many locals were unfamiliar with its name (the Verzanobres). After an exceptional downpour in October 1969, the Rio de Las Yeguas broke its banks and flooded the small village of La Roda de Andalusia with mud one metre deep. In the eastern Pyrenees, no less than 6 or 7 rainfalls, each dropping 200–400 mm within a few days, have occurred in the last 15 years.

Animals may also suffer in these storms. In June 1997, a violent one-day storm (*c.* 300 mm rainfall) was so stressful for birds breeding on the island of Corsica that up to 21 of 57 nest-boxes occupied by blue tits (*Parus caeruleus*) were deserted by the adults and the young left to die. In coastal lagoons where sterns and waders breed in large colonies, hundreds of nests may be flooded by sudden rise of water levels.

Unusually prolonged droughts also happen. The summers of 1976 and 1989 were exceptionally dry throughout the region and a large number of catastrophic fires broke out. At long intervals, severe cold spells may strike too, for example the memorable winter of 1709, when the entire harbour of Marseilles and the Rhône River froze over. Nevertheless the Mediterranean is above all a land of abundant sunshine, with nowhere less than 2300 hours per year, and more than 3000 hours in the eastern and southern parts of the Basin. This sunshine, combined with nearly 250 rain-free days per year, allows the cultivation of that famous trio of Mediterranean agriculture: spring wheat, olives, and grapes.

Climatic extremes and erosion

Sudden rises in river levels may be all the more devastating when enormous quantities of land material are carried downstream, aggravating the effects of flooding. Erosion is a major problem in many Mediterranean territories because of steep slopes and deep and narrow valleys. Moreover, most of the mountain ranges are built not of solid resistant rock, but rather are accumulations of soft, crumbly materials very sensitive to erosion. The violence of the rivers and streams during rare flash floods is all too often exacerbated by the greatly increased surface runoff resulting from uncontrolled deforestation of entire watersheds, particularly on the steepest hillsides. More recently, the sharp increase in the use of asphalt and macadam, as well as soil compaction resulting from the use of heavy machinery and the excessive use of pesticides, tends to destroy or reduce the biological activity of soils and thereby to increase the risks of flooding after exceptionally heavy rains.

Vaudour (1979) estimated that soils are being eroded at a rate of one metre per millennium in the region of Madrid. The rate of soil ablation due to erosion can be as high as 1.4 mm per year (1.4 m per millennium) in many catchments of Medierranean Europe (Dufaure 1984), and much higher rates may occur in North Africa. Sari (1977) cites several examples in the Ouarsenis region of Algeria where catchments lose 1000 to 2000 tons/km^2/year! Especially in the eastern and southern parts of the Basin, most of the region's landscapes are so highly dissected, complex and unstable, with steep slopes and shallow rocky soils, that the former protective covering of vegetation has been destroyed. Once exposed, the shallow underlying mantle of soil quickly falls prey to erosion. In the more degraded regions of the eastern half of the Basin, and most of North Africa as well, there is a mosaic of heavily degraded, eroded sites, with only a very few well-maintained sites remaining. Gully erosion and landslides of mountain slopes result in 'badlands', of no use to humans and also very poor in biological diversity. In the second half of the nineteenth century, France, in an attempt to reverse the devastation caused by flooding, began the immense task of replanting mountains (RTM, 'Restauration des Terrains en Montagne'). Some of the most famous examples are those of Mont Aigoual and Mont Ventoux (see Chapter 10). Similar efforts have been undertaken in this century in many other parts of the Mediterranean.

Climatic vicissitudes during the Pleistocene

Before closing this chapter and moving on to an analysis of the origins and development of the present diversity of life in the Mediterranean, we must summarize the salient climatic sequences that have marked the past 2.5 million years. These climatic variations have had a decisive influence in shaping the composition and distribution of Mediterranean species and communities.

Although the mediterranean-type climate was established some 3.2 Myr ago, a large series of fluctuations occurred during the Pleistocene. Twenty years ago it was recognized that changes in the Earth's orbit were the fundamental causes of the Quaternary climatic oscillations (Hays *et al.* 1976). The first massive ice sheets of the Northern Hemisphere started to grow about 2.5 Myr BP, and major climatic oscillations occurred during most of the Pleistocene, with a dominant 100 000-year cycle (Webb and Bartlein 1992). For these ice-dominated regions, the Quaternary should probably be viewed as 'a cold epoch interrupted periodically by catastrophic warm events—the brief interglacials with climate similar to that of today' (Davis 1976). Therefore, the relative climatic stability recorded during the last 8000 years seems to be the exception rather than the rule. Present-day biogeographical patterns are those of an interglacial period of the kind that has been rather rare in Europe's and the Near East's history.

North–south shifts

During the entire Pleistocene, continuous remodelling took place in the structure, functioning, and composition of both local and regional floras and faunas as a result of these shifts and upheavals. Any attempt to interpret biogeographical patterns requires an investigation of past dynamics of the biota at appropriate scales of space and time. At time-scales of 10^3–10^5 years, the distribution and composition of living systems have changed continuously with each glacial–interglacial alternation, throughout the Pleistocene, such that present biota share only a short history—a few thousand years at most (McGlone 1996). Much insight has been provided in recent years by palaeobotanical studies that reveal the transitory nature of environments, and the profound changes wrought over geological time in ecological communities.

Pollen grains are especially useful in establishing a site or region's particular history. Pollen cells possess a hard epidermis that is finely and elaborately decorated, and each plant species (or at least genus) is characterized by a unique pollen design clearly visible under the microscope. These hard cells preserve well as fossils and pollen layers accumulate through time wherever soil conditions are favourable, especially in moist and soft soils such as bogs, riverbeds, and marshes. Using data provided by fossil pollen combined with radiocarbon dating, palaeobotanists can draw 'isopoll maps' (maps with lines of equal relative pollen abundance for a taxon) showing the dynamic distribution of various species over time. These maps indicate the long-term turn-over of plant communities in a given region over time.

For example, the related Holarctic genera hazelnut (*Corylus*) and oak (*Quercus*) have progressively migrated to the north from their southern refugia* in the course of the last 12 000 years (Fig. 1.8). A similar shift in distribution occurred repeatedly during the Pleistocene, each time that a glaciation event forced biota to 'escape' to the south.

During the most severe phases of the glacial periods, almost no arboreal vegetation survived north of the mountain chains that delimit the Mediterranean

Corylus

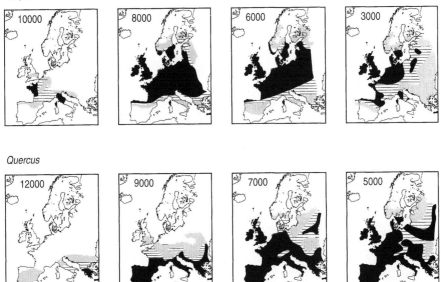

Quercus

Fig. 1.8 Isopoll maps illustrating the northward expansion of oaks (*Quercus* spp.) and hazelnut (*Corylus* sp.) over the last 12 000 years and 10 000 years respectively. Three classes of pollen density are shown: stipple 2–5%, horizontal stripe 5–10%, black fill > 10%. (Reproduced with permission, after Huntley 1988.)

region in southern Europe, that is, the Pyrenees, Alps, and the Carpathian Mountains. During these periods many Mediterranean landscapes were characterized by extensive steppe formations with the same shrub species, for example wormwood (*Artemisia herba-alba* complex), as occur in Near Eastern, Turkish, and North African steppe areas today.

Nevertheless, all the forest floras of the western Palaearctic, and their associated faunas, had to survive within the geographical limits of the Mediterranean Basin during glacial episodes since the Sahara desert and, for some groups the sea, prevented them from going further south. They concentrated around the Mediterranean Sea, especially in the larger peninsulas. During the coldest phases average sea water temperatures must have been about 8 °C lower than prevailing temperatures today. The snow limit was at 1200 m in Galicia, north-western Spain, and Liguria, northern Italy, and at 2000 m in the Atlas Mountains of Morocco.

Analyses of fossil pollen deposits have shown that even during the most extreme glacial periods, many refugia existed for plants and animals on certain mountain slopes, in the large peninsulas (Iberian, Italian, Aegean, Anatolian), on islands, and in the valleys of large rivers. The Mediterranean area was much larger, and even more ecologically complex than it is today, as a result of the

PRESENT

WÜRM

| | Tundra | | Conifers | | Deciduous | | Steppe | | Mediterranean |

Fig. 1.9 Schematic representation of the main vegetation belts in Europe and North Africa at the present time ;and during the most severe phase of the Würm glaciation (30 000 yr BP). (After Brown and Gibson 1983, reproduced with permission from Blondel 1995.)

worldwide sea level being some 100 to 150 m lower. Many opportunities thus arose for species to differentiate as repeated fragmentation of initial areas of distribution took place over time. During glacial episodes, faunas must have been a mixture of many different communities, without any clear geographic delimitation between Mediterranean and non-Mediterranean communities. For example, fossil faunas in deposits of southern France dating back to the Würm glaciation include disparate species assemblages with both 'nordic' species such as the reindeer (*Rangifer tarandus*) and the snowy owl (*Nyctea scandiaca*), and Mediterranean thermophilous species such as the Hermann's tortoise (*Testudo hermanni*). This indicates that climatic conditions allowed for the persistence of both boreal and Mediterranean species. Taxa of primarily Euro-Siberian distribution, such as the whortleberries (*Vaccinium* spp.) and several species of willow (*Salix* spp.), the tundra-dwelling snowy owl, or the reindeer, all occurred widely in the Mediterranean during glacial periods. On time scales of tens of millennia, it is perfectly acceptable to use the term migration, *sensu stricto*, in relation to these species in just the same way that we refer to the seasonal migration of animals. As birds track annual climate change, so migrating trees and their associated plants and animals track long-term climate changes.

Many of the huge colonies of seabirds that breed in northern Europe as far north as Spitzberg had to find refuge in the south during glacial times. Many cliffs and islands of the Mediterranean Sea were populated by the large colonies of seabirds that are so characteristic of steep cliffs in the northern part of the Atlantic Ocean today. A number of fossil sites have provided evidence that gulls (*Larus* spp.), auks (*Uria* spp.), and gannets (*Sula bassana*) used to breed along the coasts of Iberia, France, and Italy during glacial episodes of the Pleistocene. The flightless and now extinct great auk (*Pinguinus impennis*) was widespread in the Mediterranean during most of the Quaternary. Fossil records of this species exist from as far east as Calabria, Italy (see Box 2.1 in Chapter 2). The last pair of Great Auks was hunted out in Iceland in 1844.

During interglacial periods, the plants and animals spread north again without leaving the Mediterranean region. Thus, the survival of species of boreal origin in these refugia during the glacial periods added many species of northern origin to our contemporary Mediterranean biota. Species with very large modern distributional ranges, such as the chaffinch (*Fringilla coelebs*) or the chiffchaff (*Phylloscopus collybita*), expanded north during interglacials, just as deciduous oaks (*Quercus* spp.) and their seed-dispersing companion bird, the jay (*Garrulus glandarius*), did in southern Europe during both glacial and interglacial periods, reaching most of Europe during interglacials. They tended to alternate between small isolated populations during unfavourable glacial periods, and large, widespread populations during the kinder interglacials. Many species differentiated to some extent when they were split among several Mediterranean refugia, as we will show in the next chapter.

The balance between forest and shrubland

During interglacial periods, such as the one we are presently enjoying, large parts of the Mediterranean were covered with forests, the exceptions being high mountains and certain high plateaux in Iberia and Anatolia. However, since the end of the last glaciation period, some 12 000 years ago, deciduous forests have spread progressively from the Mediterranean area north at rates of about 10^2–10^3 m per year (Huntley and Birks 1983). Thus, the Quaternary history of the most important European forest belts and their associated faunas has been one of a series of massive migrations back and forth across the western Palaearctic. Migrations from the east (especially the Irano-Turanian region) and the south concurrently affected the Near Eastern and African portions of the Mediterranean area.

However, although the Basin was extensively forested until quite recently, it appears that shrublands and heathlands did occur spontaneously in some places, long before modern humans brought their fires, goats, and sheep. Fossil pollen analyses demonstrate that low-growing shrub formations occurred throughout the last two million years, most probably in quite patchy spatial distribution. Likely regions for the persistence of shrubby formations were coastal sands and inland areas where microclimate and/or limited soil reserves did not allow the development of taller forest. These areas appear to have been dominated by some of the same sclerophyllous shrubs that occur today, such as lentisk (*Pistacia lentiscus*), *Phillyrea*, junipers, and the remarkable rope-like shrub *Thymelaea hirsuta* (Reille *et al.* 1980). We know that these species did not occur simply as an understorey in these localities, because no pollen has been found there of any of the taller-growing trees, such as oaks, beech, or firs, known to be present in the region at that time.

Summary

The birth of the Mediterranean has been a long story characterized by a series of violent events, both geologic and geographic, which occurred on a wide range of space and time scales. This, combined with the geographical location of the Mediterranean at the intersection of two major land masses of the Eastern Hemisphere, resulted in the emergence of a very particular climatic and biological context. All this has been propitious for a complex, dynamic development of floras and faunas. The next chapter will examine more closely the processes and patterns of arrival and settlement of modern-day biota.

2 Determinants of present-day biodiversity

The Mediterranean as a melting pot and crossroads

Extant faunas and floras of the Mediterranean are a complex mixture of elements; some derive from *in situ* evolution, whereas others have colonized the region from adjacent or far-distant regions in various periods in the past. The colonizers constitute the vast majority of present-day species. Although Mediterranean ecosystems can be considered 'young' because of the relatively recent appearance of a mediterranean climate, they are composed of species originating in almost all known biogeographic realms of the world. From fossil remains dating back to the upper Tertiary to the Holocene, we can gain valuable information and insight about the turnover of Mediterranean biota over the past few million years. Four factors make the Mediterranean area exceptionally rich in biological diversity: biogeography, geology, ecology, and history. We will look at each in the following pages.

Biogeography

Located at the crossroads between Europe, Asia, and Africa, the Mediterranean Basin has served as a melting pot and meeting ground for species of varying origins. Many elements in the course of history have colonized the Basin thanks to the high number of geographic and climatic events that periodically occurred in this part of the world. As we saw in Chapter 1, the region's physical environment and climate have changed radically since the Mesozoic, with the result that biological composition of the different regions of the Basin and migration routes of invading species have changed repeatedly. Opportunities for invasion and secondary speciation have been continually renewed. As a result, one can find species originating from such different biogeographical realms as Siberia, South Africa, and even some relics of the Antarctic continent for several components of the soil fauna. The Mediterranean thus might be considered as a huge 'tension zone' (Raven 1964) lying amid the temperate, arid, and tropical biogeographical regions which surround it, a zone where intricate interpenetration and speciation has been particularly favoured and fostered as compared to the more homogeneous regions to the north and south.

One striking feature of the Mediterranean flora and fauna is the relatively low number of elements of Afro-tropical origin (except North African mammals). Something prevented them from colonizing the Basin with the same success as many Euro-Siberian and Irano-Turanian species achieved. Even the large Saharo-

Arabian province 'donated' fewer plant species to the Mediterranean than either the Euro-Siberian or Irano-Turanian regions, whereas the Mediterranean area apparently contributed a number of taxa to this province (Quézel 1978, 1985). Clearly, the massive Afro-Arabian deserts, dating back at least to the Miocene–Pliocene boundary, and the east–west orientation of seas and mountain chains, must have seriously impeded south–north biological exchanges.

One way to see this is to compare the biodiversity trends in temperate forests and woodlands of Europe, including the southern part of the Mediterranean, with those of the twin warm temperate regions of the Northern Hemisphere, eastern North America and eastern Asia. In sharp contrast to Europe, the biota of these two regions were never entirely cut off from the potential source and refuge areas to their south, at least not since the lower Tertiary. Accordingly, both eastern North American and eastern Asian biota have many more tropical-derived families than do the ecologically equivalent European ones. Comparing the diversity of tree species between Europe, eastern North America, and eastern Asia, Latham and Ricklefs (1993) reported three times more tree species in mesic* forests of eastern Asia (729) than in eastern North America (253), and 6 times more than in European forests (124). Huntley (1993) attributed the low taxonomic diversity of tree species in European forests, as compared to North America, to the limited area available to forest taxa during glacial episodes. Presumably, many forest taxa that occur today in the Nearctic and in eastern Eurasia became extinct in Europe during Pleistocene glaciations. If differential extinction rates of plant species occurred on a larger scale in Europe than in the two other regions, one would also expect more differential extinction rates in animals, which would explain for example why European forest bird faunas are poorer in species than their Nearctic counterparts (Mönkkönen 1994). Most of the tropical floras and faunas that had to leave Europe at the end of the Pliocene as a result of climatic deterioration were prevented from returning when the climate improved by the Saharan arid belt and the Mediterranean and Black Seas, which acted as barriers to dispersal. As a result, all the major groups of extant biota in the Mediterranean Basin are much more heavily derived from Holarctic 'stocks' than from the vast tropical lands to the south.

Geological history

The repeated splitting and joining of land masses has had a decisive influence on the evolution of plants and animals in the Mediterranean Basin throughout the Tertiary. The many cases of differentiation, vicariance, and endemism in various groups of plants and animals are a legacy of the long and violent evolution of the Mediterranean since the Mesozoic. Many speciation events in the Basin are apparently much more ancient than was formerly thought, and may be associated with the complex dynamics of splitting and reassociation of micro-plates during the Palaeogene* and lower Neogene,* that is roughly 65 to 30 Myr BP. Repeatedly, during these periods, isolation of pieces of land within the Tethys Sea trapped species, which then evolved in isolation, giving rise to the numerous palaeoendemics we find today in many groups of plants and animals. For example, Oosterbroek and Arntzen (1992) observed a recurrent pattern of differentiation clearly associated with isolation on tectonic microplates during the Tertiary in seven unrelated groups of invertebrates and vertebrates, including butterflies, flies, scorpions, frogs, and newts. However, because of the difficulty in establishing the phylogenetic* history of species and because the so-called 'molecular clock',* upon which depends the rate of evolutionary change, varies so widely among phylogenetic groups, there is still much uncertainty about when extant taxa differentiated. We can only state that in times in the far past and at various places in the Basin, geographical isolation on 'palaeo-islands' provided opportunities for differentiation of plant and animal species. It is in the north-eastern quadrant that the greatest richness in geological 'tapestries' is found, as we will illustrate briefly in Chapter 4.

Habitat diversity

The astonishing diversity of regional ecology is immediately perceived in the 'mosaic effect' so typical of Mediterranean landscapes. This factor obviously plays a critical role in generating and maintaining species diversity. Biologists note that as a consequence of its kaleidoscopic topographical, climatic, and geo-pedological complexity, the Mediterranean is exceptionally rich in regional or local plant and animal endemics at the levels of genera, species, and subspecies. Along with the long narrow peninsulas and isolated mountains, the huge Mediterranean archipelago is an outstanding framework for speciation to occur in populations isolated by imposing geographic and ecological barriers. Almost every island in the Mediterranean has its own set of native species. Ecologically similar to islands, high mountains are conducive to speciation, and Mediterra-nean mountains recurrently show up to 42% endemism among higher plants (Médail and Verlaque 1997).

Human history

The fourth factor underlying Mediterranean biodiversity is historical and anthropological. Variations in human land use patterns and site-specific histories of resource management, which often resulted in overexploitation and resource depletion, have had profound impact on living systems throughout the Basin. Among plants especially, evolutionary consequences of this factor can be seen in the structure and composition of the vegetation, and the life history traits of many species. Both vegetation structure and individual species show a wide array of adaptations to human perturbations that include fire-setting, clear-cutting, heavy browsing and grazing by herds of domestic livestock, and ploughing. The exceptional richness of annual, or even more ephemeral plant species in the Mediterranean flora is also to a large extent the result of long-standing human activities (see Chapter 7).

In the next chapter we will examine the extant biodiversity of some major groups of plants and animals. In this chapter we will discuss when and how species first arrived, became established, and subsequently evolved in the Basin. For this purpose, we will take examples from the two most extensively studied groups of organisms: vascular plants and vertebrates. Many historical factors have helped shape the diversity and distribution patterns in both groups, including long-distance dispersal, colonization, extinction, sorting processes on biota of various origins, as well as local or regional evolution arising from isolation and geographical and topographical heterogeneity.

In the first section devoted to flora, and in the succeeding ones devoted to fauna, we shall employ two important concepts that may need to be defined. These concepts are simple but of great analytical value when considering Mediterranean biodiversity in three or four dimensions, and will recur throughout this book, especially in Chapters 4 to 7. *Turnover* refers to successive or successional* processes that lead species, genera, and families of organisms, as well as the corresponding communities and habitats, to change in space and in time. These changes are related to environmental changes that occur as a result of disturbance events or with a steep rise in altitude or latitude, for example, or else when the time scale considered is broadened to 5000 or 500 000 years.

Sorting processes is the term employed in an evolutionary context to define any processes whereby lineages characterized by given sets of life-history traits are differently represented in regional species assemblages as a result of differences in regional responses to selection pressures. In an ecological context, a readily recognizable example of a sorting process is the differential colonization and extinction rates taking place among species in a given landscape or region, as a function of the varying life-history traits, especially overall competitive ability and reproductive effort. The impact of these forces is obviously direct and strong on ecosystem structure and landscape dynamics. We shall now turn to the vast and complex Mediterranean flora, for which a fair amount of raw and regional data are available, but very few Basin-wide, or even quadrant-wide syntheses have been attempted.

Processes that define the Mediterranean flora

The extant Mediterranean flora is a complex mixture of taxa whose biogeographic origins, respective age, and evolutionary histories vary enormously. Each region has had its own unique turnover sequences and interplay among biogeographical elements, but analyses and comparisons are simplified by the widely accepted designation of five main groups dominating the Basin's flora. These five groups differ in their biogeographical origins, which the student can often guess by comparing suites of life-history traits that often recur within a group. The five groups are Afro-tropical, which includes several different sub-groups, Holarctic, which corresponds to the 'Euro-Siberian Region' in Fig. 1.6, Irano-Turanian, Saharo-Arabian, and indigenous, corresponding to species that have apparently differentiated *in situ* in the Basin, including both ancient ('*palaeoendemic*') and recent taxa ('*neoendemic*').

Tropical components

This first historical group comprises plants that differentiated in the dry tropics of continental Africa and adjacent regions in the era of the Tethys Sea, before continental drift had separated the Americas from Eurasia. Related taxa, especially among hard-leaved evergreen species can be seen in central and southern California and other dry parts of North and South America. Axelrod (1975) has termed such links 'Madrean-Tethyan', and examples include the evergreen oaks and cypress, but also a large number of annuals as well.

However, the Mediterranean Basin as we know it today has been isolated from the dry Afro-tropical regions by the vast Saharo-Arabian desert regions since about the Miocene–Pliocene juncture, 5–6 Myr BP. This contrasts sharply with the ongoing interaction and interplay between flora (and fauna) in the Mediterranean area and that of the northern temperate forests.

Among the Afro-tropical elements of ancient lineage, the so-called '*palaeo-tropical*' relics, are the evergreen *Asparagus*, *Capparis*, *Ceratonia*, *Chamaerops*, *Jasminum*, *Nerium*, *Olea*, and *Phillyrea* (Quézel 1985). Other 'younger' African elements also occur, especially in small disjunct populations in montane or pre-montane areas of North Africa and the Near East. Plant families that are today distributed mostly in the tropics but which have one or a few representatives in Mediterranean forests and shrublands include Aquifoliaceae, Arecaceae (the palm family), Aristolochiaceae, legume trees and shrubs (both Caesalpinoideae and Mimosoideae), Moraceae, Myrtaceae, Salvadoreaceae, and Vitaceae. In certain periods of the Miocene–Pliocene and the Pleistocene, all of these, as well as many 'Afro-tropical' trees, shrubs, and vines, were far more common throughout the Mediterranean. Today they are limited to wet habitats in frost-free regions. The Mediterranean taxa of this group were either present in the area long before the onset of the Mediterranean-type climate in the Pliocene, about 3.2 Myr BP, or else arrived recently, since the last glaciation. If they are 'veterans', they survived the climatic upheavals of the upper Tertiary and the Quaternary, including the successive waves of prolonged Pleistocene glaciations. This scenario is confirmed by fossil records of such 'palaeotropical' species found in southern European floras of the upper Eocene (e.g. *Chamaerops*, *Smilax*), the Oligocene (e.g. *Olea*) and lower Miocene (e.g. *Phillyrea*) (Palamarev 1989).

Apart from these obvious tropical links, there is also evidence of relationships between the Mediterranean flora and that of the semi-arid and arid formations that occur intermittently across Africa, from Mediterranean North Africa all the way to the Cape Province of South Africa (Raven 1973). This ancient 'Rand-flora' component, as it is often called, includes the olive tree (*Olea europaea*) complex and the endemic argan tree (*Argania spinosa*) in south-west Morocco. Migration routes followed by this flora were at times restricted to the mountain ranges of Africa, which acted either as stepping stones, or as refugia for such groups as

Erica, Mesembryanthemum, Salvia, and *Helichrysum.* These and other genera are represented by endemic species on many of the high mountains throughout Africa but are most numerous in the continent's southern end (Quézel 1995). An additional sign of ancient contacts between the Cape Province and the Mediterranean is illustrated by the presence of several members of the Restionaceae and Proteaceae families in the Oligocene–Miocene fossil record of southern Europe that are entirely absent there today, but which are represented by a great many species in the Southern Hemisphere, especially South Africa, southern Australia, and southern South America.

A third subgroup among the 'tropical' components, is called 'Sudanian'. It is represented in a few southern parts of the Basin by 'tongues of penetration' in some parts of North Africa and along the Dead Sea valley shared by Israel and Jordan (Zohary 1973). The area of distribution of the so-called 'Sudanian' belt in Africa stretches from the Atlantic coast in southern Mauritania right across sub-Saharan Africa to the Red Sea coasts, and reappears in certain parts of the Indian subcontinent. Several dozen species of this biogeographical group are found both in the sub-Saharan belt of savannas and also narrow parts of the southern Mediterranean quadrants.

The fact that there is little endemism among the Sudanian elements in the Mediterranean area, and a relatively high frequency of adaptations for long-distance seed dispersal, and/or usefulness of some kind for humans, led Shmida and Aronson (1986) to argue for a recent arrival for many of them, that is, since the last glacial period. It is also notable that in many cases they occupy quite different habitats, for example in the Dead Sea area or the dry riverbeds of North Africa, from those they occupy in the Sudanian savannas south of the Sahara. Yet, to the best of our knowledge, they are the same species, showing a disjunct distribution. Problematic cases include a dozen species of the emblematic acacia trees of the African savannas, which also occur in the southern fringes of the Mediterranean, north of the Sahara, alongside trees like *Zizyphus, Balanites,* and *Salvadora persica,* and a number of vines from strictly tropical families like Menispermaceae and Asclepiadaceae. The date palm (*Phoenix*) may also be considered part of this group.

Holarctic components

This second category includes many species and families of clearly 'northern' extra-tropical origin, which are mainly Holarctic. Some of these were already established in the Mediterranean by the upper Pliocene, before the first glaciations, and have persisted mostly in the colder, wetter life zones of the north-western and especially north-eastern quadrants. In this group belong the Oriental plane tree (*Platanus orientalis*), walnut (*Juglans regia*), hazelnut (*Corylus avellana*), and beech (*Fagus sylvatica*). Many of the plants of boreal or Holarctic origin also participate in the flora of the temperate zone that constitutes most of the vegetation found today in western Eurasia. This group includes deciduous broad-leaved tree genera such as *Acer, Alnus, Betula, Fagus, Quercus,* and *Ulmus,* but also many herbaceous taxa such as *Aquilegia, Doronicum,* and *Gentiana.* The mountain ranges of the northern Mediterranean played a prominent role in the survival of these taxa during glaciations as they served as refugia and allowed many species to persist by providing ecological conditions temporarily absent elsewhere in Eurasia.

Quite a number of these Holarctic elements (e.g. various species of *Epimedium, Rhododendron, Pterocarya,* and *Zelkova*) are most common or even endemic in the north-east quadrant of the Basin, where prevailing climatic conditions are most suitable for them. Indeed, at the northern frontiers of this quadrant both the Pontic and 'Hyrcanian' provinces of the Euro-Siberian region (see Fig. 1.6) harbour a large number of taxa and formations that intermingle with and contribute to contemporary Mediterranean vegetation, especially in areas disturbed by humans. The Pontic is very rich in evergreen genera such as fir (*Abies*), pine (*Pinus*), spruce (*Picea*), and rhododendron. The Hyrcanian, by contrast, lacks these elements but is rich in deciduous trees, among which is the monotypic ironwood (*Parottia persica*) of the witch-hazel family (Hammamelidaceae). This Arcto-Tertiary relict formerly occurred widely in various life zones throughout the eastern Mediterranean. Today, it can be seen as a pioneering colonizer in mid-altitude slopes of the southern Caspian region, along with hornbeam, *Zelkova,* various oaks, and *Rhamnus.* Together these small trees and shrubs invade and occupy large areas of badly degraded beech forests. Here is an example of taxa from a more xeric flora (i.e. the Mediterranean basal zone) replacing a more mesic one after the montane habitat of the beech woods has been destroyed by humans. The alpine flora of the Pontic and Hyrcanian provinces also shows many links with the Oro-Mediterranean life zone vegetation in the various eastern Mediterranean mountains.

Irano-Turanian components

The third group is part of the Arcto-Tertiary Mesogean flora and provided many 'old invaders' to our area. It includes hundreds of Irano-Turanian elements such as *Artemisia, Ephedra, Haloxylon, Pistacia, Salsola,* and *Suaeda* whose centres of diversity and, no doubt, of origin are located in the semi-arid steppes of central Asia where summers are exceptionally hot, and winters exceptionally cold and dry. So-called 'forest-steppes' have occured in this vast region since lower Tertiary times or even before, but they are now badly degraded through long centuries of land mismanagement. As a result, the area is mostly characterized today by vast steppes, artificially deprived of most of their trees and large shrubs, and also of most of their topsoil.

Some deciduous arboreal elements of this Irano-Turanian flora do survive in patches throughout the Middle East however, and provide another insight into the deciduous components of the Mediterranean flora, especially in the eastern parts. Examples include the Judas tree (*Cercis siliquastrum*) and the storax tree (*Styrax officinalis*), both of which do also occur in the western quadrants, but only sparingly. By contrast, in the north-east quadrant both trees are abundant

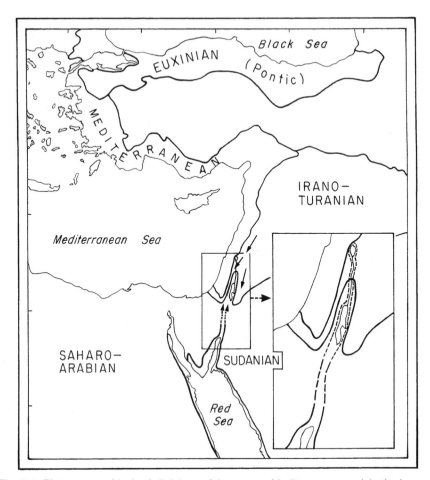

Fig. 2.1 Phytogeographical subdivisions of the eastern Mediterranean and, in the box, of Lebanon, Israel, and north-west Jordan (after Zohary 1973; Shmida and Aronson 1986), showing Irano-Turanian and Sudanian vegetative 'tongues of penetration' into the Mediterranean area.

enough to be used as bioindicators of the thermo-Mediterranean life zone. Prominent among this group of deciduous trees is a large number of species of the apple family (Rosaceae); for the Middle East as a whole this family includes about 60 species in 14 genera (Zohary 1973).

Zohary considered that there must have been waves of interpluvial penetrations by Irano-Turanian elements into the Mediterranean region since the upper Tertiary. In the Holocene, when human powers of perturbation became gradually more important, this pattern is thought to have intensified. This view is based partly on the frequently recognized ecological trend that, following human transformation of ecosystems, plants from more xeric habitats tend to invade more mesic ones rather than vice versa. We shall call this 'Zohary's law', though Zohary himself (1962) wrote in terms of the 'expansion drive of desert plants'.

The example given above of the ironwood and other shrubby Mediterranean invaders in the disturbed forests of the Pontic and Hyrcanian provinces shows that this 'law' applies to Mediterranean as well as to steppic Irano-Turanian elements, and perhaps far better than to strictly 'desert' flora.

'Zohary's law' not only explains why the Irano-Turanian region has contributed vastly more taxa to the Mediterranean region than the reverse, it also probably explains why the abrupt 'tongues of penetration' pattern is observed at their borderlands (see Fig. 2.1). However, a human factor must also be considered in the intrusion of Irano-Turanian elements into the Mediterranean flora: almost all the region's cultivated fruit and nut trees are of Irano-Turanian origin, as we will discuss in Chapter 8.

It is also worth noting that there are a far greater number of deciduous oak species in the eastern Mediterranean than in the western part (see Chapter 6). They all show strong affinities with the cluster of oaks found further east, in the Irano-Turanian region as in the foothills of the Himalayas, where some 20 species of deciduous oaks occur. In this context, it is striking that pines are totally absent from the Irano-Turanian region, while some 16 species and subspecies of *Pinus* occur in the Mediterranean region and play prominent roles in many vegetation types, usually in association with oaks. Thus, the pines may be considered as a valuable bioindicator group of the Mediterranean realm in this inter-regional context.

Saharo-Arabian components

The fourth category encompasses a few Saharo-Arabian lineages that contribute in a significant fashion to local floras and landscape diversity in the frost-free arid regions of the southern and south-eastern margins of the Mediterranean Basin. This xerophytic* desert component of the flora, along with the steppe one, appears to be quite ancient, dating back at least to the Tertiary. However, only very few Saharo-Arabian elements have succeeded in extending their distribution northwards, and establishing themselves as part of contemporary Mediterranean ecosystems. Among those which have succeeded, there is a mixture of species that apparently penetrated as a direct result of both long-term climatic fluctuations during the Pleistocene and fluctuations that occurred in more recent times. As for the Sudanian elements discussed above, it is difficult to discern between the two groups. These elements include members of the saltbush family (Chenopodiaceae), the Zygophyllaceae (*Balanites*, *Peganum*, *Tribulus*, *Zygophyllum*), a few perennial grasses, and others (Quézel 1985; Zohary 1973). Their distribution roughly coincides with the 150 mm isohyet* that marks the limit of the Mediterranean–desert borders.

Indigenous components

This category includes several thousand native elements of the so-called 'Mesogean' flora (see Chapter 1). They differentiated after the beginning of the Oligocene in the coastal regions around the Tethys Sea, especially on the many tectonic microplates which were scattered between the Gondwana land mass to the south and the Laurasian super-continent to the north (Quézel 1985). Within this Mesogean flora, the strawberry tree (*Arbutus*) and the various evergreen sclerophyll oaks are considered to be indigenous Mediterranean taxa, along with *Helianthemum*, *Lavatera*, *Salvia*, *Cupressus*, *Pinus*, and *Juniperus* (Quézel 1985).

Another important portion of this small but crucial component is the anthropogenic elements, that is, that distinct sub-flora of about 1500 segetal* and ruderal* annuals that evolved locally (Zohary 1973), and occur in varying associations in fields, pastures, and on the roadside. Surprisingly, several dozen taxa of this indigenous group actually show very limited distribution, despite their long history as weeds. As many as 200 of them are endemic to the Mediterranean and the Middle East. Among these a large number of endemic genera are represented by one or just a few species, such as the Crucifers *Bunias* and *Calepina*, the Composites *Cardopatium* and *Ridolfia*, the Umbellifers *Bifora*, *Exoacantha*, and *Smyrniopsis*, and many others.

As a result of the Basin's complicated history, one finds both palaeo- and neo-endemics side by side, some examples of which are illustrated in Fig. 2.2. Neo-endemics belong to lineages that appeared in the region by immigration, and subsequently differentiated (examples are *Amelanchier* and the many species of rock-rose as well as a wide variety of Compositae, e.g. *Centaurea* spp.).

From what has been discussed so far, we can summarize the history of the Mediterranean flora as follows. In the first part of the Tertiary, until the Oligocene, the flora was typically tropical, with a mixture of forest and savanna landscapes that were very different in structure and composition from those prevailing today. However, after the beginning of the Oligocene, and especially from the Pliocene, a mediterranean-type climate became established (3.2 Myr BP), bounded by more arid and semi-arid areas to the south and east, and more mesic ones further north. From then on the typical Mediterranean flora progressively developed and differentiated into what we know today.

From the Messinian Crisis to the end of the Pliocene and the beginning of the Pleistocene, a large turnover of floras resulted in the disappearance of most tropical species. This tropical flora was progressively replaced by modern floras that include such disparate historical elements as Saharo-Arabian taxa (e.g. *Retama*, *Lygos*), steppe species (*Artemisia*, various Chenopodiaceae), the many native Mediterranean (*Arbutus*, *Ceratonia*, *Quercus*, *Pistacia*, *Myrtus*, and *Cistus*), and boreal elements (*Alnus*, *Fraxinus*, *Tilia*, *Ulmus*). About 2.3 Myr BP a drying trend led to the development and expansion of steppe associations (e.g. *Artemisia*, *Ephedra*) as well as those of typical Mediterranean xerophytes such as *Phillyrea*, *Olea*, *Cistus*, *Quercus ilex*, and *Pistacia lentiscus* (Suc 1980). Subsequently, the large and long-lasting cyclic climatic oscillations of the Pleistocene had a decisive influence on the shaping of modern plant assemblages, as explained in the previous chapter.

One interesting and biologically important exception in the large-scale extinction crisis that resulted from the Pleistocene glaciations are the floras of

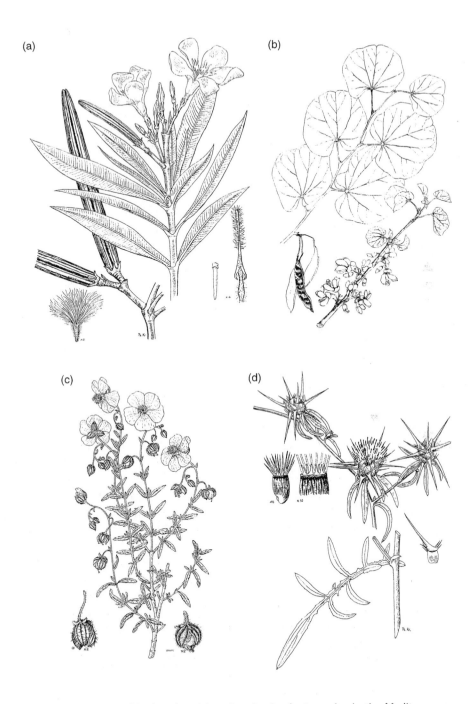

Fig. 2.2 Examples of 'palaeo-' and 'neo-' endemic plant species in the Mediterranean. Palaeo-endemics: (a) *Nerium oleander* and (b) *Cercis siliquastrum*; neo-endemics: (c) *Helianthemum vesicarium* and (d) *Centaurea iberica*. (Reproduced with permission from Zohary and Feinbrun 1966–86.)

the Macaronesian archipelago (Canary Islands and Madeira) and, to a lesser extent, those of some of the larger Mediterranean islands (e.g. Crete and Cyprus). As mentioned in Chapter 1, the Canary Islands acted as a repository of the palaeo-floras and vegetation formations that had prevailed in much of the Mediterranean Basin during the Miocene but have since mostly disappeared. This special flora includes taxa, and notably evergreen trees, in many families of palaeotropical origin (see Chapter 5). Moreover, there are sizeable patches of mixed forests on several islands that give an idea of what many lowland Mediterranean forests probably looked like in the Miocene and Pliocene eras (Suc 1984). They are therefore a unique window on a long-gone past.

The mammal fauna from the Pleistocene to the present

Contemporary mammal faunas of the Mediterranean Basin are a legacy of repeated waves of immigration and extinction that go back as far as the upper Oligocene–lower Miocene. Their history is intimately associated with the appearance and disappearance of land bridges connecting and then disconnect-ing Asia, Europe, and Africa across the Tethys and the Paratethys seas. In particular, the 'Levantine corridor' allowed many faunal exchanges during the Neogene and Quaternary (Tchernov 1992). On account of its geographical location, the Levantine region has been used many times as a land bridge between Eurasia and Africa, especially in relation to the Pliocene–Pleistocene climatic fluctuations. The first important faunal interchange in this region occurred after the closing of the Tethys at the eastern part of the Mediterranean in the lower Miocene, 20 Myr BP, which explains why many rodents in arid and semi-arid habitats of North Africa are modern representatives of ancient Asian lineages. Although the geography of the region has stabilized since the Miocene, there was also an extensive turnover in the faunas of large mammals associated with the Pliocene–Pleistocene climatic changes.

Pleistocene assemblages

The ancient large mammal fauna of the upper Pliocene, 2.8 Myr BP, which persisted into the lower Pleistocene on the northern bank of the Basin, included both tropical and boreal elements. Tropical species included a cheetah (*Acinonyx*), several large felids (*Homotherium*, *Meganthereon*), a panda (*Para-ilurus*), a raccoon-dog (*Nyctereutes*), a tapir (*Tapirus*), and several gazelles and antelopes. But the end of the Pliocene is characterized by the appearance of 'temperate' faunas including a very large bovid (*Leptobos*), the first true horse (*Equus*), the first mammoth (*Mammuthus*), and several large carnivores including the wolf (*Canis lupus*). Climatic degradation of the beginning of the Pleistocene resulted in a progressive decline of the tropical species so that at the end of the lower Pleistocene (1.5 Myr BP), many of them went extinct.

In the middle Pleistocene (1 Myr BP), a major faunal turnover, associated with large climatic cycles with long-lasting cold episodes, was characterized by the disappearance of all tropical species, the extinction of many ancient boreal species, the arrival of 'cold faunas' of boreal origin, and finally the settling-in of modern faunas. The first true bison (*Bison schoetensacki*) replaced *Eobison* and several new cervids appeared: the reindeer (*Rangifer tarandus*), the red deer (*Cervus elaphus*) and the roe deer (*Capreolus*). Besides the reindeer, mammals of 'cold origin' included the woolly rhino (*Coelodonta antiquitatis*) and the musk ox (*Ovibos pallantis*). The primitive boar *Sus strozzi* was replaced by the modern wild boar (*Sus scrofa*), and the mammoth *Mammuthus meridionalis* by a more evolved species (*M. trogontherii*), which co-occurred with the elephant *Palaeoxodon antiquus*. Large carnivores included the Etruscan bear (*Ursus etruscus*), which subsequently evolved into the well known cave bear (*Ursus spelaeus*), and the modern wolf (*Canis lupus*).

Testimonials to this magnificent upper Pleistocene mammal fauna are provided by dozens of fossil sites in southern Europe (Table 2.1) as well as by the superb wall paintings in many ornate caves of southern France and northern Spain. Beautiful examples are those of the Cosquer cave and the Chauvet cave (see Box 2.1). Some survivors of this ancient fauna, which included a surprisingly large number of large predators, disappeared only recently. The lion and elephant survived until historic times in Greece and Syria, respectively, while the porcupine and the Barbary macaque are still present in southern Italy, and southern Spain and North Africa, respectively.

In the eastern Mediterranean, many species of Eurasian origin took advantage of the gradual improvement in climate during the Riss–Würm interglacial period (*c.* 110 000–70 000 yr BP) to expand their areas of distribution (Tchernov 1984), and colonize North Africa through the 'Levantine corridor'. Examples are close relatives of the fallow deer, red deer, roe deer, wild boar, auroch, goat, and ibex. Then the Würm glaciation, the last cold episode of the Pleistocene, forced many mammals into southern Mediterranean refugia. Some, such as the weasel, roe deer, and fallow deer survived in the Levant at least until the first millennium BC, that is, several millennia after the major warming at the end of the last glacial (Dayan 1996). The current presence of the weasel in Egypt could be a relict of its widespread distribution in the eastern Mediterranean during the Holocene, perhaps because it became commensal with the dense human population in the Nile delta. The roe and fallow deer both disappeared early this century, probably as a result of hunting (Yom-Tov and Mendelssohn 1988). The progressive drying out of the eastern Mediterranean during the Holocene was the main factor in the extinction of many Afro-tropical and Palaearctic faunal elements. This resulted in an increased faunal isolation between tropical Africa and Eurasia.

In North Africa, too, a large turnover of the mammalian fauna occurred during the Pleistocene. After a short period of direct contact with Europe during the Messinian Crisis, which did not result in massive faunal interchange between the African and Eurasian land masses, the mammal fauna of the Maghreb evolved in relative isolation. This fauna was clearly African in character during the lower

Table 2.1 List of fossil remains of large mammals found in various middle and upper Pleistocene deposits of southern France (+: present; O: absent). These deposits correspond to both glacial and inter-glacial periods.

Families and species	Middle	Upper	Families and species	Middle	Upper
Canidae			Equidae		
Canis lupus	+	+	Equus sp.	+	+
Vulpes vulpes	+	+	Equus germanicus	+	0
Alopex lagopus	0	+	Suidae		
Cuon alpinus	0	+	Sus scrofa	+	+
Ursidae			Cervidae		
Ursus thibetanus	+	0	Rangifer tarandus	+	+
Ursus arctos	+	+	Megaceros giganteus	0	+
Ursus spelaeus	+	+	Cervus elaphus	+	+
Hyaenidae			Capreolus capreolus	+	+
Crocuta spelaea	0	+	Bovidae		
Felidae			Bos primigenius	+	+
Panthera (Leo) spelaea	+	+	Bison priscus	+	0
Panthera pardus	+	+	Hemitragus sp.	+	0
Lynx spelaea	+	+	Hemitragus cedrensis	+	0
Felis silvestris	+	+	Capra sp.	+	+
Mustelidae			Capra ibex	+	+
Gulo spelaeus	0	+	Rupicapra rupicapra	0	+
Meles meles	+	0	Rupicapra sp.	0	+
Proboscidae			Sciuridae		
Palaeoloxodon antiquus	+	0	Marmotta marmotta	+	+
Mammuthus primigenius	+	0	Castor fiber	+	+
Rhinocerotidae					
Coelodonta antiquitatis	+	+			
Dicerorhinus hemitoechus	+	+			

After Defleur *et al.* 1994

and middle Pleistocene, with several species of antelope, an elephant, and many species of rodents (e.g. *Ellobius*, *Meriones*, *Arvicanthis*, *Gerbillus*). Savanna-like mammal groupings of African character, including goats, antelopes, elephants, white rhinos, hares, jerboas, and jackals, were enriched during glacial periods of the upper Pleistocene by Eurasian species that colonized North Africa using the 'eastern route', the narrow belt of Mediterranean habitats that stretched along the seashore prior to the northward extension of the Sahara desert in what is now Libya and Egypt. These Palaearctic species included the brown bear (*Ursus arctos*), aurochs (*Bos primigenius*), and deers (*Megaloceros algericus*, *Cervus elaphus*). The number of large-hoofed mammals and carnivores increased from 17–20 species to 29, between the late Riss (110 000 yr BP) and the late Würm

Box 2.1 The wonderful bestiary of ornate caves

The Chauvet cave was discovered in 1994 in the Ardèche region of southern France and the Cosquer in 1996 along the coast near Marseilles. The Chauvet cave contains some of the oldest (about 30 000 years old) and most beautiful cave paintings ever found in the Mediterranean Basin (Chauvet et al. 1995). A vast bestiary is portrayed, including three hundred or more different animals such as bison, horse, bear, deer, mammoth, hyena, panther, lion, rhino, reindeer, giant deer, auroch, ibex, and the only known representation in palaeolithic art of large birds such as the eagle owl. In the underwater Cosquer cave there are paintings of many mammals such as horses, bison, and ibexes, together with extraordinarily detailed renderings of the great auk, dating back 20 000 years BP. To get an idea of how much has changed by the shores of the Mediterranean since palaeolithic times note that the entrance of the Cosquer cave is now 36 m below sea level!

(14 000 yr BP) ice ages. As on the northern shores of the Mediterranean, the middle-Pleistocene was characterized by an impressive number of large carnivores: dogs (two species), lycaon, fox, brown bear, genet, hyenas (two species), cats (two species), lynx (two species), lion, and panther (two species).

Human-induced extinctions and introductions

As in many other parts of the world, the former extraordinary rich 'megafauna' of the Mediterranean Basin, including many of the large carnivores mentioned above, was drastically reduced in Pleistocene–Holocene times through the combined effects of a changing climate and the various types of pressure exerted by prehistoric humans. The 'overkill hypothesis' suggested by Martin (1984), whereby prehistoric humans are held largely responsible for the mass extinction crisis of large mammals in the upper Pleistocene worldwide, is probably also true in the Mediterranean, including in islands (see Chapter 9).

Throughout the Mediterranean Basin, uninterrupted forest clearing, burning, hunting, persecution, and finally, both deliberate and accidental introduction of exotic species, have all combined to alter the pre-existing faunas. In the Near East, as in southern Europe, relentless human pressure from the beginning of the Holocene (c. 10 000 yr BP) has led to the extinction of the majority of large species, and especially ungulates (Tchernov 1984). The last European wild horses (*Equus caballus*) were slaughtered in the middle of the nineteenth century in the Ukraine. The aurochs occurred in the vast forests of eastern Europe until the beginning of the seventeenth century, but the last survivor died in 1627 in Poland.

The Mediterranean was unique in harbouring populations of both species of modern elephants, the Asian (*Elephas maximus*) and the African (*Loxodonta africana*). The last herd of Asian elephants was killed by an Assyrian king about 2800 years ago in what is now Syria, and the African species, which had been semi-domesticated by the Romans and occurred in southern Morocco until the eleventh century, disappeared much more recently.

Many species of the rich mammal fauna of North Africa survived in the Holocene well into the Neolithic period but progressively became extinct, mostly as a result of human influence (Kowalski and Rzebik-Kowalska 1991). For example, from the 6 species of gazelle that occurred during the middle or upper Pleistocene in North Africa, only 2 survive today. In Algeria, the long litany of species of large terrestrial mammals that have disappeared since antiquity include the serval (*Felis serval*), the lion (*Panthera leo*), the panther (*Panthera pardus*), the elephant (*Loxodonta africana*), the African horse (*Equus africanus*), the African ass (*Equus asinus*), doomed to extinction in 1925, the onager (*E. onager*), the bubal antelope (*Alcephalus busephalus*), and the red gazelle (*Gazella rufinas*). Only 5 large-hoofed mammals still survive today: the wild boar, 2 species of gazelle (*Gazella dorcas* and *G. cuvieri*), the mouflon (*Ammotragus lervia*), and the red deer (*Cervus elaphus*). Among large carnivores only some individuals of the serval (*Felis serval*), and the panther (*Panthera pardus*) still remain in remote forested areas of Algeria and Morocco, respectively (Cuzin 1996).

Concurrently with this wave of extinction, the Basin experienced many post-Pleistocene invasions by mammals. Some species, such as the marbled polecat (*Vormela peregusna*), the jackal (*Canis aureus*), and Guenther's vole (*Microtus guentheri*), invaded Mediterranean Europe from the Middle East whereas others came from North Africa, for example the mongoose (*Herpestes ichneumon*) and the genet (*Genetta genetta*). It is possible, however, that these last two were introduced into Europe by the Arabs after the collapse of the Roman Empire. In the Near East, the post-glacial immigration from south-western Asia of several species of rodents and the desert hedgehog (*Hemiechus auritus*) accompanied a deterioration of Mediterranean habitats (Tchernov 1984).

Finally, humans have intentionally introduced a number of species, for example the rabbit (*Oryctolagus cuniculus*) and red deer from Europe to North Africa and the Middle East, while several species benefited from human migrations to expand in the Mediterranean Basin as 'camp-followers'. This is the case for three commensal species: the house mouse (*Mus musculus*), the black rat (*Rattus rattus*) and, much later, in the Middle Ages, the Norway rat (*Rattus norvegicus*). The first two invaded the Near East from Asia, making use of human settlements as stepping stones. We will come back to this well-documented case study in Chapter 9. Finally, several species, for example the coypu (*Myocastor coypus*), the muskrat (*Ondatra zibethicus*), and the cottontail (*Sylvilagus floridanus*) were intentionally introduced from the Americas for their fur or for hunting. In most cases, they do not seem to have become pests, except perhaps the muskrat, which in some places causes damage to the dikes of canals and to fishing or hunting ponds.

A history of the bird fauna, or why are there so few indigenous species?

Compared to the more homogeneous bird fauna of northern and central Europe, the Mediterranean avifauna is a surprising collection of cold boreal, semi-arid steppe, and indigenous Mediterranean elements, resulting in a fascinating kaleidoscope of originally disparate faunas. Of the 366 breeding species of birds found in the Mediterranean, only 64 can be considered as being indigenous, that is, having evolved within the limits of the region (Covas and Blondel 1998). Most of these are species occurring in shrublands rather than in forests.

Bird communities in mature forests: a biogeographical paradox

Paradoxically, the taller and more architecturally complex the plants in a given Mediterranean habitat, the fewer bird species of Mediterranean origin are found there (Blondel and Farré 1988). In a long-established forest dominated by the typically Mediterranean holm-oak, there is not a single bird species that is not also found in the forests of central and northern Europe. In fact, most European forest-dwelling bird species occur more or less uniformly across the continent, including the islands of the Mediterranean and the forested parts of North Africa. Such a high degree of uniformity in the avian species of mature forests of the western Palaearctic has led to the prediction that bird communities all over this vast region will resemble each other more and more closely as they move up vegetation gradients towards maximum vegetation size and complexity. Evidence supporting this hypothesis has recently been obtained (Blondel 1995) by comparing avian species along five different vegetation gradients in the western Palaearctic: three in the Mediterranean region (Provence, Corsica, and Algeria), and two further north, in temperate climate areas of Europe (Burgundy and Poland). In each of the five regions compared, the species present tend to be native in areas where the vegetation is relatively undeveloped, with many species of definitely Mediterranean origin in Provence, Corsica, and Algeria. However, as areas of greater and greater vegetative complexity are examined, so does the degree of convergence in the composition of the bird communities increase (Fig. 2.3). All include very similar bird assemblages composed of forest-dwelling species of boreal origin. A similar pattern apparently occurs in other groups of animals and plants but few detailed studies have been carried out to verify this.

Bird communities and Pleistocene glaciations

Given the remarkable diversity of habitats and the many barriers to immigration and gene exchange among populations that should have given rise to new species, one is led to wonder why there are not more endemic bird species in the mature forests of the Basin. Endemic forest-dwelling species in the Mediterranean include the sombre tit (*Parus lugubris*), the Syrian woodpecker (*Dendrocopos syriacus*), three nuthatches (Corsican nuthatch (*Sitta whiteheadi*), Algerian nuthatch (*S. ledanti*), and Kruper's nuthatch (*S. krueperi*), and two pigeons

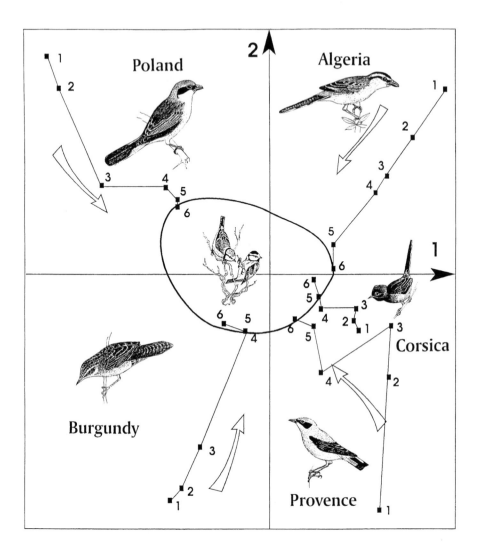

Fig. 2.3 Similarities of bird communities in mature forests as compared with those in habitats with less developed vegetation in five regions of the western Palaearctic. Numbers (1–6) on the black lines correspond to gradients of increasing complexity of vegetation, with 6 being a full-grown forest. Two of these regions are non-Mediterranean: Poland and Burgundy, and three are located in the Mediterranean area: Algeria, Corsica, and Provence. Location of the different values in the figure has been determined by multivariate statistics (correspondence analysis). (Reproduced with permission from Blondel 1995.)

(*Columba bollii* and *C. junionae*). The very low number of forest birds of obvious Mediterranean origin can be explained by the Pliocene–Pleistocene history of vegetation belts and their associated biota, briefly summarized in Chapter 1. Three main points stand out:

1. Contrary to what has been long believed, Mediterranean biota did not survive during recent glacial periods in what is today the Sahara desert. Instead, most forest biota of Europe survived in refugia within the Mediterranean area itself during these inclement periods. Forest communities of birds in the Mediterranean Basin have therefore never been geographically isolated from those at higher latitudes, which would have been a prerequisite for local differentiation according to the widely accepted model of allopatric* speciation.

2. During the relatively brief inter-glacial periods of the Pleistocene, the lowlands and the mid-altitude slopes of the Mediterranean Basin were covered with deciduous oak forests (e.g. downy oak (*Quercus humilis*) in the west and tabor oak (*Q. ithaburensis*) in Turkey, Syria, and Lebanon), while evergreen species such as the holm-oak were restricted to a limited number of relatively dry habitat types (Pons 1981). The bird faunas of these regions no doubt had a central European character quite different from what one expects to encounter now in Mediterranean shrublands. Therefore, it is not surprising that the many floristic and faunistic species of boreal origin that found refugia in the Mediterranean during glacial episodes did not leave the area when the climate improved, and thus are today part of the Mediterranean forest biota.

3. The study of fossil pollen has shown that more or less isolated shrubland formations of various size occurred throughout the whole Pleistocene in many parts of the Basin where climatic and/or soil conditions did not allow forest to develop. Such longstanding shrubland formations have been localized, for instance, in the Baetic Cordillera, Spain, and in Corsica, as well as in many parts of the Near East. Because these areas of shrubland were much smaller than is the case today, and occurred in a wide range of climatic conditions, active differentiation processes were likely to occur among avian taxa adapted to this type of habitat. The best example of speciation among shrubland birds is probably that of the warblers (genus *Sylvia*) which will be summarized in the next section.

Geographical and climatic events

In this chapter we have been emphasizing the paramount importance of the Mediterranean Basin as a crossroads for immigration and movements back and forth across continents, evidenced by the fact that a major proportion of the extant plant and animal species originated in many different regions of the Old World. But we have also shown that many species did evolve within the Basin in different epochs, giving rise to what we called 'palaeo-endemics' and 'neo-endemics'. Such differentiation was made possible by the large number of geographic and climatic events which punctuated the history of the Basin. These events led to repeated isolation of biota, and provided opportunities for evolutionary divergence and speciation. The many tectonic microplates present in the Tethys Sea during the Tertiary were examples of biological isolates, looked at over long scales of space and time. At much shorter scales, the southern refugia localized in the main peninsulas of the Mediterranean allowed biota of Europe to survive glacial periods during the Pleistocene. Most species were split into several disjunct populations, thus providing opportunities for differentiation. Depending on the plant and animal groups, as well as the duration of the isolation, this led to many opportunities for regional differentiation at various levels of organization, ranging from the genetic structure of whole populations to subspecies and species, and even to genera. There are no documented examples of vertebrates that differentiated in the ancient microplates of the Tethys although several candidates exist among archaic lineages of reptiles as well as among a great many endemic plant species. Nevertheless, a large body of evidence from palaeontological and molecular studies shows that major climatic changes occurring as far back as the Miocene–Pliocene produced biological isolates favourable for species differentiation. We will now look briefly at what befell some vertebrates, in different epochs ranging from the far past to the more recent glacial times of the Pleistocene.

Speciation from the Miocene to the Pliocene

Although it has long been thought that most modern species of vertebrates are recent and evolved during the Pleistocene, recent palaeontological findings as well as molecular systematics strongly support the idea that many species are much more ancient. For example the closely related marbled (*Triturus marmoratus*) and crested newts (*T. cristatus*) were long considered to have diverged recently. They are indeed largely separated geographically, the former occurring in the Iberian peninsula and western France, and the latter in the rest of Europe (Fig. 2.4). Moreover, where they do co-exist, as in western France, reproductive incompatibility prevents their hybridization. However, based on fossils and a variety of biochemical and molecular data, it now appears that the two species must have differentiated earlier, between 7 and 12 Myr BP.

Another example of ancient differentiation among closely related species is that of the 'Mesogean' nuthatches. Three species of narrowly endemic nuthatches occur in the Basin, the Corsican nuthatch (*Sitta whiteheadi*) in Corsica, the Algerian nuthatch (*S. ledanti*) in a small region of Algeria, and Kruper's nuthatch (*S. krueperi*) in Turkey. Since the three species are very similar, one might expect that they differentiated recently, that is, during the Pleistocene. However, molecular studies have revealed that in fact they belong to quite separate lineages that diverged at the beginning of the Pliocene *c.* 5 Myr BP (Pasquet 1998).

Still another example is that of the partridges of the genus *Alectoris*. Four species occur in the Basin and are largely allopatric with some cases of hybridization. Randi (1996) provided evidence from molecular data (mitochondrial DNA, cytochrome b) that speciation events among these partridges occurred between 6 and 2 Myr BP, probably as a consequence of lineage dispersal and isolation of allopatric populations in relation to climatic changes in the Pliocene.

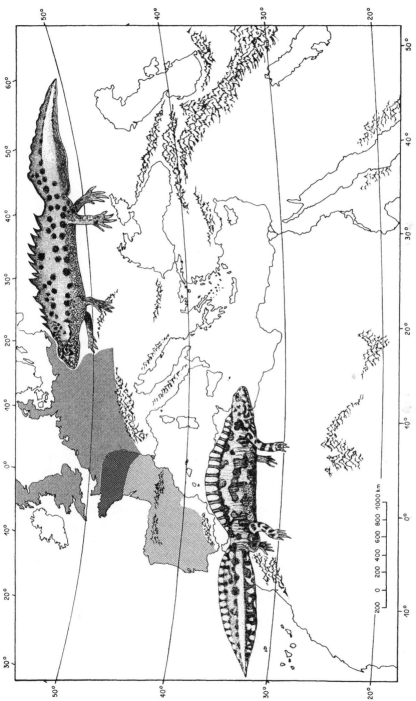

Fig. 2.4 Isolation of populations in the Miocene–Pliocene gave rise to differentiation at the species level in the crested newt (upper) and marbled newt (lower). (After Macgregor et al. 1990.)

Pleistocene differentiation

In fact, there are few examples of Mediterranean vertebrate species that differentiated at the species level during the Pleistocene. One that did is the warbler genus *Sylvia*. These small birds are abundant and characteristic in Mediterranean shrublands; indeed 14 of 19 species in this genus are Mediterranean endemics. Molecular phylogenetic techniques have suggested that the differentiation of these warblers started as far back as 6.3–6.8 Myr BP, with only the radiation of the most closely related species, that is, those that are tightly linked to Mediterranean shrublands, having occurred in the Pleistocene, between 2.5 and 0.4 Myr BP. Three principal centres of speciation have been proposed for this genus on the basis of an analysis combining a biogeographical approach and the molecular phylogeny. The first is located in the western part of the Mediterranean (three species), the second lies in the Aegean peninsula (six species), and the third in the Near East (five species) (Fig. 2.5). The hypothesis that a series of separate speciation events may have occurred in shrubland habitats is supported by palaeobotanical analyses which show that the spatial extent of these shrublands has varied with fluctuating climatic conditions (Pons 1981).

For the majority of Mediterranean vertebrates, however, Pleistocene differentiation has mostly been restricted to the subspecific level. In birds, a burst of recent studies using molecular systematics (mitochondrial DNA) provides evidence that many species are much more ancient, dating back to the Pliocene, as we saw in the previous section. This does not mean, however, that Pleistocene environmental changes had little impact on extant avian diversity. Repeated changes in selection pressures on small populations that were isolated in refugia during full-glacial times resulted in substantial microevolutioary genetic diversification (Avise and Walker 1998). Reconstructing patterns of genetic differentiation and past colonization routes across continents is possible from molecular techniques, a discipline called phylogeography.* Many species display significant geographically oriented phylogeographic units that are genetically distinct. Pleistocene scenarios of differentiation in glacial refugia and subsequent range expansions of two or more phylogeographic units have been proposed for explaining these patterns in birds as well as in other groups of animals and plants (Taberlet *et al.* 1998). These interesting findings suggest that speciation events are gradual processes, with species originating from suites of diverging phylogeographic units over long time spans, rather than at a point event in history. This sheds new light on the importance of Pleistocene events in accelerating speciation processes that had been initiated much earlier in the Pliocene.

An example of this among mammals is the brown bear (*Ursus arctos*). Significant genetic differentiation occurred during glacial periods when the species was split into several isolated populations in each of the main Mediterranean peninsulas (Fig. 2.6). Biochemical studies have allowed us to reconstruct the pattern of European recolonization by these bears as climate improved during the Holocene. The Swedish brown bear population is genetically closer to the population in the Cantabrian Mountains of northern Spain than either one of

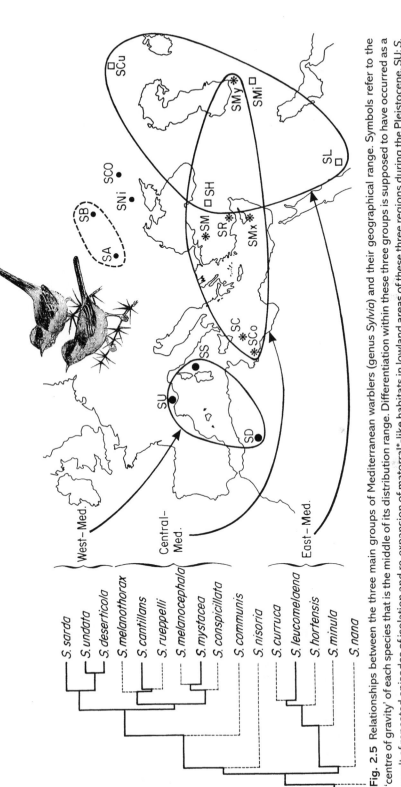

Fig. 2.5 Relationships between the three main groups of Mediterranean warblers (genus *Sylvia*) and their geographical range. Symbols refer to the 'centre of gravity' of each species that is the middle of its distribution range. Differentiation within these three groups is supposed to have occurred as a result of repeated episodes of isolation and re-expansion of matorral*-like habitats in lowland areas of these three regions during the Pleistocene. SU: *S. undata*, SS: *S. sarda*, SD: *S. deserticola*, SC: *S. cantillans*, Sco: *S. conspicillata*, SM: *S. melanocephala*, SR: *S. rueppelli*, SMx: *S. melanothorax*, SH: *S. hortensis*, SL: *S. leucomelaena*, Smy: *S. mystacea*, Smi: *S. minula*, Scu: *S. curruca*. The 'centres of gravity' of the four mid-European species (SA: *Sylvia atricapilla*, SB: *S. borin*, SNi: *S. nisoria*, SCO: *S. communis*) are also shown. The dashed lines in the phylogenetic tree refer to relationships that entail some uncertainty. (Reproduced with permission from Blondel *et al.* 1996).

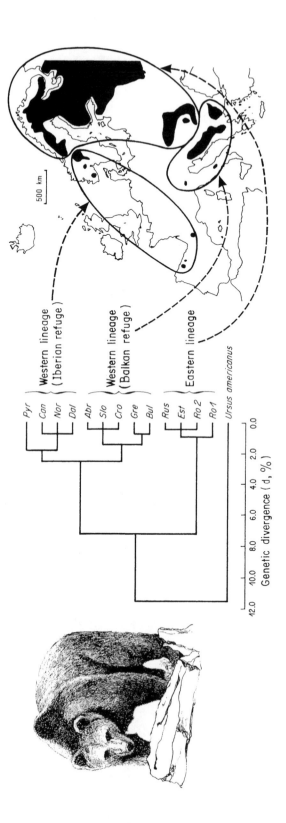

Fig. 2.6 Isolation of brown bear (*Ursus arctos*) populations during glacial periods gave rise to the differentiation of three main lineages that evolved in isolation in the main Mediterranean peninsulas and in eastern Europe. Each lineage includes well defined haplotypes (*Pyr*: Pyrenees, *Can*: Cantabric cordillera, *Nor*: Norway, *Dal*: Dalmatia, *Abr*: Abruzzo, *Slo*: Slovenia, *Cro*: Croatia, *Gre*: Greece, *Bul*: Bulgaria, *Rus*: Russia, *Est*: Estonia, *Ro 1*, *Ro 2*: Romania. The American bear (*Ursus americanus*) is used as an out-group that is a close relative species in which the phylogeny of the brown bear populations is 'rooted'. (After Taberlet and Bouvet 1994 in Blondel 1995.)

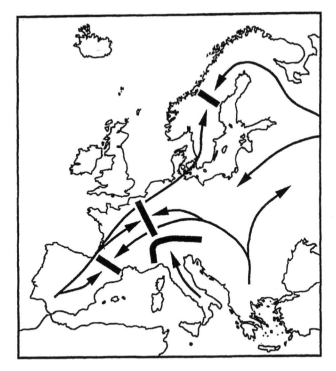

Fig. 2.7 Post-glacial colonization routes of some European animals and plants. Arrows indicate direction of northward expansion routes taken by populations. The thick black lines indicate contact zones between previously separated populations emerging from their respective refugia. (Reproduced with permission from Taberlet *et al.* 1998.)

them is to the Italian population in the Apennines. This strongly suggests that the extant Scandinavian populations are derived from an Iberian refuge stock. Based on similar evidence, the brown bear populations of central Europe are probably derived from the Italian Pleistocene refuge. Nevertheless, all European brown bear populations clearly belong to the same species.

Extending these new molecular techniques to other species, Taberlet *et al.* (1998) have identified degrees of genetic differentiation (distinct phylogeographic units) in several plant and animal species, and have tracked their migration routes as they recolonized Europe from their glacial Mediterranean refugia (Fig. 2.7).

Summary

Major geographic and climatic changes that repeatedly punctuated the Basin's history since the beginning of the Tertiary generated a large part of the biological diversity we observe today, not only in the Mediterranean, but also over the rest of the western Palaearctic. Thus, much of the intraspecific and interspecific genetic diversity we observe today across Europe is a legacy of the differentiation processes that occurred when biota were restricted and split in the many Mediterranean isolates. In the Mediterranean itself, this resulted in the exceptionally high levels of endemism that characterize to this day the various 'hot spots' in the Basin such as the three main peninsulas of southern Europe. The 'Ibero-Mauritanian' plate that resulted in present day Spain and the northern part of Morocco in particular has been a fertile matrix for differentiation in all groups of plants and animals. In the more recent past, movements resulted in a genetic 'signature' which is still visible today throughout Europe, especially in species that found refuge in the Mediterranean Basin during the glacial episodes of the Pleistocene.

3 Present-day biodiversity: plants and animals

Introduction

The extant biodiversity of the Mediterranean is a legacy of the many processes of immigration, extinction, sorting processes, and regional differentiation we summarized in the previous chapter. It would be beyond our scope, and rather tedious, to give a detailed account of the thousands of species that occur in the Basin. Instead, we will focus on particular species diversity patterns in some selected groups, pointing out features that are typical of Mediterranean biota.

How many species of animals and plants are there in the Mediterranean area today? Our knowledge of species distribution in the Basin is relatively good for some groups of vascular plants and vertebrates, and on large spatial scales. Even so, several new species of vertebrates have been discovered in the past few years. Examples are a shrew on the small island of Gozo (Malta archipelago), two frogs (*Alytes muletensis* on Mallorca, *Discoglossus montalentii* on Corsica), a nuthatch (*Sitta ledanti* in Algeria) and a land tortoise (*Testudo weissingeri*) in Peloponnese, Greece. Even the house mouse appears to be, in fact, a 'complex' of three or four species according to recent biochemical studies, and more detailed systematic analyses of other small mammals using biochemical techniques will certainly lead to the discovery of new species, as is expected to happen in the voles of the genus *Pitymys*. Cytogenetic studies and experimental crosses have also revealed similar complexity in a number of plant groups, such as the succulent *Euphorbia* species of Spain, Morocco, and Macaronesia (Molero and Rovira 1998). In most groups of invertebrates, our knowledge of the extant species diversity in the Mediterranean is still far from complete, and in some cases only just developing. One striking character of most groups of plants and animals in the Mediterranean is the high level of endemism. Indeed, local differentiation giving rise to endemic taxa was especially likely in the Basin because of its exceedingly complex history. The region has been, and continues to be, dissected and re-dissected into thousands of biological isolates in islands, peninsulas, and mountain ranges, and the level of endemism greatly exceeds that found in any other part of Europe. Table 3.1 summarizes data from various sources to give a rough idea of species richness and levels of endemism in seven major groups of organisms.

The Mediterranean Basin has been recognized by Myers (1990) as one of the 18 world 'hotspots' where exceptional concentrations of biodiversity occur. This is consistent with the general observation that species richness increases with decreasing latitude, such that one finds more species of plants and animals in the Mediterranean than further north. Many explanations have been given for such

Table 3.1 Numbers of species, levels of endemism, and percent of the world species richness for vascular plants and several groups of animals in the Mediterranean Basin. Estimated world number of species from Hammond (1995).

Group	No. of species	Endemism (%)	Proportion of world number of species (%)	Source
Vascular plants	25000	50	7.8	Quézel (1985)
Freshwater fishes[1]	300	44	–	After Crivelli and Maitland (1995)
Reptiles[1]	165	68.5	2.5	Cheylan and Poitevin (1998)
Amphibians[1]	63	58.7	1.5	Cheylan and Poitevin (1998)
Mammals	197	25	4.2	Cheylan (1991)
Birds	366	17	3.8	Covas and Blondel (1998)
Insects	150000[2]	–	1.9	Baletto and Casale (1991)
Butterflies	321	46	–	After Higgins and Riley (1988)

[1] For the northern bank of the Mediterranean Basin
[2] Tentative figure, probably largely underestimated

gradients, but one of the most respected is 'Rapoport's rule', which states that the geographic ranges of species tend to decrease as latitude decreases. Many people have also argued that the areas over which species are spread are smaller in regions of high topographic relief than in regions with more homogeneous relief, perhaps due to the combination of topographic diversity and high species colonization rates throughout history. This pattern is obvious for many groups in the Mediterranean, including plants, mammals, fishes, reptiles, and birds.

Flora

The flora of the Mediterranean area is one of the richest in the Old World. It includes more than 25 000 species of flowering plants (Quézel 1985), or about 30 000 species and subspecies (Greuter 1991), as well as some 160 or more species of ferns. This is about 10% of all known plant species on Earth, a figure estimated at between 238 000 and 260 000 (Greuter 1994), although the dry land area of the Mediterranean Basin represents only 1.5% of Earth dry land. Compare this regional richness to the mere 6000 species of higher plants found in Europe outside the Mediterranean Basin, an area three to four times greater in size! At least 100 tree species contribute to the various forest types in the Mediterranean Basin as compared to about 30 in all of central Europe. This difference can be attributed in part to differential extinction of tree species of the northern latitudes during glaciations (Latham and Ricklefs 1993).

Endemism

The main reason for the Mediterranean's richness in plant species is not so much the variety of species in any given area as the remarkable number of endemics, many of which are restricted to a single or a few localities in sandy areas, islands, geological 'islands' of unusual soil or rock type, or isolated mountain ranges. More than half the plant species are endemic, and 80% of all European plant endemics are Mediterranean (Gomez-Campo 1985). To take another point of reference: as pointed out by Médail and Quézel (1997), the Mediterranean Basin is nearly as rich in endemics as all of tropical Africa, even though it is about four times larger and harbours about the same number of vascular plant species. Thus, the Mediterranean area is an important reservoir of plant diversity, most comparable in fact to California and the three parts of the Southern Hemisphere which also have mediterranean-type conditions: the Cape Province of South Africa, central Chile, and two parts of south-west Australia (see the last section in this chapter, on convergence and non-convergence among mediterranean-type ecosystems).

Most regions of especially high endemism within the Basin have been isolated in the past, for instance the many 'tectonic microplates' (e.g. the Ibero-Mauritanian and Corsico-Sardinian) that were scattered in the Tethys during the lower Tertiary, or the many islands and upland areas that served as refugia during the various glacial periods of the Quaternary. Several authors (e.g. Bocquet et al. 1978) suggested that the Palaeogene 'palaeo-flora' has been exterminated during the Messinian Crisis and developed later, in the Pliocene, from elements of Asiatic origin. But the biased representation of elements of Asiatic origin and the large number of palaeo-endemics indicate that most of them evolved in situ during the Palaeogene (Gamisans and Marzocchi 1996). During the Pleistocene climatic fluctuations, many islands and upland regions acted both as 'conservatories' of species diversity and as centres where geographically isolated populations could differentiate, and veritable 'species flocks' could arise. Ten regions in the Basin with particularly high numbers of species and large percentages of endemics are shown in Fig. 3.1. The Mediterranean peninsulas protruding southward from Europe are all especially rich in endemic species and show very high species–area curves. As species migrated south or north in these peninsulas in response to the alternation of full-glacial and interglacial periods, they often found themselves in a 'cul-de-sac', or in a completely isolated 'island' where no further migration was possible. When these conditions were combined with high relief, speciation proceeded quickly and led to the appearance of endemic species.

On average, levels of endemism increase as altitude increases. In Mediterranean mountain ranges, whether continental (Atlas, Taurus, Lebanon, Anti-Lebanon) or insular (Corsica, Sardinia, Crete), the percentage of endemic species is very high, albeit quite variable. For example, among the 400 endemic plant taxa in Andalusia, 125 (31%) are restricted to the mountains, and levels of endemism can reach 50% in certain Spanish mountain ranges of the Baetic Cordillera, the Sierra

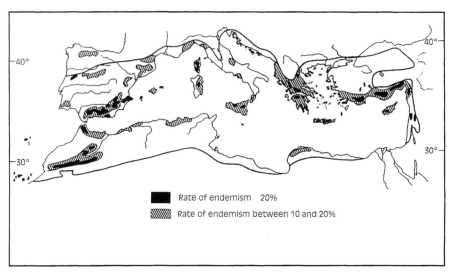

Fig. 3.1 Hotspot areas for plant species diversity in the Mediterranean Basin, including Macaronesia. (Reproduced with permission from Quézel and Médail 1995.)

Nevada, and the Serrania de Ronda (see Fig. 3.1). The largely upland Iberian peninsula as a whole harbours more than 1200 endemic species and subspecies of vascular plants (Gomez-Campo and Herranz-Sanz 1993) which, given a total of 4839 species listed for Portugal and Spain in the Med-Check-List (Greuter *et al.* 1984), amounts to an extraordinary 24.8% endemism rate for the entire peninsula.

As one might expect, Mediterranean island floras typically show high percentages of endemism, 10–13% or even greater (Médail and Verlaque 1997). Examples include Corsica with 11% (240 endemics from a total of 2150 species), as compared to only 7.2% for the nearby continental area of south-eastern France. Similarly, Crete has about 11.7% endemics among plants (200/1710), Sicily 9.7% (255/2402) (di Martino and Raimondo 1979), and the three larger Balearic Islands each about 12%. The seven Canary Islands exhibit still higher levels, ranging from 12.3 to 55.4%, and averaging around 30% (Shmida and Werger 1992).

Unusual geological substrates on islands and elsewhere are also particularly rich in endemics. Notable examples are species whose ranges are restricted to gypsum or dolomite substrates, as well as those found on ultrabasic (serpentine) formations in Cyprus, including four species of the Mediterranean crucifer genus *Alyssum* that occur here as low, often spiny shrubs, even though elsewhere this genus is represented almost exclusively by strictly annual species. An additional oddity of this genus is that some 50 species, mostly of the eastern Mediterranean region and Turkey, show the capacity to take up unusually large quantities of nickel and other heavy metals from the soil. This is probably, not surprisingly, an

Table 3.2 Endemic firs in the Mediterranean Basin.

Species	Region of endemism
Abies maroccana	Rif (Morocco)
A. pinsapo	Southern Spain
A. numidica	Algeria (Babor mountain range)
A. nebrodensis	Sicily
A. cephalonica	Greece (mostly in Peloponnese)
A. boris-regii	Northern Greece and Bulgaria
A. equi-trojani	Turkey
A. bornemulleriana	Pontic region, northern Turkey
A. cilicica	Southern Turkey, Lebanon

Source: Lemée 1967

adaptation to unusual soil conditions, but its ecological and evolutionary significance is still unclear. The genus *Alyssum* does serve as a good example of one of the many genera that have produced exceptionally large numbers of species in the Mediterranean area. In these phyletic groups, geographic speciation and adaptive radiation were clearly favoured by the exceptional degree of environmental heterogeneity provided by the Mediterranean area.

Patterns of limited species distributions and varying species–genus ratios are common within the Mediterranean flora as can be illustrated with three examples. The first is that of the fir (*Abies* spp.), of which nine endemic species occur, each in localized mountain ranges throughout the Basin (Table 3.2).

The second example is that of the Barbary thuja (*Tetraclinis articulata*), a monotypic conifer whose surviving populations are found in the thermo-mediterranean life zone of North Africa and a few southern Spanish mountains. This long-lived, fire-resistant tree is a palaeo-relict whose closest living relatives are the 14 species of the genus *Callitris* found in subtropical forests of south-east Australia and nearby islands such as New Caledonia.

The third example concerns the genus *Arbutus*, represented in the Mediterranean by two relatively widespread species that may well be a vicariant pair, and by two geographically isolated species. Strawberry trees, as they are called in English because of their edible red-orange and mottled fruit, generally occur on non-calcareous soils, like all the numerous heaths and heathers in this same family (Ericaceae). In addition to the most widespread species, *Arbutus unedo*, absent only in Libya and Egypt, there is a closely related but largely allopatric eastern species, *A. andrachne*, and two endemic species, one in the Canary Islands (*A. canariensis*) and one in the Djebel Akhdar of Cyrenaica, Libya (*A. parvarii*) (Fig. 3.2). The other 10–15 members of this genus are found in western North America, especially in subtropical woodlands of Mexico.

In part, the situation of the Mediterranean *Arbutus* seems similar to that of the Mediterranean firs and the Barbary thuja, in that early and successful adaptation

Fig. 3.2 Distribution of the four Mediterranean species of strawberry trees (*Arbutus*). Two are widespread and two are highly restricted. (After Sealy 1949.)

to rather special and difficult habitats, especially in mountains or islands with strong geographical boundaries, apparently led to a low level of adaptive radiation or speciation. However, two of the four strawberry tree species are widespread (and interfertile, since horticultural hybrids of the two have been repeatedly produced). In this context, it is likely that people have intentionally selected, modified, and introduced the strawberry tree in various parts of the Mediterranean area. In ex-Yugoslavia, for example, superior, large-fruited cultivars of A. unedo are well known and have long been widely propagated for commercial plantations. It could well be that the geographical range of A. unedo has been artificially expanded by people. In the case of the Barbary thuja, by contrast, human exploitation has been focused on the wood and attractive burl used for furniture-making (see Chapter 8) rather than for the annually renewed fruit crop.

In all three cases cited, very few species have evolved, sometimes only one, even over very long geological periods. Our purpose here has been simply to illustrate that while adaptive radiation and 'explosive' speciation have taken place, mostly in short-lived Mediterranean plant taxa, the opposite trend has also occurred, primarily in long-lived species of mountain or island habitats. In fact, an unbroken continuum exists between these two extremes.

Mediterranean characteristics

Some features that are typical of (but not exclusive to) Mediterranean flora are worth mentioning. Annual species in general, and ruderals and segetals in particular, are especially well represented in the Basin's flora, as this life-history strategy is highly adapted to various kinds of perturbations, including the stress of two or as many as five to six months of absolute summer drought.

In addition, an impressive number of taxa have shown particular success in adaptive radiation and speciation, such as *Alyssum* mentioned above, or the genus *Centaurea*. This Asteraceae (sunflower family) genus has an estimated 450 species in the Mediterranean area and the Near East. That high figure includes *c.* 170 Turkish species, including 105 endemics, as well as 44 endemic species in Greece and 78 in Iberia (Quézel and Médail 1995). A remarkably wide range of growth forms, life-history strategies, and other ecological adaptations also occur in this genus, no doubt correlated to the evolutionary potential under favourable conditions. Additional examples, amongst others, include the *Astragalus* and *Euphorbia* genera.

A further interesting feature is that the distribution of endemic plant species among so-called life forms (or growth forms) varies greatly according to regions. For instance, there is only 5.6% endemism among the annual plants of the Canary Islands, as compared to more than 70% among the Islands' trees and shrubs (Shmida and Werger 1992). This discrepancy may be related to the small representation of annuals on oceanic islands in general, a result of their generally low capacity for long-distance dispersal. In contrast, a very large proportion of the endemic plant species in both Israel and Sicily are annuals: 37.7%

(Shmida 1984) and approximately 42% (after di Martino and Raimondo 1979), respectively.

Invertebrates

It would be beyond the scope of this book to review in detail the diversity of invertebrate faunas in the Mediterranean, especially those that live under the ground. In fact, species diversities of soil organisms in the Mediterranean appear to be higher than anywhere else in the world, including the tropics (di Castri and di Castri 1981). The abundance, diversity, and high level of endemism amongst invertebrate faunas is indicated by the large number of enthusiast, and too-often collection-oriented, entomologists who stream into the varied and changing Mediterranean environments in spring and summer.

Species richness and endemism

In invertebrates, as in plants, the Mediterranean area is the richest in Europe in terms of species diversity; 75% of the total European insect fauna are found in the Basin (Baletto and Casale 1991). However, it harbours only a limited fraction (2%) of the world's estimated 8 million insect species (Hammond 1995), a figure that could be overestimated. On the European scale, the very high species diversity for most groups of insects in the Mediterranean fits the latitudinal trend of increasing diversity as one moves from boreal to tropical regions, which was mentioned above. However, a reverse trend occurs in some Mediterranean peninsulas, for example in the case of butterflies. This is probably due to the so-called 'peninsular effect' a biogeographic pattern found also in many other parts of the world.

Baletto and Casale (1991) tentatively estimated at 150 000 the number of insect species in the Basin, only 70% of which have been described and named. Figures for most groups of invertebrates increase rapidly as scores of new species are being described each year. Dafni and O'Toole (1994) estimated that there are 3000–4000 species of bee, which makes the Basin a prominent centre of diversity for this group. As many as 1500–2000 species of bees occur in Israel alone (O'Toole and Raw 1991). In Chironomids, a large family of Dipterans, 703 species have been reported in the Basin, 97 (14%) of which are exclusive to this area (Laville and Reiss 1992).

Levels of endemism, too, are high for most groups of insects. In some isolated mountains and larger islands, endemics may account for 15–20% of the insect fauna, a figure that may rise to 90% in some caves. As for most other groups, classic refuge areas are mountain chains such as the Kabyle, Atlas, and Rif chains in North Africa, the central cordilleras of Iberia, the Pyrenees, the Alps, the Balkans, and the Taurus Mountains of Turkey. In remote mountain ranges, extremely limited populations of 'odd' species are sometimes discovered. For example, several very archaic Carabid beetles have recently been found in moist relictual habitats of the High Atlas in Morocco (e.g. *Relictocarabus meurguesae*) and in the high mountains of Kurdistan. Localized in moist habitats amidst arid landscapes, these species could be relicts of ancient faunas that occurred under different ecological conditions but which survived changes in the Mediterranean climate during the Pliocene.

A study of overall patterns of species richness in butterflies reveals that regions with particularly high richness are often those that straddle mountain ranges in southern Europe, with peak diversity values occurring in the southern Alps, and decreasing throughout the Italian and Iberian peninsulas. This pattern suggests that much differentiation took place in southern Europe during the Pleistocene. In addition, populations of all species that live north of the main mountain chains were subjected to periodic devastation during the Pleistocene glaciations. Interestingly however, systematic studies have shown that during glacial–interglacial cycles, conditions remained stable enough in southern Europe, including the Alps, for evolutionary changes to occur in at least some groups (Dennis *et al.* 1991).

Peaks in species richness vary geographically among families of butterflies. For example, high numbers of species of Pieridae and Hesperiidae occur in the Atlas mountains of North Africa whereas Satyridae and Nymphalidae are particularly well represented in the southern Alps. *Erebia* (Satyridae) is an example of a genus that has had an explosive adaptive radiation in the mountain ranges of southern Europe. The majority of species in this genus currently occupy narrow subalpine habitats scattered at high altitudes of 1200–1500 m. Evolutionary divergence is more likely to occur in groups such as *Erebia*, whose habitats are 'three-dimensional mountain' regions with finely dissected landscapes, than in groups that are distributed in more uniform areas, such as the Lycaenidae in Iberian upland regions. In the families Papilionidae, Lycaenidae, and Hesperiidae, peak values of species richness are found in the Balkans and Iberia. On the other hand, particularly low species numbers are found on Mediterranean islands, which probably reflects the influence of island area and isolation.

Percentages and regions of endemism vary greatly among groups of insects according to their particular histories and evolutionary potential. Endemism is generally fairly high in beetles, stoneflies, many families of flies and butterflies, and several groups of spiders that evolved locally from an ancient Tertiary fauna, especially on islands and high mountains. The vast majority of endemic butterfly species are restricted to areas south of the northern edge of the large mountain ranges extending from the Cantabrian chain in northern Spain eastward to the Carpathians. Especially high concentrations of species are found in the Atlas ranges, the Pyrenees, the southern Alps, including the Dinaric chain, and the Apennines. Dennis *et al.* (1995) noticed that the distribution of the major groups of butterfly species in southern Europe coincides with major geographical divisions, for example peninsulas and montane zones. The Mediterranean countries with the highest percentages of endemism in butterflies are Spain, with 16 species (7.5%), and Greece and Italy, with 13 species (9%) each. This pattern supports the contention that the presence of large peninsulas has had an important role in the development of endemic insect taxa in the Mediterranean Basin as was also the case for plants and many other groups of animals.

By contrast, several groups of insects other than butterflies show a surprisingly low level of endemism, even on islands. It is possible that many insects were introduced to islands in recent times by humans, as was the case for mammals and plants. For example, in Corsica only 3 of the 83 species (3.6%) of ants are considered endemic to the island and only 23 (27.7%) are considered to be Mediterranean in their distribution. These figures mean that well over half of the species present come from very distant centres of origin. Not a single species is of Afro-tropical origin.

Tropical influences

A recurrent general trend in almost all groups of Mediterranean animals, except mammals, is their predominantly and unambiguously boreal character. Thus, invertebrate faunas of the Basin show much closer affinities with Palaearctic or Holarctic faunas than with those of any other biogeographical region (Casevitz-Weulersse 1992). Many boreal species and groups colonized the Mediterranean during glacial periods, partially replacing the existing palaeotropical fauna. Good examples are provided by butterflies in the genera *Parnassus*, *Colias*, and *Pieris* that live in montane and lowland habitats. Many forest-dwelling species that are usually plant-eating are also boreal species which followed their host trees whose distribution ranges repeatedly shifted north and south in response to glacial cycles. Examples are the beetle *Rosalia alpina* and the moth *Aglia tau*, both of which are tightly linked to beech. Many other groups of boreal origin are to be found among beetles, dragonflies, Diptera, and other invertebrate groups.

As in all other groups of terrestrial animals except mammals, invertebrates species of tropical origin are scarce in the Mediterranean and have only marginal biogeographical importance. For example, not a single species of Mediterranean ant is of tropical origin and most Chironomids (81%, Laville and Reiss 1992) and butterflies (90%, Larsen 1986) have a Palaearctic distribution. Among the 321 butterfly species, only 24 (7.5%) have tropical affinities. The few non-Palaearctic elements represent either the last remnants of original tropical faunas or else have entered the Basin in various epochs. Some examples of the former group may be found in Lepidoptera, for example the beautiful *Graellsia isabelae* or the two-tailed pasha (*Charaxes jasius*) (Fig. 4.6). Examples exist among Carabid beetles and Diptera (Psychodidae, Chironomidae) as well. An example of a butterfly species that entered the Basin recently is the palaeotropical migrant *Danaus chrysippus* (the monarch). This butterfly has resident populations in the Canary Islands and probably in the Jordan valley. It is a regular visitor to the eastern Mediterranean and may sometimes be seen in the north-west quadrant of the Basin, in Italy, France, and the Iberian peninsula (Larsen 1986).

Among the 213 species of butterflies occurring in the Iberian peninsula, only 5 species of Anthocarinae and some species of Polyommatinae are considered of African origin, along with the two-tailed pasha mentioned above (Martin and

Gurrea 1990). Many Mediterranean butterfly species of tropical origin share certain characteristics. Most are very mobile migrants and occur only sporadically in the Basin, fluctuating widely in the numbers present from one year to the next. They are often limited to very special ecological conditions at the southern limits of the Basin and do not differ, even at the subspecific level, from other populations in the tropics proper (Larsen 1986). Other pre-Quaternary elements in the families Carabidae, Staphylinidae, Buprestidae, Chrysomelidae, and Scarabaeidae are forest dwellers that were initially tied to subtropical forests, mainly laurisilvas. They are found today in soils and forest litter, as well as in caves. Distribution patterns of many species (e.g. Chironomids and butterflies) strongly suggest that the tiny fraction of species that are of tropical origin reached the Mediterranean in various epochs, either by an eastern route following the Nile River valley, or through a western route along the shores of the Atlantic Ocean. Examples of the latter group are some very rare species of Chironomids of Afrotropical origin (*Dicrotendipes collarti, Paratendipes striatus*) that occur only in the 'khettaras' of Marrakech, Morocco, traditional irrigation systems in arid areas where ground water is brought to the surface and kept at a constant temperature of *c.* 19–23 °C (Laville and Reiss 1992).

Freshwater fishes

The number, diversity, and geographic isolation of watersheds in the Mediterranean Basin, an area highly dissected and divided by mountain ranges, has favoured a remarkably high species richness among freshwater fishes. In the northern part of the Mediterranean, from Iberia to Turkey, as many as 131 of the 300 (44%) species are regional or local endemics, 31 of them being limited to a single water catchment (Crivelli and Maitland 1995). Most of these endemics (63%) belong to the Cyprinids, but endemic species also occur in other families such as the Cobitidae (11%), Gobiidae (8%), Salmonidae (6%), and Cyprinodontidae (5%). To these must be added two endemic freshwater species of lampreys (Petromyzontidae) and one of sturgeon (Acipenseridae). As for many other groups, levels of endemism are especially high in the large Iberian, Italian, and Aegean peninsulas, as well as in Turkey (which is peninsula-like), presumably because these regions acted as refugia during glacial periods (Fig. 3.3).

With 110 species, Greece harbours the largest number of fish species of any region in the Basin (Economidis 1991). Of these, 21 euryhaline* species occur in freshwater and brackish water lagoons. Excluding 11 introduced species, 78 (71% of the total) are indigenous and 37 are endemic (47.4%). However, as in many other groups of plants and animals, regional endemism levels for Mediterranean fishes are highest in the Iberian peninsula, as 25 of the 29 species and subspecies of indigenous Cyprinidae, Cyprinodontidae, and Cobitidae that occur there are endemics (Blanco and Gonzalez 1992). Most of them live in springs, mountain torrents, lakes, and lowland rivers, while a few are restricted to marshes and coastal lagoons. Fish communities of the more arid regions of the

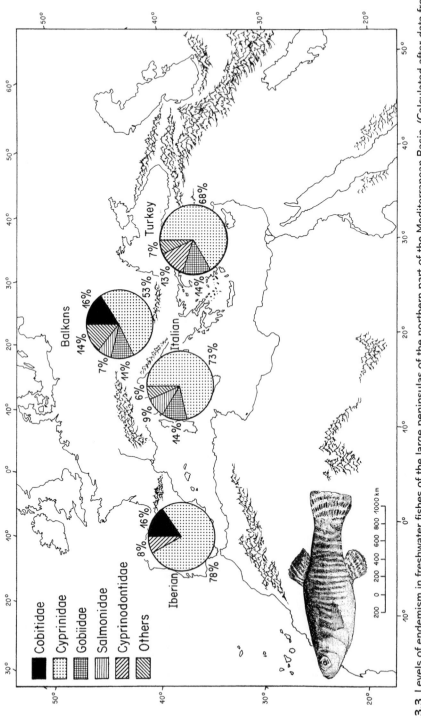

Fig. 3.3 Levels of endemism in freshwater fishes of the large peninsulas of the northern part of the Mediterranean Basin. (Calculated after data from Crivelli and Maitland 1995.)

Basin are much less rich in species. For example, the native freshwater fish fauna of Israel includes only 28 species and subspecies (Ben-Tuvia 1981) and that of Tunisia no more than 12 species, 4 of which have been introduced recently (Kraiem 1983). Some endemic species are restricted to very small areas. Examples are several species of the genera *Noemacheilus* (Cobitidae) and *Aphanius* (Cyprinodontidae), found in very small springs, streams, and ponds near the Dead Sea, and three species of Cichlids in the oases around the Chott el Djerid, Tunisia. Although most species of the Mediterranean fish fauna belong to Palaearctic lineages, some are of Afro-tropical origin (Cichlidae, Clariidae), or are relicts of the Tethys Sea (several species of Cyprinodontidae). The Cyprinid species that succeeded in colonizing North Africa presumably did so from Iberia in the Miocene when the two land masses were connected by a narrow isthmus.

Phylogenetic uncertainties

Although the phylogenetic status and time of differentiation of fish species in the Mediterranean Basin are in general poorly known, there are some exceptions. For example the barbels of the genus *Barbus* (Cyprinidae) include many species and subspecies, the systematic status of which has recently been clarified through molecular studies (Tsigenopoulos 1996). The genus includes two well-defined groups of species. The first includes large species (the *B. barbus* type) that live in the large rivers and lakes of central and southern Europe. The second group includes smaller species that live in small mountain rivers of the Mediterranean (the *meridionalis* type). Species of the two groups at times co-occur in the same river and then hybridize, for example *B. barbus* and *B. meridionalis* in France, *B. barbus* and *B. meridionalis petenyi* in Slovakia, and *B. barbus* and *B. meridionalis* subsp. *peloponnesius* in Greece. Although one might have expected that all these taxa, which at first sight seem very closely related, would derive from a common ancestor, molecular studies have shown that in fact they derive from distinct lineages and therefore do not belong to the same species complex.

Several other very interesting endemic fish species of the Mediterranean are found in the trout complex, but their origin and systematic status remain largely conjectural. Examples are the endemic Corsican trout (*Salmo trutta macrostigma*), which is restricted to the upper parts of streams, and the Prespa trout (*S. trutta peristericus*) in the Prespa lakes of western Greece and Albania. Another endemic species is the marble trout (*S. trutta marmoratus*), an endangered freshwater fish of the Adriatic basin, including the catchments of the river Po, its alpine tributaries, and some rivers of the Dinaric Alps that empty into the Adriatic (Fig 3.4). Although the current status and biology of the various forms of *Salmo trutta* have recently been summarized by Povz *et al.* (1996), there is still much uncertainty concerning the taxonomy of the marble trout. Some scientists classify it as a species, others as a subspecies of the brown trout (although it differs from the latter in colour and shape). Apart from the huchen (*Hucho hucho*) this beautiful fish is the largest trout in Europe; individuals may weigh up to 30 kg and be 140 cm long. Diagnostic characters include the absence

Fig. 3.4 Three trout taxa of the Mediterranean. From bottom: brown trout (*Salmo trutta fario*), Corsican trout (*Salmo trutta macrostigma*), and marble trout (*Salmo trutta marmoratus*).

of the red spots or blotches that characterize the brown trout, the whitish or yellowish belly, and the marble patterns of its skin, which is olive-brown to olive-green with a copper-red tint. Populations of the marble trout show little genetic variation, with heterozygosity ranging from 0–1% as compared to 5–7% in other Mediterranean trout species. This fish is restricted to upland streams where summer water temperatures do not exceed 15 °C and winter water temperatures are 2–3 °C. All these trout species are threatened because of hybridization with the brown trout (*Salmo trutta fario*), which has been repeatedly introduced in their range for sport fishing. Pure marble and Corsican trout can today only be found in some upper reaches of rivers in Slovenia and Corsica.

Reptiles and amphibians

Contrasting patterns of species richness and distribution

Reptiles are at home in the dry, warm Mediterranean area, and thus are much more abundant and diverse than amphibians (Table 3.3). Reflecting the contrasted ecology and physiology of the two groups, the diversity of reptile species increases from north to south (and from west to east), paralleling aridity gradients. For example, there are 20 species of reptiles on a surface area of 141 500 km^2 in Italy as compared to 25 species in Cyrenaica, a region of 27 000 km^2, and 21 species in Cyprus (9250 km^2). On the other hand, amphibian species richness increases from south to north (and from east to west). Since the north–south and east–west aridity gradients favour reptiles in the southern and eastern parts of the Basin, whereas the more humid climates in the northern and western parts favour amphibians, regions that are rich in reptiles tend to be poor in amphibians, and vice versa. In all, 165 species of reptiles in 15 families are found in the Mediterranean Basin, as compared to only 63 species of amphibians in 9 families (Delaugerre and Cheylan 1992; Cheylan and Poitevin, 1998 and personal communication). The many factors that have moulded the Mediterranean Basin and its biota are reflected in levels of endemism which are very high in the two groups (Table 3.3), but regional levels of endemism sharply contrast between them. In reptiles, levels of endemism amount to 44% in North Africa, 31% in the Near East, and 24% in the Iberian peninsula, whereas in amphibians, endemism is the highest in Italy (35%), then decreasing in Iberia (32%), the Balkans (17%), North Africa (17%), and the Near East (13%). The reptile fauna as a whole includes snakes, lizards, tortoises (see Box 3.1), and even tropical relicts such as two species of chameleon, *Chamaeleo chamaeleon* in North Africa and southern Iberia, and *Chamaeleo africanus* in Greece, Crete, Cyprus, and the Near East. Other tropical relicts include a tortoise (*Tropnyx triunguis*) and the desert monitor (*Varanus griseus*) in Anatolia and the Near East, as well as several species of snakes. The crocodile (*Crocodilus niloticus*) occurred in southern Morocco until the 1950s. Many neo-endemic species of reptiles in the genera *Podarcis*, *Lacerta*, *Chalcides*, and *Vipera* evolved in the Basin as a result of intensive adaptive

Box 3.1 The European pond terrapin (cistude)

Among the six tortoises that occur in the Mediterranean Basin, three are terrestrial: the spur-thighed tortoise (*T. graeca*), the Hermann's tortoise (*T. hermanni*), and the marginated tortoise (*T. marginata*), and three are mostly aquatic: the Spanish terrapin (*Mauremys leprosa*), the European pond terrapin (*Emys orbicularis*), and the stripe-necked terrapin (*Mauremys caspica*).

The European pond terrapin (*Emys orbicularis*), also called cistude, occurs across a large part of Europe but is much more common in Mediterranean countries than further north. A secretive animal, unknown to most people, it spends most of its life at the bottom of shallow ponds, marshes, and slow-running rivers and canals. Adults are 20–30 cm long with a blackish shell sprinkled with yellow spots and stripes. Males are much smaller than females. The cistude becomes active at water temperatures above *c.* 28 °C, walking slowly along the bottom in search of food, surfacing now and then to breathe. It can also swim quickly using its flat palmate legs as powerful oars. On warm and sunny days, when the air temperature rises to several degrees above that of the water, the cistude likes to climb onto any handy floating object that will support its weight, for example dead trees or stumps, piles of dead reeds, even an ancient muskrat hut, to sun itself. Up to 30 animals may sit together basking in the sun. Extremely shy, they all splash noisily into the water when disturbed. This carnivorous animal feeds mostly on aquatic insects and molluscs, but also on small fishes, newts, and carrion. Six large eggs are buried in spring, 10–12 cm deep, in fields or pastures, sometimes as far as 600 m from bodies of water. After 3–4 months incubation the young tortoises, which are very small at birth (less than 2 cm long) spend their first winter in the burrow. Their first trip to the nearest body of water in early spring is risky and around 50% are taken by predators. The shell of this slow-growing animal becomes hard only after 4–5 years but life expectancy is exceptionally high and some individuals may live for 40–50 years. When ponds and small rivers dry up in summer, the cistudes bury themselves in the ground until water returns in autumn. They spend the winter, from October to March, sunk 20–30 cm deep in the mud at the bottom of the pond.

radiation in localized areas. In the Lacertidae, the genera *Algyroides* and *Psammodromus* (four species in each of them) are typical relict Mediterranean endemics.

In amphibians most endemic species belong to archaic lineages that have remained relatively unchanged morphologically since their first appearance in the Eocene (55 Myr BP). Examples are toads of the genera *Pelobates* (three species

Table 3.3 Total numbers of species and numbers of endemic species (percentage in parentheses) of reptiles and amphibians in various regions of the Mediterranean area.

Region	Reptiles		Amphibians	
	Total	Endemics	Total	Endemics
Mediterranean area	165	113 (68)	63	37 (59)
Mainland regions				
Iberia	33	8 (24)	22	7 (32)
Italy	20	0	17	6 (35)
Balkans	45	11 (24)	17	4 (24)
Near East	84	26 (31)	15	2 (13)
Cyrenaica	25	0	2	0
Maghreb	59	26 (44)	12	2 (17)
Insular regions				
Balearic Islands	10	2 (20)	4	1 (25)
Corsica	11	3 (27)	7	2 (29)
Sardinia	16	3 (19)	8	5 (63)
Sicily	18	1 (6)	7	0
Crete	12	0	3	0
Cyprus	21	1 (5)	3	0

Source: from Cheylan and Poitevin 1998

in the Basin) and *Discoglossus* (five species), and salamanders of the genus *Euproctus*. The nearly eyeless *Proteus anguinus* of the Dinaric karsts and caves is another example of an ancient lineage of tropical origin. This species is the only representative in Europe of a family (Proteidae) that has five species in North America (Box 3–2).

From a biogeographical perspective, the Mediterranean reptile and amphibian faunas became established as early as the upper Eocene to mid-Miocene (38–15 Myr BP), from several biogeographical regions including the Euro-Siberian, Saharo-Arabian, and Turano-Caucasian. Biochemical studies have shown that divergence among major groups of lizards (e.g. Lacertidae) apparently took place during the Oligocene–lower Miocene. Thereafter, Pliocene–Pleistocene climatic fluctuations remodelled this fauna through extinctions and new waves of speciation. Although certain reptiles, like the ringed snake (*Natrix natrix*) or the Schokar sand snake (*Psammophis schokari*), are of central European and Saharo-Arabian origin, respectively, most Mediterranean reptiles originated in western Asia, notably the Caucasian region which is a hotspot for this group (Meliadou and Troumbis 1997), and in North Africa. Indeed, the desert belt that limits the Mediterranean in the south favoured the differentiation of many groups

Box 3.2 The blind cave salamander, *Proteus anguinus*

This is one of the most fascinating Mediterranean amphibians, an almost eyeless and highly specialized salamander-like animal whose distribution is restricted to the caves and underground rivers of the Dinaric karsts, from the border of north-eastern Italy to the south of Herzegovina. The species is patchily distributed in more than 250 localities where genetic isolation has led to a close matching between ecotypic adaptations and habitats (Sket 1997). It probably derives from an ancient lineage that arose in the region when tropical climates prevailed during the Oligocene–Miocene and then colonized the Dinaric area as karstic formations progressively developed.

Proteus is large for an amphibian, reaching 25–50 cm in length, and bears large feathery gills whose pink colour contrasts dramatically with the animal's pigment-free, bleach-white skin. This neotenic* animal, one of the six species in the family Proteidae, has only two toes on each hind leg, and reproduces without ever reaching an 'adult' stage of body development. *Proteus* lives in large slow-running, underground rivers with temperatures at a nearly constant 8 °C, and in deep caves. Its metabolic rate is very low, in keeping with the fact that organic matter and especially animal prey are scarce in this lightless habitat. By contrast, its hearing is extraordinarily acute, and it also finds prey through chemical sensory means.

of reptiles, for example *Acanthodactylus*, which penetrated the Mediterranean area to some extent.

The ecophysiology of amphibians helps explain why regional species richness is so uneven in the Basin. Mediterranean North Africa has only 12 species of amphibians (Lescure 1992), several of which colonized the area from the north after the closing of the Strait of Gibraltar at the end of the Miocene, about 6 Myr BP. Examples of 'recent' immigrants include the fire salamander (*Salamandra salamandra*), the common toad (*Bufo bufo*), the midwife toad (*Alytes obstetricans*), and probably also the stripeless tree frog (*Hyla meridionalis*) (Lescure 1992).

In North Africa, the number of amphibian species declines from west to east, with 11 species in Morocco, 7 in Tunisia, and only 2 in Tripolitania, western Libya, namely the green toad (*Bufo viridis*) and the marsh frog (*Rana ridibunda*), both species that have very wide distributional ranges. Differential distribution patterns between reptiles and amphibians suggest that historical effects differed greatly between the two groups. From this arose various regional specificities and several cases of vicariance with east–west species replacements. Examples of such species pairs include the marbled and crested newts (Fig. 2.4), the palmate (*Triturus helveticus*) and common newt (*T. vulgaris*), the Iberian (*Pelobates cultripes*) and common spadefoot (*P. fuscus*), and the stripeless (*Hyla meridionalis*) and European (*H. arborea*) tree-frogs.

Impoverished faunas

By comparison with mainland areas of similar size, species impoverishment in Mediterranean islands is 43% and 60% in reptiles and amphibians, respectively. However, patterns of distribution have been perturbed by introductions and extinctions of endemics by humans since the beginning of the Holocene. This may explain why there are, on average, few endemic reptiles and amphibians in Mediterranean islands, especially in those of the eastern half of the Basin. As in mainland areas, there are more amphibian species in islands of the western half of the Basin than in those of the eastern half: seven species in Corsica, eight in Sardinia, and seven in Sicily, as compared to only three each in Crete and Cyprus. Examples of highly endemic species in the western islands are two species of salamanders (*Euproctus montanus* in Corsica and *E. platycephalus* in Sardinia). These are very particular archaic species which, contrary to the closely related newts, do not have lungs. They breathe through the mouth and skin. Moreover, their breeding biology is unique for their family. After having laid her eggs under a stone in a small river, the female takes great care of them, actively watching them until they hatch. In the Balearic Islands, the recently discovered toad *Alytes muletensis* is the last survivor of an endemic fauna that has been enriched by nine species of reptiles and three species of amphibians, all introduced by humans (Delaugerre and Cheylan 1992).

Although based on comparable habitats elsewhere one would expect to find several endemic reptiles and amphibians in Cyprus, there are only one endemic species of snake, *Coluber cypriensis*, and ten subspecies of reptiles (Böhme and Wiedl 1994). This level of endemism is very low for an island of volcanic origin that has been geographically separated since the Pliocene. One possible explanation is that repeated colonization events on natural rafts prevented local evolution of these taxa. But a more likely explanation is that several species, intentionally or inadvertently introduced by humans, have directly or indirectly pushed the endemic species to extinction. Thus, differences in species richness and endemism on islands between the western and the eastern half of the Basin could be a result of the islands of the western Mediterranean having been colonized by humans much later than those of the eastern Mediterranean.

Birds

Judging from the great numbers of bird-watchers that visit the Mediterranean each spring and summer, the diversity of their quarry must be very high. Indeed, there may be as many as 366 species, compared to only 500 for all of Europe (Covas and Blondel 1998). Following the progressive disappearance of tropical taxa from the western Palaearctic during the cooling and drying trends of the Oligocene and Miocene, the extant Mediterranean bird fauna became established during the Pliocene–Pleistocene, primarily from elements that colonized the Basin from as many as nine different biogeographical areas. However, the relative contribution of each varies greatly and three groups are clearly dominant. The largest one includes 144 species of northern origin that are characteristic of forests, freshwater marshes, and rivers all over western Eurasia. The second group consists of 94 steppe species, most of which presumably evolved in the margins of the current Mediterranean area, notably in the 'Eremic*' Saharo-Arabian region which extends from Mauritania—where the Sahara meets the Atlantic Ocean—eastwards across Africa, the Red Sea, the Arabian peninsula, and on to the semi-deserts of southern Asia. This belt has almost always (at least since modern species evolved) isolated the Palaearctic from the Afro-tropical and Oriental realms. The importance of this faunal element in the Mediterranean Basin has been secondarily favoured, both in geographical distribution and in population sizes, by the generalized human-induced retreat of forest cover since the Neolithic. As a result, many species that were formerly rare are now wide-spread and common throughout the Basin.

The third group encompasses all the species that are more or less linked to shrubland habitats, the so-called matorrals (see Chapter 5). Good examples are the partridges (*Alectoris* spp.) and the many species of warblers (*Sylvia* spp., *Hippolais* spp.). Given the extent and high diversity of shrubland formations in the Mediterranean Basin the number of species in this group is surprisingly small (42 species, or 11.5% of the regional bird fauna).

To these three principal elements of the contemporary regional bird fauna should be added two smaller groups of great biogeographical and ecological interest. The first is the rock-dwelling species found on steep cliffs, screes, and rock outcroppings of hills or mountainous Mediterranean regions and, indeed, throughout the southern Palaearctic as a whole. Examples are the lammergeier (*Gypaetus barbatus*), the blue rock thrush (*Monticola solitarius*), and the chough (*Pyrrhocorax pyrrhocorax*), which attract bird-watchers from all over the world. The second small group, known as 'sarmatic'*, includes bird species inhabiting lagoons and coastal swamps. They include the flamingo (*Phoenicop-terus ruber*), white-headed duck (*Oxyura leucocephala*), marbled duck (*Anas angustirostris*), slender-billed gull (*Larus genei*), and Mediterranean gull (*Larus melanocephalus*).

As in most other groups of animals, only very few Mediterranean species are of Afro-tropical origin. Examples of birds that presumably reached the Atlantic shores of Morocco from tropical latitudes are a goshawk (*Melierax metabates*), the cape owl (*Asio capensis*) and the guineafowl (*Numida meleagris*).

Few endemics

The bird fauna of the Mediterranean differs from that of other groups of animals and from the flora in two main points, which were raised in Chapter 2. First, despite their disparate biogeographical origins, many species and groups of birds are rather homogeneously distributed across the Basin, so that regional variation in the composition of local bird assemblages is not very marked. Of course, there are several endemic species and regional variations but, on the whole, the vast

Fig. 3.5 Some examples of endemic bird species of the Mediterranean Basin. Clockwise from upper left: Moussier's redstart (*Phoenicurus moussieri*) (North Africa), Cyprus wheatear (*Oenanthe cypriaca*) (Cyprus), Cyprus warbler (*Sylvia melanothorax*) (Cyprus), and Corsican nuthatch (*Sitta whiteheadi*) (Corsica).

majority of species are widespread throughout the Basin. This is especially true for forest-dwelling birds. From statistical tests devised to detect regional differences in species composition, Covas and Blondel (1998) were able to discriminate only some of the most conspicuous differences, for example between the south-east and the north-west quadrants. This homogeneity of regional bird faunas results partly from the long distances birds are able to cover.

Secondly, the low level of endemism is a surprising feature. Only 64 species (17% of the total) appear to have evolved within the geographic limits of the Mediterranean Basin (Table 3.4). One wonders why more speciation events did not occur among birds in the highly fragmented, heterogeneous landscapes and regions of the Basin. One possible explanation is that the present extension of scrub and shrublands is secondary, resulting directly from human activities and perturbations, and therefore species have not yet had time to differentiate. However, palaeobotanical data have shown that more or less isolated patches of matorral have existed in the area at least since the Pliocene. Accordingly,

Table 3.4 Numbers and examples of bird species that presumably evolved in forest, steppe, shrubland and other habitats of the Mediterranean area.

Habitat type	Forest	Steppe	Shrubland	Other (sea, freshwater, rocks)
No. of species	16	19	20	9
Examples	Laurel pigeon	Barbara partridge	Moussier's redstart	Marbled teal
	Corsican nuthatch	Red-necked nightjar	Cetti's warbler	Slender-billed gull
	Sombre tit	Dupont's lark	Sardinian warbler	Audouin's gull
	Spotless starling	Berthelot's pipit	Marmora's warbler	Pallid swift
	Syrian woodpecker	Black-eared wheatear	Black-headed bunting	Rock nuthatch

successive episodes of expansion and contraction of these shrublands have presumably favoured differentiation in at least a few groups (see p. 47).

However, although the Mediterranean bird fauna includes few endemic species, a much higher degree of subspecific variation has been reported there than in any other part of the Palaearctic. Such high levels of intraspecific variation results from the high geographical diversity of the Basin with its many islands, peninsulas, and other geographic and ecological barriers to dispersal. Bird species that occur as breeders in the area are represented by an average of 5.4 subspecies per species on a worldwide distribution scale, by 2.3 subspecies per species on the Palaearctic scale, and by 2 subspecies per species in the Mediterranean Basin itself. Weighting these figures by the sizes of surface areas involved in order to obtain an index of regional subspecific variation, the proportion rises from 0.56 for the entire Palaearctic realm to 6.7 in the Mediterranean Basin (Blondel 1985a). This represents a truly remarkable range compared to other parts of the world. Examples of resident birds that exhibit a large degree of subspecific variation are the jay (*Garrulus glandarius*) and the blue tit (*Parus caeruleus*).

Terrestrial mammals

Approximately 197 species of mammals occur, 25% of which are endemic to the area (Cheylan 1991) (see Table 3.5). Three main factors influence the composition of the non-flying mammal fauna (i.e. excluding bats):

(1) multiple biogeographic origins, due to the proximity of and contribution from three continental land masses;

(2) repeated faunal turnover provoked by climatic variation during the Pliocene-Pleistocene period, including numerous intercontinental exchanges; and

(3) more so than for any other group of animals, species richness and distribution have been influenced by local human history, especially persecution and hunting, since the early or mid-palaeolithic.

One characteristic of the Mediterranean mammal fauna is the large number of browsers such as asses, goats, gazelles, hamsters, gerbils and jerboas.

Regional specificities

In contrast to birds, the mammal faunas of the four quadrants of the Basin differ sharply. The eastern part of the Mediterranean is richest in mammals (106 species), with as many as 23 species of Asian origin that do not occur elsewhere in the Basin. The second richest region is North Africa with 84 species, followed by the Aegean (80), and finally western Mediterranean Europe, with 72 and 77 species in Italy and Iberia, respectively (see Table 3.5).

Non-flying mammals are of course more sensitive to physical barriers than bats or birds, which means that the mammal faunas of southern Europe, the Middle

Table 3.5 Numbers of mammal species in different regions of the Mediterranean Basin (excluding islands). Figures in parentheses in first column are the numbers of families and species in each group for the entire Basin.

Group	North Africa	Iberia	Italy	Balkans	Anatolia	Near East
Hedgehogs (1, 4)	1	2	1	1	1	1
Shrews (1, 14)	6	7	8	7	4	1
Moles (1, 4)	0	4	3	2	1	1
Bats (7, 39)	25	26	27	25	34	9
Primates (1, 1)	1	1	0	0	0	0
Canids (1, 3)	2	2	2	3	3	3
Ursids (1, 1)	0	1	1	1	1	0
Mustelids (1, 9)	5	5	6	5	6	7
Genet, mongooses (2, 3)	2	2	0	0	2	1
Cats, felids (1, 7)	4	2	1	2	5	5
Hyraxes (1, 1)	0	0	0	0	1	0
Pigs (1, 1)	1	1	1	1	1	1
Deers (1, 3)	1	2	2	2	2	3
Goats, sheep, gazelles (1, 9)	3	2	2	2	5	3
Squirrels (1, 5)	2	1	1	2	1	1
Glirids (1, 5)	1	2	4	5	5	1
Porcupines (1, 2)	1	0	1	0	1	1
Mole–rats (1, 3)	0	0	0	1	2	0
Mice, rats (1, 16)	8	6	6	10	11	8
Gerbils, hamsters (1, 26)	15	0	0	2	9	9
Voles (1, 19)	0	8	5	8	7	4
Jerboas (1, 4)	2	0	0	0	3	2
Gundis (1, 1)	1	0	0	0	0	0
Hares, rabbits (1, 3)	2	3	1	1	1	1
Elephant–shrews (1, 1)	1	0	0	0	0	0
Total	84	77	72	80	106	62

Source: after Cheylan 1991

East, and especially North Africa are quite distinct. Although only 14 km wide, the Strait of Gibraltar has effectively isolated Europe from Africa for non-volant mammals since its opening after the Messinian Crisis. As a consequence, mammal faunas on the northern side of the Sea are basically Euro-Siberian in origin with wild boar (*Sus scrofa*), deer (*Cervus, Capreolus*), and the brown bear (*Ursus arctos*) as typical elements. Apart from a few exceptions, such as the porcupine (*Hystrix cristata*) in southern Italy, the Barbary macaque (*Macaca* sylvanus) in southernmost Spain, and some rodents and shrews (*Mastomys*, *Xerus*, and *Crocidura*), mammal species of tropical origin were eradicated from the Euro-Mediterranean regions by the beginning of the Pleistocene.

Although many mammals of North Africa are of Palaearctic origin (41%), in contrast to most other groups of animals and plants a large number of species are Afro-tropical or Saharo-Sahelian. Colonization of North Africa by typical Afro-tropical elements such as large felids, elephants or rhinos presumably occurred long ago when the Sahara region was more humid than it is today. More recently, at both the western and eastern extremities of the desert, along the shores of the Atlantic Ocean and through the Nile River valley, various mammals of tropical origin also entered the Mediterranean. At the western end of the desert, small shrews and rodents (*Crocidura*, *Mellivora*, *Xerus*, *Mastomys*, and *Acomys*) colonized western North Africa, mainly Morocco, from the south. Examples of species of tropical origin that colonized the Near East through the Nile valley are the genet (*Genetta genetta*), a mongoose (*Herpestes*), *Procavia*, the bubal antelope (*Alcephalus busephalus*), *Acomys*, a large fruit-eating bat (*Rousettus aegyptiacus*), and the ghost bat (*Taphozous nudiventris*). The macaque is the only native primate of the Basin, living in disjunct populations in Algeria and Morocco. Occasional vagrants in North Africa of species living in the Saharo-Arabian region include the hunting dog (*Lycaon pictus*), the guepard (*Acynonyx jubatus*), and several gazelles (*Addax nasomaculatus*, *Gazella dama*, and *Oryx damah*).

In eras when the Sahara desert did not reach the shores of the Mediterranean Sea, that is, in the last glacial episodes of the Pleistocene, immigration of Palaearctic elements contributed to the mammal fauna of North Africa through the Levantine corridor as we saw in Chapter 2. During those periods, a belt of Mediterranean-like habitats, the so-called 'eastern route', extended along the shores of the Sea, allowing the colonization of North Africa as far west as Morocco by species of boreal and south-east Asian origin. Later, when the Sahara desert extended northward to the Sea, this belt disappeared and was replaced by arid habitats that forged a link between North Africa and the Near East. Examples of species adapted to such habitats, and co-occurring in both regions thanks to this ancient link, are asses, antelopes, gerbils, and jerboas.

Today, the mammals of the Mediterranean Basin include few local or regional endemics; notable amongst these are the amazing Etruscan shrew (Box 3.3), the Spanish hare (*Lepus granatensis*), and the spiny mouses (*Acomys minous* in Crete, and *A. cilicicus* in Turkey). In Mediterranean islands, because of the mass extinction of the endemic mammal fauna that started in the upper Pleistocene, only two endemic species of shrews are left, one in Sicily (*Crocidura sicula*), and one in Crete (*C. zimmermani*). Endemic rodents include several gerbils and no fewer than eight species of voles (*Pitymys* spp.). This last group, of boreal origin, provides us with the best documented example of mammal speciation in Mediterranean refugia (Chaline 1974).

Box 3.3 The world's smallest mammal

Among many other oddities, the Mediterranean region harbours the Etruscan shrew (*Suncus etruscus*), which grows to be no more than 3.6–5.2 cm long and 1.6–2.4 g in weight (Jürgens *et al.* 1996). This typically Mediterranean species, with very long hair for a shrew, especially on the tail, lives only within the limits of the thermo- and meso-mediterranean life zones and is widespread wherever it finds suitable habitats warm enough to allow for survival through the winter. Suitable habitats include fallow terraces, stone walls, stone piles, ruins, and traditional human settlements, all of which provide it with shelter, food, and protection against predators (Fons 1975). A buffered microclimate between or under stones makes this kind of habitat particularly suitable for this mammal, as well as for two other shrews, *Crocidura suaveolens* and *C. russula*, which occur in the same type of environment. Although the species is active all year round, it is in summer months, from early June to late September, that numbers are greatest as a result of high reproductive rates. In spite of its small size, the Etruscan shrew is a fearsome predator for any living prey, provided it is small enough to be tackled: insects, worms, myriapods, and snails are all eaten voraciously because its survival requires incessant feeding. As expected for such a small endothermic organism, this species also exhibits the most outstanding physiological records of all mammals. Electrocardiogram recordings have shown that the mean heart rate of resting animals at ambient temperature (22 °C, which is the mean temperature experienced by this species while resting in its nest within stone walls), is *c.* 835 min^{-1}, a value that can reach 1093 min^{-1} in active animals. The mean resting respiratory rate is 661 min^{-1} and the highest value recorded was 894 min^{-1}. This species also has the highest mass-specific energy consumption of all mammals. At ambient temperatures, the Etruscan shrew consumes 267 ml O_2 kg^{-1} min^{-1}, 67 times as much as resting humans (Jürgens *et al.* 1996)!

Marine biodiversity

Although this book is primarily devoted to land biota, we will briefly consider marine biodiversity. Here, as in all terrestrial groups, biological diversity of marine plants and animals is extremely high with *c.* 50% of the species being endemic to the Basin (Tortonese 1985). The Mediterranean Sea is one of the richest in the world for biodiversity, harbouring 7.5% of the world marine animal taxa and 18% of the world marine flora in an area covering only 0.82% of the 'world ocean' (Bianchi 1996). The number of species inhabiting the Mediterranean Sea ranges from 2.2% of world species richness for Echinoderms to 18.4% for sea mammals. However, these percentages may be overestimated because the Mediterranean Sea has been studied more than many other seas and oceans. Summarizing data from

Table 3.6 Species richness of some groups of aquatic plants and animals in the Mediterranean Sea.

Group	No. of species	Proportion of world species richness (%)	Group	No. of species	Proportion of world species richness (%)
Plants			Invertebrates		
Red algae	870	17.4	Sponges	600	10.9
Brown algae	265	17.7	Cnidarians	450	4.1
Green algae	214	17.8	Bryozoans	500	10.0
Sea grasses	5	10.0	Annelids	777	9.7
			Molluscs	1376	4.3
Vertebrates			Arthropods	1935	5.8
Cartilaginous fishes	81	9.5	Echinoderms	143	2.3
			Tunicates	244	18.1
Bony fishes	532	4.1	Other invertebrates	c. 550	4.1
Reptiles	5	8.6			
Mammals	21	18.4			

Source: after Bianchi 1996

various sources, Bianchi (1996) estimated at roughly 8500 the number of macroscopic species existing in the Mediterranean Sea today (Table 3.6).

As in terrestrial biota, Mediterranean marine biota are composed of species with many different biogeographic origins: Atlanto-Mediterranean, pan-oceanic, palaeo-endemic of Tethyan origin, neo-endemic, and subtropical. An additional group invaded the eastern part of the Sea when the newly dug Suez Canal connected the Mediterranean Sea to the Red Sea at the end of the nineteenth century (see Chapter 9). Thus, the Sea is a crossroads for marine life forms, just as the lands surrounding it are for land biota.

Several historical factors have contributed to the high marine diversity. These include the variety of climatic and hydrologic conditions in the western half of the Sea relative to the south-east quadrant. In the former, primarily temperate-zone biota are found, while in the latter there are many subtropical species. However, although the Mediterranean Sea is a remnant of the warm equatorial Tethys Sea, the Messinian Salinity Crisis, which led to a nearly complete desiccation of the sea, resulted in a mass extinction of the former marine biota and the disappearance of many palaeotropical elements. It was not until the re-opening of a connection with the Atlantic Ocean, about 5 Myr BP (see Chapter 1), that the Mediterranean was repopulated by biota of Atlantic origin. Thus, from a biogeographic point of view, the Mediterranean Sea is fundamentally an 'Atlantic Province' (Briggs 1974).

Despite the very high species richness of most groups of marine organisms (Table 3.6), including sea turtles (Box 3.4), population sizes are usually low in Mediterranean Sea biota. One never finds the huge, spectacular seabird colonies that are so characteristic of steep cliffs in the North Atlantic. The low productivity

Box 3.4 Sea turtles

Five species occur more or less commonly throughout the Mediterranean Sea, among which the caouan (*Caretta caretta*) and the green sea turtle (*Chelonia mydas*) are the most abundant. They feed in deep waters upon a variety of animals, mostly deep sea invertebrates (salps) and jellyfish, but also venture into shallower waters where they prey upon crabs, urchins, and molluscs. The large lute turtle (*Dermochelys coriacea*), which can weigh up to 500 kg, formerly bred in the Mediterranean, but is only a rare visitor there today. Two other rare visitors are *Eretmochelys imbricata* and the Kemp turtle (*Lepidochelys kempii*). The caouan still regularly breeds in the eastern Mediterranean, in the Ionian islands, Peloponnese, southern Turkey, Cyprus, and Israel as well as in Libya and perhaps Tunisia. Since the beginning of this century, many breeding sea turtle sites have been deserted or destroyed, for example in Malta, Sicily, Sardinia, and Corsica. Only the green sea turtle still breeds in fairly large numbers in Turkish waters and to a lesser extent in Israel and Cyprus. However, for both the caouan and green sea turtle the destruction of breeding sites resulting from tourist and industrial encroachment combined with fishing accidents has led to a sharp decline in populations (Delaugerre 1988).

of the Mediterranean Sea also explains why most of the Basin's fisheries are relatively minor as compared to those of the Atlantic.

Convergence and non-convergence among mediterranean-type ecosystems

A long-standing hypothesis upheld by many scientists is that similarity of bioclimates in different parts of the world should result in similar adaptations arising among unrelated organisms. However, in the case of the Mediterranean biota, the specific and unique history that all groups of organisms have undergone has marked indelibly their evolutionary trajectories and ecological attributes. Therefore, the relative merits of the arguments for and against convergence are worth considering briefly.

Apart from the Mediterranean Basin, four other areas in the world are characterized by 'mediterranean-type' climate and ecosystems: central Chile, southern and central California, the Cape Province of South Africa, and disjunct parts of south and south-west Australia (Fig. 3.6). As a result of general atmospheric circulation patterns and cold off-shore ocean currents, these areas are almost all located on the western shores of continents, between 30–35° and 40–43° latitude, and have an adjacent arid region to the south, north, or east of them.

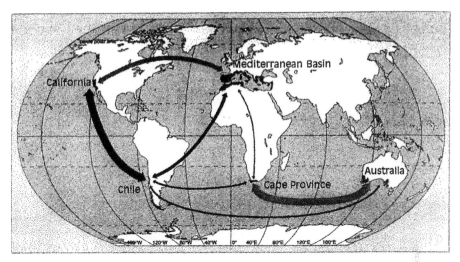

Fig. 3.6 The five areas of the world with a mediterranean bioclimate. Relative thickness of arrows between pairs of regions indicates floristic similarities, which appear to follow overall biodiversity trends for several groups of animals as well. (After F. di Castri, personal communication.)

If we compare the floristic diversity of the Mediterranean Basin with that of these other four areas, taking into account the much smaller surface areas involved there, then the Mediterranean appears as the poorest in number of plant species per unit area (Table 3.7). But it is third to Australia and only to the Cape Province of South Africa in the percentage of endemic species. Despite their considerable differences in size, the floras of the two mediterranean-climate areas of the Northern Hemisphere (the Mediterranean region and California) have lower proportions (50% and 35%, respectively) of endemic plant species than those of both the Australian and South African regions, which are considerably smaller in size (Table 3.7).

Furthermore, the histories of humans in these five regions differ greatly, as well as their impact on local floras and associated vegetation. Table 3.8 compares the relative duration of human occupation in the various areas. One prominent feature of the Mediterranean Basin, as compared to these other regions, is the considerably longer period of human occupation and intensive exploitation of resources. Consequently, human as opposed to non-human determinants of biodiversity must have been of paramount importance in the Mediterranean Basin. This idea can be presented in the form of an 'impact factor' (right hand column of Table 3.8).

Convergence

Given the particular constraints of the mediterranean bioclimate, major patterns of ecosystem structure and function have been thought to involve different kinds of species and communities that independently acquire similar sets of adaptations in these different areas. The number of publications on evolutionary convergence

Table 3.7 Comparison of plant species diversity in the five mediterranean-climate areas of the world.

Region	Area (km² × 10³)	No. of species	Species/area ratio	No. of endemic species	Endemics in total no. of species (%)	Endemism as function of surface area
Mediterranean	2300	25000	10.9	12500	50.0	5.43
California	320	4300	13.4	1505	35.0	4.70
Cape Province	90	8550	95.0	5814	68.0	64.60
Central Chile	140	2400	17.1	648	27.0	4.63
S + SW Australia	310	8000	25.8	6000	75.0	19.35

Sources: after Cowling *et al.* 1996; Hopper 1992; Arroyo *et al.* 1994; Keeley and Swift 1995

Table 3.8 Human occupation histories of the major mediterranean-climate regions.

Region	Arrival of indigenous people	Indigenous agriculture, livestock husbandry	First European settlement	Introduction of cultivated cereals and non-native livestock	Impact factor (time + habitat change)
Mediterranean	> 500000[1]	10000	–	10000	++++++
California	10000	Scarce	229	140	+++
Cape Province	> 200000	Scarce	300	250	+++
Central Chile	10000	Scarce	465	460	+++
S + SW Australia	40000	Nil	122	100	++

[1] All figures are in years BP.
Source: after Fox and Fox 1986

between and among taxa and species assemblages is enormous (e.g. Cody and Mooney 1978; di Castri *et al.* 1981; Schluter and Ricklefs 1993; Hobbs *et al.* 1995). Convergence in form and function has been unequivocally demonstrated in different mediterranean-type phyla, especially plants, reptiles, birds, and mammals. Convergence at the community-wide level is more difficult to demonstrate.

Perhaps the two most striking resemblances among the different mediterranean-type ecosystems are their remarkable floristic richness in relation to their geographic size and the preponderance of evergreen sclerophyllous trees and shrubs belonging to unrelated plant families. This leaf type and the generally low shrubby vegetation structure usually associated with it, are arguably a logical outcome of the bimodal mediterranean climate, with the dry period falling at the hottest season of the year. Yet, as will be explained in Chapter 7, there are several ways to consider the evolutionary significance of sclerophylly. Many studies support the view that the morphology (life forms) of plants and the overall physiognomy of plant communities are quite similar in the five regions with mediterranean bioclimates (di Castri *et al.* 1981).

However, other factors must also be considered, such as different histories of the biota, and the effects of fire and grazing, which vary from one region to another. For example, natural fire is a critical ecological factor in some but not all of the regions. It is much less common in California and the Mediterranean Basin than in South Africa or south-west Australia and uncommon in central Chile (Cowling *et al.* 1996). Similarly, grazing pressure over the millennia has not had the same importance in the five regions. Transformation of native ecosystems and biota has proceeded exceedingly fast in Chile, California, South Africa, and Australia since European-style agriculture and livestock husbandry were introduced 100 to 500 years ago. Yet the ecological and evolutionary impact of livestock grazing and browsing has been incomparably more profound in the Mediterranean Basin. Both the fires and grazing imposed over 10 millennia by pastoralists and agriculturalists have certainly contributed in important ways to the floristic diversity and shrubby vegetation structures of the Mediterranean Basin. The same cannot be said for the other mediterranean-climate areas.

If a certain degree of convergence can be seen for the dominant evergreen trees and shrubs, no such trend is found for the understorey plants, for which soil types appear to play a more important role than climate. In South Africa and south-west Australia, small coastal strips of sedimentary limestone occur in a larger matrix of metamorphic parent rocks. In central Chile and California there is hardly any limestone at all but in the Mediterranean Basin limestone is by far the major type of parent rock. These pedological* and lithological* differences among the different mediterranean-climate regions are reflected in vegetative differences. For example, the soils of the Australian and South African regions are extremely poor in phosphorus and other essential elements. As a result, plants there have evolved varied mechanisms, both in their root systems and in their diverse mutualistic relationships with fungi and bacteria, to facilitate the absorption of nutrients (Lamont 1982).

The end of a myth?

In spite of the enthusiastic impetus given to studies of convergence, many authors have contested the generalization of convergent evolution in regions with mediterranean climates, pointing out the many *divergences* among the regions (e.g. Shmida and Whittaker 1984; Blondel *et al.* 1984). As shown above, these differences may be related to factors such as soil fertility, fire, topographic and climatic heterogeneity, and human occupation histories. Two non-human historical determinants must also be considered: sorting processes (defined in Chapter 2), and phylogenetic constraints (i.e. the conservation of ancient traits that may have lost their adaptive significance). Two examples follow to illustrate the importance of these historical effects in explaining composition patterns and community structure in mediterranean-type ecosystems.

1. Sorting processes in woody plant taxa

Basically, the idea of convergence of living systems among the different mediterranean-type regions of the world leads one to predict that organisms would evolve characters that are adapted to a mediterranean climate, whatever their origin and history. The convergence hypothesis predicts that the large variety of growth forms and life-history traits that characterize extant mediterranean-type floras result from evolutionary responses to a mediterranean-type bioclimate. Alternatively, if phenotypes* and life-history traits of plant taxa that evolved *before* the establishment of a mediterranean climate still persist *after* the large climatic and ecological changes associated with the appearance of this climate, then historical factors must be considered of prime importance. Extant Mediterranean assemblages of plants include species that originated at different geological times and, for many of them, under tropical conditions. The question thus arises: do these co-occurring taxa sets share morphologies and life-history traits specifically adapted to present-day Mediterranean conditions? If so, extant plant assemblages would be the result of sorting processes, keeping only mediterranean-adapted species among plants of different origins.

Analysing variations in life-history traits (e.g. summergreenness,* spiLLnescence,* sclerophylly, sexual reproductive systems, and seed dispersal) among the woody plant flora of Andalusia, southern Spain, Herrera (1992) found no evidence of differential extinction events among pre-Mediterranean genera in the regional flora he analysed. This means that certain traits observed today already existed among the set of woody plants present in the area when the climate was tropical in the upper Miocene and lower Pliocene. Life-history traits that evolved under tropical conditions have largely survived as 'ecological phantoms' despite a dramatic shift in overall climatic regime. In contrast, all 'new' taxa that differentiated after the establishment of a mediterranean climate clearly evolved adaptations to mediterranean climatic conditions. This example demonstrates that observed traits are not always adaptive and do not necessarily fit the 'adaptationist programme', as pointed out by Gould and Lewontin (1979). The conservation of ancient traits in modern plant species casts a doubt on the hypothesis of convergence as an evolutionary response to similar bioclimatic conditions.

2. Convergence among bird communities

Evolutionary convergence among bird communities of different areas with a mediterranean bioclimate was studied by Cody (1975) for California and Chile, and by Blondel *et al.* (1984) for Mediterranean France as compared to California and Chile. Blondel and colleagues compared ecomorphological* configurations of the bird species, in particular their shape, among different mediterranean regions. The basic assumption in this study was that ecomorphology reflects the different means by which a given species utilizes resources. Comparisons were made among mediterranean bird communities along matched habitat gradients of increasingly complex vegetation structure, from shrubland to forest. These data, involving 31 species in France, 31 in California, and 38 in Chile, were then compared to a non-mediterranean gradient in France (Burgundy, 42 species) that was chosen as a control group. Convergence would be demonstrated if there were more overlapping or similarities within the ecomorphological 'spaces' of mediterranean communities than between any of them and those of non-mediterranean Burgundy. Statistical analyses revealed considerable overlap among all four regional 'spaces'. This result was interpreted to mean that the mediterranean communities do not resemble each other any more than the non-mediterranean control. Only the hummingbirds of Chile and California scored differently on the morphological 'space' but this is not surprising as this group has no equivalent in western Europe. The lack of convergence was attributed to differences in the phylogenetic origin and biogeographic history of the different sets of species. Just as was the case for woody plants in Andalusia, morphological, physiological, and behavioural constraints on these bird lineages of different origins and history presumably had more influence on species assemblages and species-specific habitat requirements than their sharing a similar type of environment.

However, conclusions from tests of convergence depend on the level of similarity and the choice of variables used. Clearly, convergence may exist for some community attributes, for instance the total numbers of species, but not for others such as those that involve large evolutionary changes in species and genera. Although convergence in morphology, structure and, presumably, ecological function, is most likely to take place among groups of organisms that depend strongly on climatic variation and seasonal patterns of nutrient cycling (e.g. plants, invertebrates, and lizards), it is less likely for homeotherm vertebrates that rely more on the structural attributes of the ecosystems in which they live. Thus, convergence in such attributes as species richness and community structure may be a by-product of similar patterns of resource-sharing in relation to the structure of habitats. As for plants, historical and phylogenetic constraints may limit adaptation such that some animals do not always evolve life-history traits tightly adapted to the particular environment in which they now occur.

4   Scales of observation

The regional tapestry

At almost any spatial scale, from satellite images taken 200 km above the surface of the Earth down to a single metre of ground viewed from a standing position, the Mediterranean Basin shows striking patterns of biological patchiness that reflect its topographic, climatic, geological, and edaphic* heterogeneity. This patchiness derives to a large degree from the profound 'tinkering' with biota and landscapes that farmers, herders, and woodcutters have practised here over the past millennia. In this chapter we present several approaches for bringing Mediterranean biodiversity into focus and evaluate the components of biodiversity at various scales, especially habitat- and landscape-scales. Firstly, we will explain how habitats and communities in the Basin are organized in well-defined 'life-zones' that range along gradients of altitude and/or latitude as a response to climatic variation. Secondly, to give an insight into the spatial turnover of habitats and communities, we will go for a stroll across landscapes, along 'transects' ranging from sea level to the summit of mountain ranges that limit the Mediterranean. Transects have been chosen in contrasting regions: two in the north-west quadrant (France), and two in the south-east quadrant (Near East). Finally, we will explore the dynamics of biological diversity on still smaller scales.

A succession of life zones

Because terrain in the Basin is mountainous almost everywhere, 'life zones' are elevational/latitudinal belts of plants and animals that tend to share ecological affinities and to occur together. Fundamentally, the term refers to each band within the succession of species and communities found along gradients of altitude (and latitude) (Ozenda 1975; Quézel 1985). The most convenient method for characterizing life zones and recognizing them in the field is to note the two or three dominant tree and shrub species that serve as bioindicators of their particular life zone. Although such communities or 'associations' are not sharply defined in nature, the life zone concept is quite powerful for purposes of field orientation. However, microclimatic conditions can pull certain species or even assemblages far from the expected limits of their distribution.

Eight readily recognizable life zones are apparent along the temperature gradients that occur as one moves up mountain slopes, or traverses the Basin from south to north (Fig. 4.1). The first one, at the lowest altitude/latitude, in the warmest regions, is the 'infra-Mediterranean' life zone, found only in frost-free

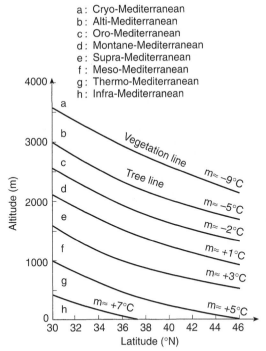

a : Cryo-Mediterranean
b : Alti-Mediterranean
c : Oro-Mediterranean
d : Montane-Mediterranean
e : Supra-Mediterranean
f : Meso-Mediterranean
g : Thermo-Mediterranean
h : Infra-Mediterranean

Fig. 4.1 Altitudinal/latitudinal gradients showing the zonation of the various vegetation belts, or life zones, in the western Mediterranean area; m = average of the minima of the coldest month. (Reproduced with permission after Le Houérou 1990 *in* Blondel and Aronson 1995.)

parts of south-west Morocco and Macaronesia. In south-west Morocco its indicator species are the argan tree (*Argania spinosa*), of the tropical family Sapotaceae, and an endemic acacia, *A. gummifera*. This is one of the very few acacias, of the 120 or so species found in Africa, that is considered native to a Mediterranean region. In the Canary Islands and Madeira, this life zone is far more complex, as will be briefly discussed in Chapter 5. Notably it includes a large number of tropical plant species no longer found elsewhere in the Basin.

The second life zone, called 'thermo-Mediterranean', is found at low altitudes in all the warmer parts of the Basin, especially in North Africa and the Near East, where it is characterized by dense coastal woodlands of the wild olive tree (*Olea europaea* subsp. *oleaster*) and carob or Saint John's bread tree (*Ceratonia siliqua*). Other common components are the lentisk (*Pistacia lentiscus*), false olive (*Phillyrea media*), laurel (*Laurus nobilis*), and Barbary thuja (*Tetraclinis articulata*), which is a North African and southern Spanish relative of the cypress. In some parts of the western Mediterranean, this life zone also contains the cork oak (*Quercus suber*), cluster pine (*Pinus pinaster*), and Mediterranean dwarf palm (*Chamaerops humilis*) (Fig. 4.2). Not surprisingly, the plant communities found in this life zone are all heavily imprinted with changes wrought by humans. Many areas have been planted with stone pine (*Pinus pinea*) in this century. One prominent feature of this life zone is that nearly all woody plant species are

Fig. 4.2 The Mediterranean dwarf palm (*Chamaerops humilis*). It occurs in rocky hillsides near the coast where it is frequently dug out for use in gardens. In Morocco it is often considered a troublesome weed in cultivated fields.

evergreen and sclerophyllous (see Chapters 3 and 7 for discussion of this feature).

Next in increasing altitude (and latitude) comes the 'meso-Mediterranean' life zone, the most familiar one. Here one or sometimes two species of evergreen oaks dominate a wide variety of woodlands and shrublands. The evergreen holm-oak is the dominant tree species in the western and central parts of the Basin, and its vicariant, *Q. calliprinos*, dominates in the east. Aleppo pine (*Pinus halepensis*) and Calabrian pine (*P. brutia*) occupy large parts of this life zone in the western and eastern parts of the Basin, respectively. In many regions, formerly oak-dominated formations have been more or less replaced by artificial plantations of the fast-growing Aleppo and Calabrian pines over several millions of hectares. In other areas one finds a remarkably monotonous anthropogenic community dominated by a dwarf, prickly form of the kermes oak (*Quercus coccifera*) in the western Mediterranean, or by spiny legume shrubs such as *Calycotome villosa* and *Genista acanthoclada* in the east.

The fourth life zone in altitude is the 'supra-Mediterranean', the domain of deciduous oak forests that occur around almost all the Basin. On each mountain range of the area, a distinctive and yet readily recognizable medley of deciduous and some evergreen tree species occurs, in a broad belt from about 500 to 1000 m, depending on latitude and slope. No other mediterranean-climate area outside the Basin has a deciduous oak belt of this kind. Palynological* studies indicate that many of the dominant deciduous oak species found in this life zone today formerly occurred widely in the meso-Mediterranean life zone as well. In France and northern Spain, the downy oak (*Quercus humilis*) dominates the supra-Mediterranean life zone but other deciduous oaks do so elsewhere. Thus, from Italy to Turkey and the Levant are found the Turkey oak (*Q. cerris*) and several other species, including *Q. boissieri*, *Q. infectoria*, *Q. frainetto*, *Q. trojana*, and *Q. macedonia*. In this same belt, many other broad-leaved deciduous tree species also occur, such as hornbeams (*Ostrya carpinifolia*, *Carpinus orientalis*), hazelnuts (*Corylus* spp.), lime or linden (*Tilia* spp.), manna ash (*Fraxinus ornus*), several apple family members including mountain ash (*Sorbus* spp.) and several small-leaved Mediterranean maples (*Acer mouspessulanum*, *A. campestre*, and *A. opalus*).

In the north-east quadrant, this life zone also harbours two evergreen maples, *Acer sempervirens* in Crete and *A. obtusifolium* in the eastern Mediterranean mountains. Apparently these are the only two evergreen species of the more than 100 species of maples spread throughout the Northern Hemisphere. In North Africa and the southernmost parts of Europe, broad-leaved or semi-deciduous oak stands in this life zone are dominated by the Spanish oak (*Quercus faginea*) and the zeen oak (*Q. afares*). They have been so heavily exploited for timber, however, that only isolated relict patches subsist. Many of the formerly co-dominant tree and shrub species have been altogether eliminated by humans.

In the supra-Mediterranean life zone, various cold-sensitive Mediterranean plant species gradually disappear, first from north-facing slopes and then, higher up, from south-facing ones as well. These include lentisk, olive tree, kermes oak, honeysuckle (*Lonicera implexa*), rosemary (*Rosmarinus officinalis*) and buckthorn (*Rhamnus alaternus*). The pines of the lower life zone are gradually replaced here by other pines, especially southern varieties of the Scots pine (*P. sylvestris*) and the numerous subspecies of black pine (*Pinus nigra*).

As the altitude increases further the 'montane-Mediterranean' life zone replaces the supra-Mediterranean. It is normally dominated either by beech (*Fagus sylvatica*) or by conifers (pines, cedars, firs) although most of these forests have been so radically transformed that only a few intact remnants remain. In most countries around the Basin, a few venerable trees of great age can be seen, accompanied by an understorey sadly lacking in seedlings or saplings of the canopy species. In these montane forests, scattered centuries-old junipers and firs occur well above 3000 m in the Atlas Mountains, and up to 2600 m in the Taurus ranges of southern Turkey, bearing testimony to the majestic forests that once grew there.

The sixth life zone is the 'oro-Mediterranean', characterized by pines in the northern part of the Basin, with mountain pine (*Pinus uncinata*) being the most common species, along with the related Mugo pine (*Pinus mugo*) of central Europe. The former is usually a much bigger tree, growing up to 25 m tall, while the latter rarely exceeds 4 m in height. There are patches of white fir (*Abies alba*) and Norway spruce (*Picea abies*) in some parts of this life zone, which generally grow much further north in Europe. The rather rare *Pinus heldreichii* also occurs in high mountains on either side of the Adriatic. In peninsular Greece it is sometimes accompanied by *P. peuce* (see Fig. 4.3) and an eastern beech, *Fagus moesica*.

Continuing to climb, one moves into the seventh life zone, the 'alti-Mediterranean' or subalpine, which usually includes dwarf junipers mixed with diverse grasslands of *Bromus*, *Festuca*, *Poa*, *Phleum*, and other perennial grasses. A large range of herbaceous perennials also occurs here, many of which are endemic owing to the geographic isolation of Mediterranean mountaintops. Indeed, biogeographical features are often highly pronounced over relatively short distances in higher altitude Mediterranean life zones. This is well illustrated in southern Greece (Fig. 4.3), where vegetation in the thermo- and meso- life zones is rather uniform throughout the peninsula, but plant assemblages in higher life

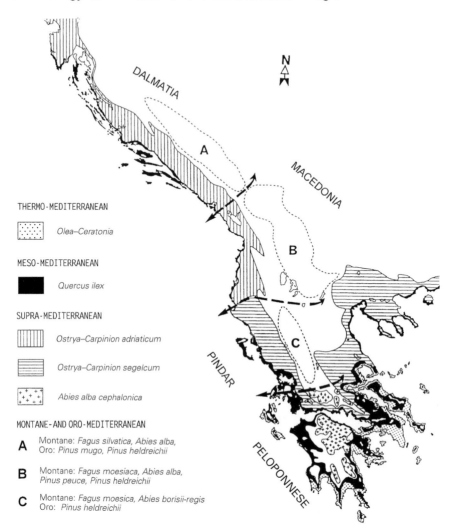

THERMO-MEDITERRANEAN

Olea–Ceratonia

MESO-MEDITERRANEAN

Quercus ilex

SUPRA-MEDITERRANEAN

Ostrya–Carpinion adriaticum

Ostrya–Carpinion segelcum

Abies alba cephalonica

MONTANE-AND ORO-MEDITERRANEAN

A   Montane: *Fagus silvatica, Abies alba,*
Oro: *Pinus mugo, Pinus heldreichii*

B   Montane: *Fagus moesiaca, Abies alba,*
Pinus peuce, Pinus heldreichii

C   Montane: *Fagus moesica, Abies borisii-regis*
Oro: *Pinus heldreichii*

Fig. 4.3 Schematic distribution of five life zones in peninsular Greece, showing regional variation especially in high mountains. (Reproduced with permission from Ozenda 1975.)

zones around each of the separate mountain ranges all differ considerably. Frequently included are various representatives of widespread alpine genera like *Viola, Androsace, Saxifraga, Linaria, Arenaria, Primula* and *Vicia*. In the Mediterranean mountains of France, Italy, and along the eastern Adriatic coast, yet another perennial legume–grass association occurs, consisting of *Onobrychis* and *Sesleria* spp. In southern Greece and western Turkey there is at these altitudes a very particular formation of daphnes (*D. oleiodes* and *D. pontica*) accompanied by a range of fescues (*Festuca* spp.), and a medley of *Primula, Soldanella* and other widespread alpine plants.

In the Taurus Mountains of southern Turkey, as in the High Atlas of Morocco and the Sierra Nevada of southern Spain, the subalpine life zone is occupied by treeless grasslands dotted with many spiny dwarf legume shrubs with evocative names like 'goat's thorn' or 'hedgehog' (especially *Astragalus, Erinacea,* and *Genista*), and a rich variety of bunch grasses of diverse genera.

Finally one reaches the eighth life zone, the cryo-Mediterranean. It is almost entirely devoid of vegetation except among rocks, scree, and gravel where a range of widespread alpine plants of the genera *Saxifraga, Androsace, Aubretia* and others do occur.

Transects

As explained in Chapter 1, Mediterranean landscapes are, as a rule, squeezed between the sea and the mountain ranges that encircle the Basin and dissect it, literally, into thousands of small, more or less isolated catchments. One convenient way to discover the extraordinary diversity of any one of the regions within the Basin is simply to make observations along a straight line, a 'transect' from the coastline where one happens to be to the top of the nearest mountain range, a distance typically of 100 km. It can thus be seen that altitude has a dramatic effect on physical environments and biota, since higher elevation almost always brings not only higher rates of precipitation but also more even distribution of rainfall throughout the year.

Having introduced the eight life zones, we will now briefly describe the succession or turnover of habitats along four transects in two highly contrasted parts of the Mediterranean, the north-west quadrant with its sub-humid climate, and the eastern Mediterranean with its semi-arid climate. Precipitation gradients contribute to differentiate habitats between the two regions. For example, habitats in the north-west quadrant receive nearly 10% of their annual precipitation during the summer, a situation that contrasts sharply with the weather patterns found in the eastern Mediterranean where no more than 2 or 3% of annual rainfall occurs during summer. For each transect, we will give details on different groups of animals, plants, or land use patterns, as appropriate.

North-west quadrant

We will look at two transects that extend from the coast to the top of Mt. Aigoual (1567 m) and Mt. Ventoux (1912 m) in southern France, represented by Figs 4.4 and 4.5, respectively. These two mountain peaks mark the limits of the Mediterranean bioclimate area. Five of the eight life zones are represented here; the infra- and thermo- do not occur due to the persistent cooling of the region by the mistral and tramontane winds, and the cryo-Mediterranean is absent due to insufficient altitude. In a day's drive one can see cave-dotted limestone hills, moonscape-like outcroppings of denuded karst, crumbling dolomitic cirques and, more sparsely, weathered granites, flints, and schists, with marble and mica intrusions. There are even basalts and spurs of bauxite and labradorite. Local microclimate and edaphic variations also play a leading role in creating the regional mosaics that are essential in maintaining biodiversity (see Chapters 5

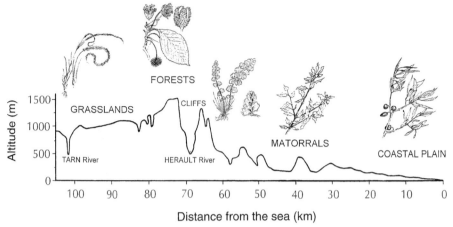

Fig. 4.4 Transect showing a cross-section of the Languedoc–Roussillon region, from the coast to Mount Aigoual and the plateaux beyond. Typical plants of each habitat shown are, from left to right, angel's hair grass (*Stipa pennata*), beech (*Fagus sylvatica*), Petrarch's fern (*Asplenium petrarchae*), holm-oak (*Quercus ilex*), and olive tree (*Olea europaea*).

and 6). Starting from the sea and moving inland along these two transects, each time we encounter a new habitat it will be described briefly.

Until the first half of this century, the coastal plain of southern France were still largely undeveloped; consequently, the fauna and flora were far richer than today. Even as late as the 1930s, in fact, the coastal marshes and brackish lagoons from Marseilles to Valencia (Spain) were considered unusable for agriculture and unsafe for human habitation because of malaria. After the Second World War, however, large drainage operations were undertaken, and the vast marine and brackish biota that once occurred here began to decline.

Just behind the coastal lagoons and marshes stretches a strip of arable land, about 13 km wide in Languedoc and highly variable in Provence, due to the presence of several large river deltas. In Provence there is a complex mosaic of wheat fields, orchards, vineyards, hay meadows, rice fields, and early fruit and vegetable plots. Each piece of land is separated from its neighbours by tall hedges of cypress trees, whose function is to break the force of the mistral wind. In Languedoc, the great bulk of cultivated land has been planted as vineyards since the end of the last century.

Here and there in the north-west quadrant one encounters dry and flat steppes. For example 'La Crau', the former delta of the Durance River in Provence, is an area of some 50 km^2, covered by large round stones. Several species of animals occur here that are typical of the semi-arid habitats of central Spain and North Africa, of which the most spectacular is the pin-tailed sandgrouse (*Pterocles alchata*). The Crau is the only breeding place in France of this species which, with no more than 100–150 breeding pairs, is among the most threatened bird species of the country. Other rare breeding birds found in this coastal strip include the lesser kestrel (*Falco naumanni*), the endangered little bustard (*Otis tetrax*), and

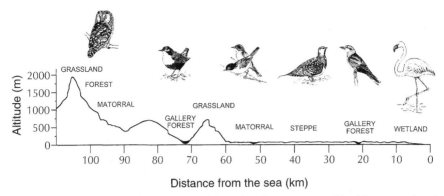

Fig. 4.5 Transect through Provence, from the Camargue to Mt. Ventoux, showing bioclimatic zones, vegetation life zones, and prominent bird species that characterize them. Birds depicted are, from left to right, tengmalm's owl (*Aegolius funereus*), dipper (*Cinclus cinclus*), Sardinian warbler (*Sylvia melanocephala*), pin-tailed sandgrouse (*Pterocles alchata*), roller (*Coracias garrulus*), and flamingo (*Phoenicopter ruber*).

several steppe species that hardly occur elsewhere in France, such as larks (*Melanocorypha calandra*, *Calandrella brachydactyla*), and the pied wheatear (*Oenanthe hispanica*).

Several times along the transects, both in Languedoc and in Provence, a number of rivers cut through this hilly region, and on leaving the coastal plains one frequently encounters narrow riverine forests. Mediterranean riverine forests used to be rich and varied ecosystems, punctuating landscapes and regions and supporting a high diversity of plants and animals. Many species of plants and animals from non-Mediterranean biota penetrate the Basin in these particular habitats thanks to the abundance of standing water present all year round. A Mediterranean note is brought to the bird assemblages of this habitat by species such as the colourful roller (*Coracias garrulus*), whose relatives are mostly found in tropical Africa, Australia, and Asia.

In the coastal hills just above the arable plain, one enters the realm of evergreen shrublands, called matorrals, whose characteristics may vary greatly depending on local conditions and land use history. Among the most degraded forms are the dwarf, depauperate, kermes oak shrublands referred to previously, which generally do not occur above 200 m altitude. Most of them are the result of millennia of exploitation and transformation of former woodlands. In some areas, however, they were intentionally managed until quite recently as a 'pasture' for the herbivorous cochineal insect *Kermococcus vermilio*, from which a strong, colour-fast red dye was obtained for dying wool and other fabrics. Indeed, the Latin name of the kermes oak, *Quercus coccifera*, means 'beetle-bearing oak'! Although the cochineal-raising practice was progressively abandoned after the invention of aniline dyes in the late 1800s, in southern France and parts of Spain the kermes oak stands have remained remarkably stable (see Box 8.4, p. 225).

A different group of matorrals is found in the 'garrigue*' region, which displays dense or open woodlands 4–8 m tall. The holm-oak is far and away the dominant tree here but it sometimes mingles with larger deciduous downy oaks (*Quercus humilis*) preserved to provide shade for livestock. Fire and charcoal production from holm-oak coppices have for centuries been the major use of this type of vegetation. In climbing up the slopes one inevitably encounters some kind of cliff or cave; these have a very particular assemblage of plants and animals that will be described in the next chapter.

One remarkable feature of the garrigues is their abundance of invertebrates, quite different from those found in northern Europe. Turning over rocks and logs will sometimes uncover the Languedocien scorpion (*Buthus occitanus*) with its large, yellow body, or the yellow-tailed black scorpion (*Euscorpius flavicaudis*). Among insects, three outstanding groups are the cicadas (see Box 4.1 and Fig. 4.7), various mantises, and the emblematic two-tailed pasha (*Charaxes jasius*) (Fig. 4.6). This large showy butterfly, with a 75 mm wingspan, is the sole Mediterranean representative of an otherwise tropical family, with its closest relatives occurring in Ethiopia and equatorial Africa. Its distribution is intimately tied to that of the strawberry tree (*Arbutus unedo*), on whose leaves it lays its eggs to assure a food source for the emerging caterpillars. The two-tailed pasha has two generations per year. Adults may be seen on the wing in May–June, and again in August–September. Adult butterflies sometimes feed on overripe figs in summer and can get completely 'drunk' from the juice!

In contrast to insects, the bird fauna of the garrigues is not very rich, as we explained in Chapter 3, but it is of great interest. The most unusual feature is the large number of Mediterranean warblers. These small birds are difficult to watch for more than a few seconds at a time because they are secretive and move rapidly among the thick bushes. They are segregated according to the overall structure of vegetation. The small dartford warbler (*Sylvia undata*) lives in the lowest scrub and the larger orphean warbler (*Sylvia hortensis*) lives in the highest shrublands. However, as many as four species may occur together in the same habitat. The sweetly singing nightingale (*Luscinia megarhynchos*) also occurs abundantly throughout this area.

Leaving the garrigues, we climb into the supra-Mediterranean life zone where mixed evergreen–deciduous forests dominate, and where an increasing proportion of plant and animal species have temperate, non-Mediterranean affinities. Dominant trees include several species of deciduous oak, maple, beech, Scots pine, and several species of mountain ash. Among plants, the most noticeable changes occurring with altitude are the appearance of evergreen plant species more common in northern forests than in the Mediterranean Basin, such as yew, holly, *Arctostaphyllos*, and *Pyrola*. Song thrush (*Turdus philomelos*), dunnock (*Prunella modularis*), and tree pipit (*Anthus trivialis*) are prominent birds here. In some places with mature mixed forests of beech and pines, one may hear the call of the black woodpecker (*Dryocopus martius*). Abandoned breeding sites of this woodpecker, in deep holes in the trunks of large trees, are sometimes colonized by Tengmalm's owl (*Aegolius funereus*), which is a typical representative of the bird fauna of the large forest blocks of the boreal belt of conifers.

In this life zone, intricate networks of smaller rivers stemming from the high

Fig. 4.6 The two-tailed pasha (>*Charaxes jasius*) adult and its caterpillar on a stem of its host plant, the strawberry tree (*Arbutus unedo*).

mountains collect waters falling on the huge karstic plateaux stretching northwards. These cold and oxygenated waters are the home of the brown trout (*Salmo trutta*) and the grayling (*Thymallus thymallus*). Bird species not yet encountered along the transects include the dipper (*Cinclus cinclus*) and the grey wagtail (*Motacilla cinerea*). Both these species are strictly associated with clear waters with a strong current, and indicate that we are not far from the northern limits of the Mediterranean area.

Upon approaching the summit of these mountains, in the oro- and alti-Mediterranean life zones, one finds a very particular habitat consisting of stony ground with some interspersed montane pines (*Pinus uncinata*) and low wind-adapted bushes of juniper (*Juniperus communis*). This habitat is home to a very

Box 4.1 Cicada life history

Cicadas are certainly the most popular insects in the Mediterranean because of their unmistakable stridulant singing in the hottest parts of the summer. Few people, however, are familiar with the extraordinary life cycle of these animals, which spend 95% of their lives underground. In July, females insert their long abdominal 'drills' into the bark of a tree where they lay as many as 300–400 eggs. Three months later, minute larvae hatch, pierce the bark to get out, and then undergo their first moult. They next drop from the tree, suspended by a delicate silk thread, and when they reach the ground dig a hole with their powerful forelegs, which resemble those of a tiny mole cricket. They then spend four years in the soil, feeding on roots. Between the egg and adult stages nearly 99% of the cicada larvae will die from predation or parasitism. On a warm June evening the few fortunate survivors will emerge from the soil through a small vertical tunnel they construct themselves. These nymphs will climb the nearest tree and seek a safe place where they can shed their skins and emerge as light green adults. Their chitin quickly gets hard and turns a blackish colour. Some hours later, the full-grown cicada takes off for its short adult lifetime, which lasts only a few weeks. This is the period when males 'sing', using the 'sawing organs' located at the base of their abdomen between the two hind legs. While 'singing', the males suck the sap of pines, olive, almond, or other species using their specialized mouth parts which they insert deeply into the tree's bark.

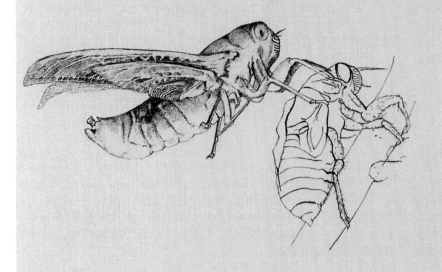

Fig. 4.7 A young cicada (*Tibicen plebejus*) emerging from its nymphal sheath.

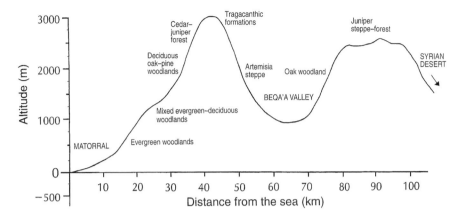

Fig. 4.8 Lebanon transect from Mediterranean coast to the Anti-Lebanon range and the Syrian desert. Indications of bioindicator plant species are given for each life zone.

particular set of bird species, like the wheatear (*Oenanthe oenanthe*) which breeds in holes under big stones, the citril finch (*Serinus citrinella*), and the crossbill (*Loxia curvirostris*), which lives in pines.

South-east quadrant

The part of the eastern Mediterranean region called Levant consists of *c.* 40 000 km² and is divided among Lebanon, northern Israel, Palestine, and north-western Jordan (Fig. 4.8). Compared to southern France, this region is far drier and more degraded. It is also the 'crossroads of crossroads', so to speak, the permanent contact zone between the Mediterranean world and the Asian continent. Anthropogenic impact is exceedingly important here (Naveh and Dan 1973; Le Houérou 1981; see Chapter 8). Human demographic expansion began about 10 000 yr BP, a figure to be compared with the 3500–2500 years of intensive human activities in the north-west quadrant.

Figures 4.8 and 4.9 provide highly simplified transects that proceed 100 km inland from the eastern Mediterranean coast at 33°N and 31°N, respectively. The first of these traverses Lebanon and ends at the edge of the vast Syrian desert. The second traverses Israel and Palestine, crosses the Dead Sea valley, the lowest place on earth at 400 m below sea level, and finally climbs to the top of the Mt. Edom range of central Jordan.

1. From the coast to Mt. Lebanon

Coastal dune areas support tall tufted grasses such as *Ammophila arenaria* and *Stipagrostis lanatus*, while cliffs and rocks near the shore harbour the silvery-leaved *Otanthus maritima* and the semi-succulent sea Samphire (*Crithmum maritimum*), whose fleshy leaves can be pickled to make a pungent condiment. Another remarkable plant in the 'spray belt' is the sea daffodil (*Pancratium maritimum*), with large white 'trumpet' flowers in early summer. This relative of

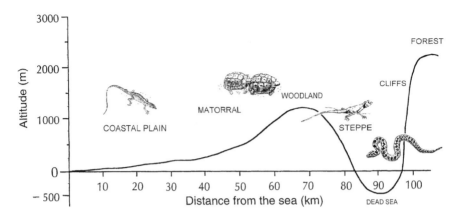

Fig. 4.9 Israel–Jordan transect, showing some prominent lizards and snakes for the various habitats: the fringe-toed sand lizard (*Acanthodactylus schreiberi*) in coastal sands; the Greek turtle (*Testudo graeca*) in the matorral zone; the Sinai agama (*Agama sinaitica*) in the Judean desert and the Dead Sea area, and the Palestine viper (*Vipera palestina*) in the Dead Sea area.

Amaryllis was formerly common along sandy seashores of the Mediterranean, the Black Sea, and the Caspian Sea, but is now quite scarce due to over-harvesting of the flowers and bulbs.

No less than 47 species and subspecies of endemic plants in 17 families are found in the coastal sand dunes of Israel (Auerbach and Shmida 1985). Reminiscent of the 'crossroads' character of the eastern Mediterranean, it is noteworthy that 29 of these taxa have strongest associations with the Mediterranean flora while 14 have strongest affinities with the Saharo-Arabian desert flora, and one grass, *Aristida sieberiana*, is a palaeo-relict of tropical origin.

A large number of snake and lizard species are found in the coastal area, including the fringe-toed sand lizard (*Acanthodactylus schreiberi*) (Fig. 4.9), which abounds on stabilized sands. Both this species and its relatives *A. pardalis* and *A. scutellatus*, of inland desert areas, have exceptionally long agile toes, which allows them to run about during daytime on extremely hot sands in search of their prey. Like most members of their family (Lacertidae), these lizards are endowed with remarkable camouflage equipment in their versatile skin colouring.

At low elevations near the coast, typical 'thermo-Mediterranean' woodlands once occurred, very similar to what is found across southern Europe to the Aegean and western Turkey. Only fragments survive today with species like the lentisk and carob, wild olive tree and the Aleppo, stone, and brutia pines. Among the oaks, *Quercus calliprinos*, the eastern vicariant of the evergreen holm-oak, is a dominant species, and wild cypress (*Cupressus sempervirens*) is also found; this latter derives its scientific name from the Greek word for Cyprus, an island where this evergreen tree was once native and extremely abundant. To see large populations of wild cypress today, one has to go to Crete, or to Mt. Elburz in northern Iran. Many wild and planted carob trees (Fig. 4.10) are also found in

Fig. 4.10 A branch of carob tree (*Ceratonia siliqua*), showing primitive inflorescence types and large fleshy pods of great value as animal fodder. As carob seeds are unusually uniform in weight they were used by the ancient Greeks as a measure of gold, hence the word 'carat', which is still used a measure of gold, silver, and diamonds. (Reproduced with permission from Zohary and Feinbrun 1966–86.)

Israel and Cyprus, dating from the days when animal husbandry was an important component of local agriculture.

At around 400–500 m elevation in the north-east quadrant, especially in Lebanon, the evergreen *Quercus calliprinos* and other typical thermo-Mediterranean elements gradually give way to a series of mixed-deciduous–coniferous woodlands typical of the meso-Mediterranean life zone. The Turkey oak (*Quercus cerris*) also occurs in Lebanon, as in Italy, the Balkans, and western Turkey, along with the Lebanon oak (*Quercus libani*). This latter species is recognizable from its very typical leaf traits, with serrate margins and coarse hairs on both sides of the leaves. Further south, *Q. boissieri* and tabor oak (*Q. ithaburensis*) are the two dominant deciduous oak species.

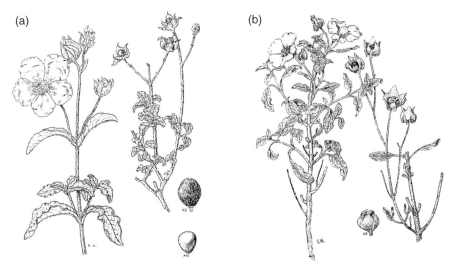

Fig. 4.11 Two eastern Mediterranean rock roses: (a) *Cistus creticus*, and (b) *C. salviifolius*, two of the most widespread species in the genus. (Reproduced with permission from Zohary and Feinbrun 1966–86.)

At these altitudes we also find *Pistacia palestinus*, the vicariant of terebinth (*Pistacia terebinthus*), and storax (*Styrax officinalis*), a deciduous tree of tropical origin with large, white, fragrant flowers. A related species of *Styrax* is the source of a valuable gum used to this day for incense and perfumes. The substance, called 'stacte', used in the sacred incense of Hebrew rites (Exodus 30: 34), may have been derived from storax (Hepper 1981). As throughout southern Europe and Turkey, rock roses form a dramatic part of the shrublands in coastal foothill areas in Lebanon and Israel, especially where grazing and fire-setting have been important in the recent past (Fig. 4.11).

Dozens of wild lily and iris family flowers occur here, especially among rocks. These include autumn crocus and saffron (*Colchicum* spp. and *Sternbergia* spp.), tulips, black iris, gladiolus, and *Bellevalia*, and a remarkable number of annual species. Several species of lizards, snakes, and turtles also abound in this zone, such as the land turtle (*Testudo graeca*), and the highly venomous and viviparous Palestine viper (*Vipera palestina*) (Fig. 4.9).

Climbing higher on the flanks of Mt. Lebanon in the supra-Mediterranean life zone, we meet several new coniferous species, including four junipers (*Juniperus drupacea, J. oxycedra, J. phoenicea*, and *J. excelsa*) and the black pine (*Pinus nigra*). A number of trees belonging to the apple family (Rosaceae) also occur in the premontane and alpine zones, for example wild pear (*Pyrus syriaca*), three species of wild almond (*Amygdalus*), and three of hawthorn (*Crataegus*). Other deciduous elements of Central Asian affinities include the deciduous gall oak (*Quercus infectoria*) and, at higher altitudes, both *Q. brantii* subsp. *look* and *Q. cedororum*, along with hornbeam and European ash.

In the montane-Mediterranean life zone of Lebanon, at 1500–1900 m, there are scattered stands of cilician fir (*Abies cilicica*) and the Lebanon cedar (*Cedrus libani*), the 'glory of Lebanon'. These handsome conifers, which can reach a height of 30 m, once covered the snow-clad mountain tops all the way from southern Lebanon to southern Turkey. Today, only two large stands of Lebanon cedar survive, in the Mt. Troödos range of eastern Cyprus, where an endemic sub-species occurs on ultrabasic rock, and in the Elmali region north of Antalya, in the eastern Taurus Mountains of southern Turkey. In Lebanon itself, only some 2700 ha of this emblematic tree survive, scattered over 14 sites above 1400 m on the western side of the Lebanon range. Apparently the species only occurs in spots where incoming fog collects in the afternoons and thus dramatically raises the relative humidity of the air. The absence of such conditions on the eastern flanks probably explains the absence of cedar groves there (Al Hallani *et al.* 1995).

Yet another stately conifer occurring at high altitudes in Lebanon is the Eastern savin (*Juniperus excelsa*) (Savin was the ancient name for Mt. Hermon). In Biblical and Greco-Roman times, these alpine conifers were prized by royal houses of the entire Near East and Eastern Mediterranean. Today, on Mt. Lebanon and Mt. Hermon, *Quercus libani* subsp. *look* (from the vernacular 'lik' of local shepherds) forms an open shrubby 'pygmy forest' mixed with isolated individuals of *Juniperus excelsa*.

On the dry eastern slopes of Mt. Lebanon's oro-Mediterranean life zone there occurs a distinctive 'hedgehog' or 'thorncushion' vegetation type, unknown in the western Mediterranean. This is the so-called 'tragacanthic' formation, named after a series of a dozen densely spiny *Astragalus* (or 'Goat's thorn') shrub species that occur in the high mountains of Lebanon, Syria, Iraq, Turkey, and Iran. When their stems and roots are pierced by a knife these shrubs yield a valuable gum used in pharmaceutical and food products. The centuries-old exploitation of these wild gum-bushes has been the principal determinant of this odd ecosystem's structure and composition.

Next on the Lebanese transect, the eastern slopes drop off precipitously to bring us to a dry range-land area in the rain-shadow of the mountains, where the semi-desert or steppe vegetation is dominated by grasses and shrubs used for grazing sheep and goats. The vegetation here is rich in Central Asian shrubs, hemi-cryptophytes*, bulbs, and annuals adapted to life under constant grazing pressure. The most prominent shrubs are wormwoods (*Artemisia herba-alba* complex), several of which yield a bitter juice referred to as wormwood in the Bible, La'ana in North Africa, and sagebrush in the western United States, where related species occur. As elsewhere in the Near and Middle East, in patches where these shrublands are badly degraded, various grasses (*Poa*, *Stipa*, etc.), saltbushes and thistles predominate.

By contrast, the fertile Beqa'a valley is almost entirely cultivated with a range of crops irrigated with water pumped from the Litani River. Climbing up the slopes of the Anti-Lebanon range, amidst fields and pastures, one once again finds woodland fragments and matorral, mixed with bath'a*. The Anti-Lebanon is almost entirely deforested, except for *Quercus boissieri* woodland fragments at mid-altitudes, where conifers are curiously absent. In the montane belt, occasional vestiges of a 'steppe forest' are found, consisting of a bizarre mixture of *Juniperus excelsa* and other trees, with low gum-species elements. Middle sized shrub layers are altogether missing, no doubt as a result of culling and over-harvesting. Descending to the semi-arid inland valleys, one finally reaches the arid Syrian desert, where no further trace of Mediterranean biota is found.

2. *Around Jerusalem: where the Mediterranean area meets the desert and the tropics*

A severe rain shadow desert also occurs on the eastern slopes of the Jerusalem hills. In this transition zone between Mediterranean, Irano-Turanian and Saharo-Arabian biota, there is a remarkable peak of species richness in many groups of plants and animals, related in part to high inter-annual rainfall variations and particularly varied human activities over past millennia (Aronson and Shmida 1992). Striking and abundant reptiles include the Sinai agame (*Agama sinaitica*) (Fig. 4.9) and the endemic snake *Ophisaurus apodus*, the only representative in Israel of the large Anguidae family.

The Dead Sea depression is part of the Pliocene Afro-Syrian Rift valley which extends from East Africa northward to Turkey, and includes the Sea of Galilee (Tiberias Sea) and the Red Sea, as well as the Beqa'a and Litani river valleys of Lebanon. Along both shores of the Dead Sea are scattered oases where freshwater springs allow the cultivation of specialty crops of tropical origins that can not be grown elsewhere in Israel or Jordan.

In the uncultivated parts of oases, and in the canyons and gorges of dry river beds, there are arid tropical trees of 'Sudanian' affinities such as wild date palm (*Phoenix dactylifera*), horseradish tree (*Moringa aptera*), and the depauperate remnants of an *Acacia–Zizyphus* woodland very similar in physiognomy to those found in the Rift Valley in Kenya or Tanzania. Outside these specialized habitats, a sparse desert vegetation composed mostly of Saharo-Arabian elements is found, concentrated in the dry watercourses called wadis*, while 90% or more of the area is bare of all vegetation.

On the Jordanian side of the Dead Sea, the valley slopes are much steeper than on the Israeli side. Along with desert vegetation on the Jordan slopes, with their numerous cliffs, canyons, and sandstone areas, there is a gradual transition into upland Mediterranean woodland, with numerous remnants from the once abundant juniper–pistacia–oak formations. These abrupt slopes are cut by 15 deep, rather inaccessible canyons where a surprising collection of elements from four different phytogeographical regions intermingle. A quick summary would include the Mediterranean evergreen oleander (*Nerium oleander*), *Thymelaea hirsuta*, *Quercus calliprinos*, buckthorn, red juniper (*Juniperus phoenicea*), and deciduous elements such as Syrian ash (*Fraxinus syriaca*) and various Rosaceae. Species like *Pistacia atlantica*, *P. palestina*, *Rhus tripartita*, and the odd-looking bladder senna (*Colutea istria*) represent the Irano-Turanian contingent, whereas *Lygos raetam*, *Ochradenus baccatus*, and *Calligonum comosum* are some of the more common Saharo-Arabian elements. Finally, *Acacia raddiana*, *Caralluma*

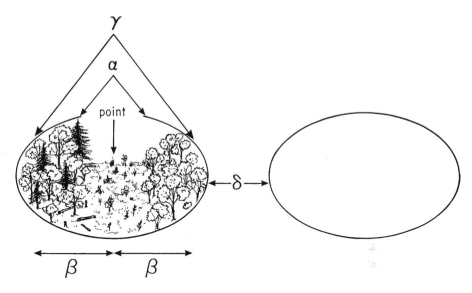

Fig. 4.12 The different components of diversity within and between communities in a landscape. The division of diversity into point, alpha (α), beta (β), and gamma (γ) components characterizes patterns of diversity and turnover on different scales. Alpha and gamma diversities pertain to the number of taxa at the local and regional scales, respectively. Gamma, or 'regional', diversity is an expression of the pool of species within which each local community will 'select' those species that will constitute each separate species assemblage. Beta diversity measures the rate of change, or turnover in diversity, between two habitats. Units of measure are dimensionless. Finally, the delta (δ) diversity is sometimes used to describe changes in diversity patterns among larger units of space such as landscapes or regions (Reproduced with permission from Blondel 1995).

spp., and *Tamarix* spp. are tropical 'Sudanian' elements in this crossroads region par excellence.

Small-scale diversity

We will now look briefly at species diversity and turnover on the spatial scale relevant to a bird or a mouse, that is, 1–1000 m². One of the simplest statistics used to make comparisons among sites in this size range is species richness in a given group of organisms within standardized plots of logarithmically increasing size.

A popular and useful tool for measuring diversity and for investigating its changes across space is to recognize several 'levels' or components of diversity (Fig. 4.12). Distinguishing the components of diversity is closely associated with quantifying the local distribution of species, similarity and dissimilarity among local species assemblages, and rate of change in species composition with respect to ecological conditions or distance. A recurrent pattern in most Mediterranean

Table 4.1 Species richness of vascular plant communities in 1, 100, and 1000 m² plots.

Locality	Vegetation	Disturbance regime or structure age since fire	Number of vascular plants in plots of		
			1 m²	100 m²	1000 m²
Israel					
Mt. Gilboa	Open grassland	Lightly grazed	29	105	179
Mt. Carmel	Open shrubland	Disturbed	20	75	119
Allonim	Dwarf shrubland	Grazed/burnt	21	88	135
Mt. Carmel	Oak woodland	Ungrazed	14	48	65
Neve Ya'ar	Oak woodland	Grazed	10	36	47
France					
Puechabon	Holm-oak coppice	1 yr after cutting	9	45	64
Grabel/Bel Air	Kermes oak shrubland	10 + yr after fire	12	40	54
Puechabon	Holm-oak coppice	10 + yr after cutting	6	33	51
St Clément	Open pine forest	1 yr after fire	7	29	30
St Clément	Kermes oak shrubland	5 yr after fire	8	23	28

Sources: after Naveh and Whittaker 1979, reprinted in Westman 1988; F. Romane and A. Shmida, unpublished (last line)

ecosystems and landscapes is that alpha diversities are not always very high but beta and gamma diversities are generally quite high because of the physical and biological heterogeneity of the systems.

Perturbations and plant species diversity

Vascular plants lend themselves readily to an analysis of the components of diversity. Table 4.1 shows vascular plant species richness data from five sites in northern Israel and five in southern France, near Montpellier. From this table several points emerge. Vascular plant species richness at all the three spatial scales we are considering is much higher in the eastern Mediterranean than in the western part, across a range of vegetation structures. Yet in each area, recent land use history greatly affects plant species diversity. As Naveh and Whittaker (1979) pointed out, moderately grazed woodlands and shrublands in the eastern Mediterranean show some of the highest plant alpha diversities in the world. They argued that the unusual species richness of these east-Mediterranean floras (especially in annual plants) is the product of relatively rapid evolution under stress by drought, fire, grazing, and cutting. They further suggested that the longer the period of human perturbations, the greater the chance that differentiation will occur. This opened a line of investigation that has still not been resolved (see Chapter 9).

The time since fire or other major disturbance has a clear effect on floristic diversity. Thus, grazed oak woodland at Allonim in northern Israel had two- to threefold greater plant species diversity than the ungrazed woodland site at Neve Ya'ar. Disturbance-adapted annuals contributed 97 of 135 total species in this site, and half to two-thirds of species richness in the Mt. Gilboa and Mt. Carmel sites as well. Similarly, four years' data from a study site in southern France (Cazarils) also revealed that moderate grazing had a significantly positive effect on plant species diversity as compared to fenced, ungrazed plots in the same landscape units (Le Floc'h *et al.* 1998). These data corroborate the widely held view that moderate disturbance regimes* result in higher species diversities in Mediterranean ecosystems than heavy disturbance or the absence of disturbance (see Chapter 6).

Data presented in Table 4.1 for the French sites of Puechabon and St Clément show that as the time since a major fire or clear-cutting increases, species diversity declines gradually. In addition to changes in species richness, the botanical composition and life form spectra change in response to decreasing or increasing intervention by humans.

Habitat and species turnover at the landscape scale

At Cazarils, a 600-ha site situated 25 km north-west of Montpellier, southern France, a long-term ecological research station was recently established to study the spatial distribution and ecological roles of biodiversity in ecosystem and landscape functioning (Le Floc'h *et al.* 1998; Aronson *et al.* 1998). The site is a representative sample of the increasingly large portion of much of southern Europe formerly occupied by agricultural and pastoral activities but now reverting to mixed evergreen–deciduous oak woodlands and/or evergreen oak shrublands.

At Cazarils, not only alpha but also beta and gamma diversities are being studied for plants (Fig. 4.13) as well as for insects and soil-borne microorganisms contributing to leaf litter decomposition and topsoil renewal. Life form spectra are also considered as an important variable, since plants of differing life forms exploit different components of available energy, water, and nutrients and, in turn, provide different resources and habitat for soil-borne and other animals. In Fig. 4.13 a distinction is therefore made between woody plants (trees and shrubs), herbaceous perennials, and annuals.

Although preliminary and representing one year only, the data presented here suggest that three distinct habitats exist in the 'landscape'. Sites 1 and 2 had much greater similarity than sites 2 to 3, or sites 3 to 4. Moreover, along a series of ten topographical gradients set out at Cazarils, each *c.* 500 m long, and similar to the one depicted schematically in Fig. 4.13, the turnover of plant species was also very high between the exposed hillsides, the abandoned fields, and the dry riverbed at the bottom of the transects. However, over five successive years, the

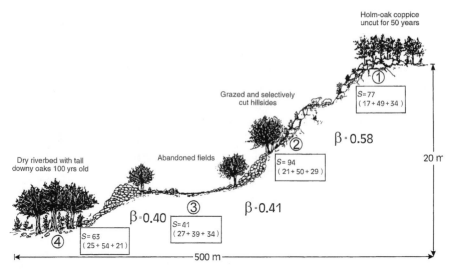

Fig. 4.13 Plant species diversity along a 500 m topographic sequence at Cazarils, southern France. Data for each site include S (total vascular plant species /100 m^2) and contribution (%) of woody plants, herbaceous perennials, and annuals to S. As determined by the Jaccard index of similarity, beta diversity (β) indicates the relative turnover rate of species composition between each pair of sites. (After Le Floc'h *et al.* 1998.)

abandoned fields were in fact highly dissimilar from the rest of the landscape in terms of their floristic makeup and microbial and soil-borne macroarthropod biota (Le Floc'h *et al.* 1998). Life-form spectra also changed along the transect, and in rather predictable fashion. The contribution of annuals was highest in the abandoned fields and lowest in the riverbed site. Woody species were most numerous in the open hillsides.

However, five years of data for the 54 sites situated along the 10 transects revealed the presence of six types of sites, differentiated in their floristic composition by relative habitat openness of the site and degree of soil degradation, both of which traits are directly related to land use history.

It thus emerges from this landscape-scale study (on about 2 km^2) that, regardless of the spatial scale chosen, the 'mosaic' effect is very much in evidence regardless of how the biodiversity in analysed, and that micro-variations in land use history heavily influence all 'spot' measurements. Inter-annual fluctuations due to rainfall variations were relatively unimportant in this study, even though in other situations, especially in drier regions of the Basin, it is known that they can lead to large fluctuations in the composition and richness of plant assemblages.

Summary

In this chapter we have endeavoured to convince the reader that the most striking feature for anyone travelling across Mediterranean landscapes is the diversity of biota, regardless of the scale of observation. From moist beech–fir forests in the northern fringe of the Mediterranean to tropical-like oases near the Dead Sea, and from one square metre to large regions, diversity of life is an amazingly recurrent theme. In the next chapter we will try to organize this diversity by describing the main families of habitat types.

5 A patchwork of habitats

The habitats and vegetation types encountered along the transects described in the previous chapter can assume a seemingly unlimited number of variations depending on local conditions and human land use histories. The combination of a great many habitats with a large number of possible pathways and stages of degradation or regeneration occurring together in separate units gives Mediterranean landscapes particular mosaic characteristics that set them apart from more northern temperate areas, or from arid and humid tropical zones. For plant and animal populations, and for communities and ecosystems, this landscape-scale patchwork pattern has profound consequences, which have been poorly analysed and documented. Moreover, the distribution of plant and animal species is also quite 'patchy'.

In this chapter we will look at the major habitat types encountered in the Mediterranean area, and organize them into broad categories. We will also introduce some of the 'special' habitats that occur in the Mediterranean, including cliffs and caves, riverine forests, steppes and grasslands, and 'old fields'. We will leave aside those areas that are so intensively cultivated or urbanized that there is little or nothing left that is *mediterranean* about them.

We can say, simplifying somewhat, that the north–south bioclimatic gradient discussed in Chapter 1 is reflected by forest and shrubland formations. In the humid and sub-humid parts of the north-west and north-east quadrants, dense woodlands and true forests are growing back since widespread rural exodus and agricultural abandonment began at the beginning of this century (see Chapter 10). In contrast, on the southern and eastern shores of the Basin, where rainfall is lower, and where human population pressure is still growing rapidly, ecosystems in all life zones are often badly depleted and degraded.

Of course there are many exceptions to this schematic picture. There are degraded shrublands and 'badlands' in almost all parts of the north-west quadrant (i.e. south-western Europe), and beautiful bits of nearly undisturbed fir or oak–beech forests still remain in a few mountain ranges in North Africa, Greece, Turkey, and several large islands.

Forest

In the Mediterranean area, the term 'forest' should not necessarily bring to mind an image of high, dense stands of trees with closed canopies. Mediterranean forests are highly diverse in their architecture, appearance, and woody plant species composition, since there are at least 40 tree species that are quite common, and more than 50 others that occur more or less sporadically, often in one region or quadrant alone. The combined figure of nearly 100 Mediterranean tree species should be compared to the corresponding figures of 12 common and 25 uncommon tree species in the far vaster forests of central and northern Europe (Quézel 1976b). Table 5.1 indicates the more common and floristically important tree species that occur in the four Mediterranean quadrants and life zones. The distribution of the oaks and pines is only partially indicated, as additional information on these crucial groups will be provided in Chapter 6. In particular, many species are noted for a single life zone, whereas in fact they can occur in three, four, or even five, in different quadrants and peninsulas.

Mediterranean forests are also highly varied in the growth forms, morphology, physiology, and phenology* of the dominant trees in each region. For example, four leaf types occur in varying combinations. Firstly, leaves may be sclerophyllous and evergreen, leathery in texture, and often spiny or prickly. A second group has laurel-like leaves (Fig. 5.1a) that are somewhat softer and shiny but still evergreen, like the foliage in many tropical forest trees. Thirdly, they may be 'semi-deciduous' and remain on their stems over winter, with reduced or terminated growth and photosynthesis. Leaves are not shed until spring when they are replaced by a new crop of leaves. Examples of such species are *Quercus faginea* and *Q. infectoria*.

The fourth group has typically deciduous leaves such as predominate in northern temperate forest trees. Many examples are found in the north-east quadrant, such as *Carpinus*, *Corylus*, *Ostrya*, and *Zelkova* species, primarily occurring in higher rainfall areas and higher altitude zones. One biogeographically interesting deciduous tree, which is restricted to a small part of the north-east quadrant, is *Parrotia persica*, the sole member of the witch hazel family (Hamamelidaceae) in the Mediterranean flora, but present at least since the Miocene. Deciduous species often show reduced leaf size, as well as other adaptations to a warm, summer-dry climate, compared to congeneric species found further north. An example is the small-leaved Montpellier maple (*Acer monspessulanum*) (Fig. 5.1b). The leaves of this and several other Mediterranean species are much smaller than in the species of maple found in northern boreal forests.

Mediterranean forests also differ in the structure or physiognomy they assume under human management. Examples include forests where all conifers have been removed, but not the oaks, or vice versa. The point to emphasize here is that it is frequently difficult to 'read' a Mediterranean forest properly in the field, unless there is background information available to supplement what is visible to the naked eye.

The largest and most diverse evergreen sclerophyllous forests in the Mediterranean area today are the 'lauriphyllous*' forests found on the wetter, north coasts of the larger Canary Islands, La Palma and Tenerife. These forests are relicts of a now virtually extinct Tertiary flora that was widespread in southern

Table 5.1 Diversity of tree species in forest canopies in six life zones of the four quadrants of the Mediterranean Basin.

Life Zones	Geographical Extension in the four quadrants	Dominant tree species	Elevation
Infra-mediterranean	SW Morocco Macaronesia	*Argania spinosa, Acacia gummifera* *Pinus canariensis*, various Lauraceae	< 250 m
Thermo-mediterranean	SW, SE, NW	*Olea, Ceratonia, Pistacia lentiscus*	< 500 m
	SE, NE	*Zelkova sicula*	
	SE, NW	*Styrax officinale*	
	SW, NW	*Pinus halepensis*	
	NE	*P. brutia*	
	SW, NW	*P. pinea* subsp. *mesogeensis*	
	SW	*Tetraclinis articulata*	
	SW, NW	*Quercus suber*	
	SW, SE	*Pistacia atlantica*	
Meso-mediterranean	NW, SW	*Acer monspessulanum*	0–600 m in the northern quadrants; 500–1000 m in the southern ones
	NW, SW	*Quercus ilex*	
	NW, SW	*Q. coccifera*	
	SE, NE	*Q. aegilops*	
	NW, SW	*Q. suber*	
	NW	*Q. faginea*	
	NE, SE	*Q. infectoria*	
	NE, SE	*Q. calliprinos*	
	SE	*Cedrus brevifolia* (Cyprus)	
		Q. alnifolia (Cyprus)	
	NW, NE	*Celtis australis*	
	NE	*C. tournefortii*	

	Laurus nobilis	NW, NE, SE
	Pinus pinea, Pinus spp.	NW, SW, NE
	Juniperus oxycedrus	NW, SW
		600–1200 m in the northern quadrants; 800–2000 m in the southern ones
Supra-mediterranean	*Quercus humilis*	NW, NE
	Q. fraínetto	NW, NE
	Q. macedonia	NE
	Q. trojano	NE
	Ostrya carpinifolia	NE
	Carpinus orientalis	NE
	Quercus faginea	SW
	Q. canariensis	SW
	Q. afares	SE
	Q. infectoria	SE, NE
	Q. cerris	NW, NE
	Abies cephalonica	NE
	A. pinsapo	SW
	A. maroccana (very scarce)	SW
	Zelkova abelicca	SE (Crete)
	Acer sempervirens	SE, NE
	Castanea sativa	NW, SW
	Cupressus sempervirens	NW, NE
Montane-mediterranean	*Fagus sylvatica*	NW
	Pinus nigra subsp. *nigra*	NW
	P. nigra subsp. *clusiana*	NW
	P. nigra subsp. *laricio*	NW
	P. nigra subsp. *salzmannii*	NW
		> 1000 m

Table 5.1 continued

Life Zones	Geographical Extension in the four quadrants	Dominant tree species	Elevation
	SE	*Cedrus brevifolia* (Cyprus)	
	SW	*P. nigra* subsp. *mauretanica*	
	NW	*P. sylvestris*	
	SE	*P. nigra* subsp. *pallasiana*	
	SE	*P. heldreichii*	
	NW, SW	*Cedrus atlantica*	
	NE	*C. libani*	
	NE	*Abies cephalonica, A. cilicia*	
	NE	*A. nebrodensis*	
	NE	*A. numidica*	
	SE	*Cupressus sempervirens*	
	SW, NW, NE	*Carpinus betulus*	
	NW, NE	*Juniperus foetidissima*	
	NE		
Oro-mediterranean	NW	*Pinus uncinata*	> 2000 m
	NE	*P. mugo*	
	NW, SW	*Juniperus communis*	
	SW	*J. thurifera*	
	NW	*J. excelsa*	
	SW	*J. turbinata*	
	NW, SW	*Arceuthos drupacea*	

Sources: after Quézel 1976b and P. Quézel, F. Romane, M. Barbéro, and A. Shmida, personal communication

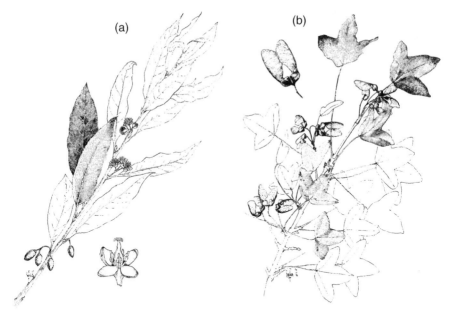

Fig. 5.1 (a) The evergreen sclerophyllous laurel tree (*Laurus nobilis*), and (b) the small-leaved, deciduous Montpellier maple (*Acer monspessulanum*).

Europe and northern Africa about 15–40 Myr BP (see Chapter 3). This forest type is named 'laurisylva*' because it includes no fewer than five species and subspecies of the tropical family Lauraceae (*Apollonia barbujana*, *Laurus azorica*, *Persea indica*, and *Ocotea foetens*), as well as several endemic broad-leaved evergreen trees such as *Arbutus canariensis*, *Myrica faya*, and *Visnea mocanera*. All of these trees share their broad-leaved, sclerophyllous leaf shape with the laurel (*Laurus nobilis*), which still occurs widely throughout the Mediterranean Basin. The highest concentration of tropical relict species is found in the wild olive–carob–holm-oak (or *Quercus calliprinos*) associations that dominate in foothills and uncultivated slopes at lower altitudes throughout the Basin. These associations can be considered as a 'lauriphyllous' vegetation type that has survived since the Tertiary, albeit with continuously changing and regionally variable botanical composition.

At low altitudes, especially in parts of southern Iberia, Turkey, and North Africa, open woodlands or park-like glades alternate with very dense, and much lower stature vegetation types. These formations are usually the result of human management schemes of some sort, involving livestock raising, woodcutting, and some regime of proscribed fire. At higher altitudes, however, open formations of conifers (e.g. *Pinus*, *Abies*, and *Cedrus*) are found with an understorey of spiny shrubs (e.g. *Astragalus* and *Genista*) on rocky outcrops. Taller, denser forests of varying composition are found in the supra-, montane- and oro-mediterranean life zones (see Table 5.1). Although heavily influenced by humans, these forests tend to reflect more closely the natural potential for vegetation provided by the local soils and climate. They often show a thorough mixture of evergreen trees, conifers, and winter-deciduous trees, shrubs, and vines. Where mean annual rainfall is greater than 500 mm, sclerophyllous elements tend to diminish in numbers, while deciduous ones gain, both in number and ecological importance.

Mediterranean forests contain a surprising number of conifers, including pines (*Pinus*), junipers (*Juniperus*), cypress (*Cupressus*), cedars, (*Cedrus*), firs (*Abies*), and the Barbary thuja (*Tetraclinis articulata*). Like *Parrotia persica*, the Barbary thuja is a monotypic* 'palaeorelict' restricted to North Africa and some bits of southern Spain. Its closest relatives are the 14 species of the genus *Callitris* found in south-east Australia and nearby islands such as New Caledonia. Conifers are still more abundant in Mediterranean mountains than in the lowlands and their contribution to overall vegetation cover is greater at higher altitudes. The 'primeval' forests at higher elevations in much of the area probably combined conifers and broad-leaved species in intricate mixtures of many species. The segregation often seen today in forest canopies in the Basin, whereby pines or evergreen oaks can form nearly pure stands, is almost always a product of human interventions, and does not reflect the natural dynamics of these forests.

Matorrals

Dozens of shrubland formations occur around the Mediterranean Basin, most of which are secondary and the direct result of some combination of human activities. Given the wide range of substrates, microclimates, and land use histories to be found, matorrals show a wide range of structural forms (Tomaselli 1981). Their diversity is so great that almost every region or country has its own name or names to designate the diverse local forms: *garrigue* and *gariga*, and *maquis* or *macchia* in France and Italy, *xerovuni* in Greece, *matorral* and *tomillares* in Spain, *choresh* or *maquis* in Israel. Comparable terms used elsewhere for similar vegetation types are *chaparral* and *coastal sage* in California, *matorral* and *jaral* in Chile, *fynbos*, *renoster veld*, *karroid shrubland*, and *strandveld*, in South Africa, and *mallee* in Australia. These names are often poorly defined and their usage varies among authors. In many countries, *maquis* (or *macchia*) is considered as the first major stage in forest degradation, followed by *garrigue*, *phrygana** or *bath'a* which are all of still lower stature and complexity than *maquis*. In France the distinction between *garrigue* and *maquis* is considered by many geographers and phytoecologists to depend on substrate, so that *garrigues* are said to occur primarily on limestone substrates and include the full range of species associated with holm-oak, while *maquis* is reserved for those formations occurring on acid, siliceous soils. In addition to the cohort of species found in nearby *garrigues*, *maquis* includes such calciphobe* marker species as the strawberry tree (*Arbutus*) and other heath-family shrubs (*Erica* and *Calluna*), as well as certain rock-roses (e.g. *Cistus ladanifer*), lavenders, and other shrubs. Yet this dichotomy has only limited value, and a third matorral type somewhat intermediate in floristic terms also occurs, and is found on dolomite substrates. In the north-east quadrant of

Table 5.2 Main growth forms recognized by Raunkiaer.

Growth form	Definition	Example
Therophyte	Annual	Poppy
Cryptophyte	Bulb	Orchids, tulips
Hemicryptophyte	Perennial herb	Alfalfa
Chamaephyte	Shrub	Thyme, lavender
Phanerophyte	Tree	Kermes oak, laurel

Source: after Raunkiaer 1934. For more details, see Orshan 1989.

the Basin, a fair number of volcanic (ultrabasic) rock types also occur in patches and certain shrubland types are primarily restricted to these as well. Following many authors, we will use the Spanish word *matorral* as a general and generic term to designate the entire class of mediterranean-type shrublands, including the most reduced in size, the *tomillares*, *phrygana*, and *bath'a*.

The most characteristic feature of matorrals, which are exceedingly attractive to animals, is that they include a fine-grained mosaic of almost all the growth forms recognized by plant ecologists. In their understorey can be found the full range of plant life forms, described by Raunkiaer in 1934 in his attempt to propose a universal system of plant growth form classification based on the position of leaf and stem renewal buds relative to ground level (Table 5.2).

Matorrals are characterized by the dominance of shrubs with 'evergreen, broad and small, stiff and thick (sclerophyllous) leaves, an overstorey of small trees sometimes being present, and with or without an understorey of annuals and herbaceous perennials' (di Castri, 1981). Zohary (1962) designated as matorral any sclerophyllous evergreen vegetation type that is dense and capable of attaining 4–6 m in height, but which are usually much lower. Prominent examples of dominant sclerophyllous species in matorrals are the various evergreen oaks, carob, and lentisk, but also various species of *Arbutus, Daphne, Laurus, Phillyrea, Myrtus, Rhamnus*, and *Viburnum*, all of which are sclerophyllous evergreen. The low and medium shrub layers of matorrals, especially in the western Mediterranean, include a number of familiar mint family members such as lavender, rosemary, thyme, and others, as well as the bright-flowering rock-roses (*Cistus, Fumana*, and *Helianthemum*). In the Near East and North Africa, where human pressure is higher, matorral understoreys tend to have fewer shrubs and more hemicryptophytes.

Even if matorrals appear predominantly evergreen, about half of their woody species are in fact winter-deciduous. Examples are the maples, most *Pistacia* species, *Cotinus*, and *Rhus*, as well as numerous deciduous oak species. The brooms (*Cytisus, Genista, Spartium* and *Teline*), notably common and rich in species and subspecies in Iberia and Morocco, have the distinction of bearing evergreen stems that are photosynthetically active all year round. Most species in the so-called retamoid* group have small deciduous leaves that fall readily during drought.

A great many geophytes* also occur in a number of monocot plant families. Among these, the Orchidaceae is represented by over a hundred species. Unlike many tropical orchids, here the family has only terrestrial species that prefer exposed, open, and warm habitats supporting dwarf-shrub communities. As many as 50 species of orchids have been found to co-exist in a single matorral area of 100 ha in Greece! Many of them are common in frequently human-disturbed habitats that are also rich in solitary bees that pollinate them. It has been suggested that the chances for successful cross-pollination in terrestrial orchids increase when they are visible to their insect-pollinators from great distances. This reliance on optical clues for specialized pollinators may be one of the selective pressures determining the distribution of most orchid species to open habitats and also contributing to the great evolutionary success of this group of plants in matorrals (Dafni and Bernhardt 1990) (see Chapter 7).

Much work has been devoted to deciphering the pathways of degradation of former Mediterranean forests and their transformation into matorrals. Although most matorrals appear to retain the ability to recover spontaneously if human intervention ceases, some of them are so badly degraded that without long-lasting intervention they remain blocked in stunted species – poor formations. A large number of possible ecosystem trajectories* exist in Mediterranean ecosystems, in the course of both degradation and recovery. The history of local land use plays a critical role and many matorrals are so far removed from the 'primeval forests' that they never return spontaneously to the stature and structural complexity of true forests, at least not without massive human intervention. Instead, they appear frozen in self-perpetuating systems that can apparently endure for centuries with little visible change. One example is the formations dominated by the dwarf kermes oak (*Quercus coccifera*), which extend unbroken over hundreds of hectares at a stretch in the north-west quadrant. Spiny legume shrubs, including *Calycotome spinosa*, brooms, and gorse (*Ulex europaeus*) also form such formations, often described as 'leopard's skin' because patches of low green 'brush' or scrub alternate with patches of bare ground and rocks.

Tomillares, Phrygana and Bath'a

The Spanish term *tomillares*, the Arabic/Hebrew *bath'a*, and the Greek *Phrygana* all refer to plant formations that are much lower than other matorrals but closely related to them floristically and historically. The tomillares of Spain are shrublands characterized by a large number of thyme (*Thymus*) species ('tomillo' in Spanish). Phrygana, in Greece and Turkey, also consists of formations dominated by dwarf shrubs typically 20–70 cm tall. Structurally similar to these formations, bath'a formations are the Near East equivalent, found intricately interspersed with higher matorral types in the higher rainfall areas near the coast as well as at middle altitudes on mountain slopes. More commonly, bath'a occupies large stretches in semi-arid, Mediterranean-desert border regions further inland. Common species here include *Calycotome villosa*, *Genista acanthoclada*, *Eryngium* spp., *Satureja* spp., and *Sarcopoterium spinosum*, as well as several bulb species, some rock-roses, and hundreds of species of annuals.

Describing these dwarf formations, Shmida (1981) noted that canopy cover may be as extensive as in matorral, but total biomass is much lower. Leaves of dominant shrubs tend to be small, soft, and pungent, emitting volatile terpenes when touched. Winter foliage is regularly replaced by even smaller, more sclerophyllous foliage during summer. Leaves and shoots die back in part or full, in a surprisingly flexible fashion that is determined by both rainfall distribution and temperature extremes (Orshan 1972). Such drought deciduous species are often spiny and hemispherical shrubs that appear to be resistant to herbivory*, fire, and prolonged drought. They occasionally include stem- or leaf-succulents, such as spurges (*Euphorbia* spp.), that are toxic to animals.

Steppe and grassland

The history of a great many plants of the Mediterranean Basin (especially in the eastern parts) can be traced to the arid and semi-arid steppes of central Asia, the so-called Irano-Turanian region described in Chapter 1. Recall that 'Zohary's law' suggests that disturbed areas are generally invaded by weedy elements coming from the drier habitats in the vicinity, rather than from the wetter ones. This tendency is reflected at another spatio-temporal scale by an apparent 'intrusion' of Irano-Turanian elements in adjacent Mediterranean vegetation zones, as we saw in Chapter 4. A great many Irano-Turanian steppe elements have also invaded disturbed sites in the western Mediterranean, and from there, most of the other climatically similar areas around the world. This biogeographic success is due to their exceptionally long period of pre-adaptation to frequently and variously disturbed sites.

In the southern Mediterranean regions, particularly at higher altitudes and in localities receiving less than 300 mm mean annual precipitation, natural vegetation is mostly made up of mixed annual and perennial grasslands, with only scattered bulbs, shrubs, and trees. The most extensive of these are the vast 'alfa steppes' of North Africa, with their dominant bunch grass, alfa (*Stipa tenacissima*), and the shrubs wormwood (*Artemisia* spp.) and *Rhanterium* (Asteraceae). In certain areas, very large and old Betoum trees (*Pistacia atlantica*) are found here and there. Where aridity combined with a high water table creates saline or alkaline flatlands, salt-tolerant Chenopodiaceae called 'salt bushes' (*Atriplex*, *Haloxylon*, *Salsola* and *Suaeda*) constitute extensive formations.

Old fields

As noted in Chapter 3, there are over 1500 (and probably close to 2500) species of annuals, biennials, and bulbous plant species in the Mediterranean flora that mostly occur in early stages of succession in abandoned crop lands and/or fallows, both of which can be called 'old fields'. These sun-tolerant and often highly competitive 'pioneers' include the ruderal and segetal species that will be discussed in Chapters 7 and 9. Numerous ecological studies have been conducted in these special habitats, one of which will be described in detail in Chapter 6 (see p. 145).

Riverine forest

Mediterranean riverine forests were once complex, biologically varied ecosystems, extending over more than 2000 km^2 in the Basin, punctuating landscapes and entire regions with a high diversity of plants and animals. Unfortunately, most of them have been definitively removed and replaced by agriculture. Only some remnants are left, for example along the Moraca River in ex-Yugoslavia, along the River Strymon (lake Kerkini, Greece), in the middle Po valley and in the lower Rhône River valley. The few patches that have been preserved give an idea of what such habitats must have once looked like. In the Nestos delta of eastern Greece, for example, about 60 ha of intermittently flooded riparian* forest, with poplars, alders, and willows remain today in a nature preserve. Thanks to particularly favourable soil conditions, these forests are dominated by deciduous trees such as oaks, poplars (*Populus alba*), elms, and willows, upon which climb luxurious vines of wild grape (*Vitis silvestris*), hops (*Humulus lupulus*), and various species of *Clematis*. In the Balkans, a dominant tree species of this habitat is *Platanus orientalis*. Ribbon-like, these forests may penetrate from temperate areas into the warmer Mediterranean ecosystems, bringing with them a series of species and life forms that would not survive outside the shadow of the trees and moist microclimate. Some surviving remnants of riverine forests, such as those found in southern France along the Rhône River and in the Camargue delta, also contribute greatly to regional biodiversity (see section on Wetlands below).

Cliffs and caves

Most landscapes in the Mediterranean are broken somewhere by cliffs and escarpments that provide highly specialized habitats for a number of plants and animals. In addition, there are tens of thousands of caves, shafts, holes, and crevices in the hills and mountains. The larger cliffs and virtually all the caves are made of compact limestone and often harbour unusual micro-habitats and rare species. In southern France only one species, the rare Petrarque fern (*Asplenium petrarchae*), is endemic to cliffs on south-facing slopes near the coast. In other parts of the Basin, cliff endemics are more common. Frequently encountered Mediterranean chasmophytes*, or cliff-dwelling plants, include *Phagnalon* spp., *Melica minuta*, *M. amethystina*, and various mustard family members. Moist crevices provide homes for a certain number of fern species (*Polypodium vulgare*, *Asplenium fontanum*, *A. trichomanes*, and *A. ceterach*). Other plant species that are found almost exclusively on north-facing cliffs are *Globularia repens*, *Potentilla caulescens*, *Silene saxifraga*, *Galium pusillum*, *Hieracium humile*, *Campanula macrorhiza*, *Teucrium aureum*, and *Hesperis laciniata*.

Cliff habitats include step-crevices, vertical faces, overhangs, pavements, and sloping cliffs. Long periods of hot sunshine and soil moisture conditions are often limiting factors to plant growth. Ecological adaptations related to a perennial life form, prolonged and conspicuous flowering, high germinability of seeds, and various long-distance seed dispersal mechanisms also characterize chasmophytes. Prominent among these species are an eastern Mediterranean contingent including *Varthemia iphniodies* and *Phagnalon rupestre* (Asteraceae), *Rosularia lineata* (Crassulaceae), and several species of *Micromeria*, *Stachys*, and *Teucrium* (Labiatae). A second group has been identified as a 'middle Mediterranean' element, containing many mustard family members such as the showy wall-flowers (*Erysimum* spp.) and the *Brassica cretica* species group. The spectacular flowering capers (*Capparis spinosa*) are visible from afar on cliffs in warmer areas.

Chamaephytes* are by far the most common life form found in cliffs but geophytes also occur, including cyclamens, ferns, succulents (*Sedum*, *Cotyledon* and *Caralluma* spp.), wild snapdragon (*Antirrhinum majus*), and *Parietaria judaica*. In some groups where most species are annuals, the cliff-dwelling species are all shrubs. Cliff faces and overhangs also harbour a number of tree species with a dwarfish appearance because of water and nutrient limitations. Some of these appear to be clinging to the cliffs like twisted survivors of a thousand storms. Among the most spectacular are Phoenician juniper, holm-oak, and the fig tree. It has been suggested by Snogerup (1971) that Mediterranean cliffs may have served as refugia for various matorral and phrygana species of trees or shrubs during cold periods of the Pleistocene.

This highly specialized habitat offers breeding sites for one of the most unique group of birds in the Mediterranean (Fig. 5.2). Several large raptors use ledges to breed in colonies. A ballet of griffon vultures (*Gyps fulvus*) is certainly one of the most unforgettable sights for a bird watcher. More secretive cliff-dwelling species include the eagle owl (*Bubo bubo*), Egyptian vulture (*Neophron percnopterus*), raven (*Corvus corax*), peregrine falcon (*Falco peregrinus*), kestrel (*Falco tinnunculus*), stock dove (*Columba oenas*) and, in the southern Mediterranean, colonies of the lesser kestrel (*Falco naumanni*). In some parts, notably the south and east, cliffs dominating the sea are inhabited by large colonies of Eleonora's falcon (*Falco eleonorae*). This species has a fascinating biology since it breeds very late in the season, that is, in July and August, taking advantage of the autumnal migration of passerine birds upon which it feeds. Quite surprisingly, the only winter ground of this falcon is Madagascar.

Looking in more detail at the cliff habitat, the observer finds many other birds as well, such as the shy blue rock thrush (*Monticola solitarius*) that builds its nest in the darkest parts of overhangs, or the black redstart (*Phoenicurus ochruros*). Colonies of alpine swifts (*Apus melba*), rock swallows (*Riparia rupestris*), and the rare Pallid swift (*Apus pallidus*) occur near the coast, along with various species of bats that breed in deeply fissured vertical rock faces.

Fig. 5.2 A typical Mediterranean cliff with some of its inhabitants. From left to right: lesser kestrel (*Falco naumanni*), Bonelli's eagle (*Hieraaëtus fasciatus*), Egyptian vulture (*Neophron percnopterus*), and eagle owl (*Bubo bubo*). Sketches of the birds are not at the same scale.

The lives of these 'cliff birds' are not always peaceful. Ravens have been observed robbing recently captured prey from nests of eagle owls, while eagle owls themselves sometimes catch and eat peregrine falcons!

Wetlands

The Mediterranean Basin includes a variety of wetlands, ranging from large inland lakes and extensive coastal lagoons to small temporary ponds. In summertime they often stand out as oases of greenery in the dry and yellowish countryside. Britton and Crivelli (1993) estimated at 21 000 km^2 the area covered by wetlands in the Basin, of which 4700 km^2 are coastal lagoons, 2800 km^2 are freshwater lakes and marshes, and 11 600 km^2 are temporary salt lakes, found mostly in North Africa. Except for wetlands that are connected to large permanent rivers and some inland freshwater lakes, the main characteristics of Mediterranean wetlands are the fluctuations in water levels and salinity, which reflect the large variation in rainfall both within and between years. Five main categories of wetlands are recognizable in the Mediterranean, described below.

Freshwater lakes

Except for the many humans-made reservoirs built to increase water supply, most inland freshwater lakes in the Mediterranean are of glacial origin. They are

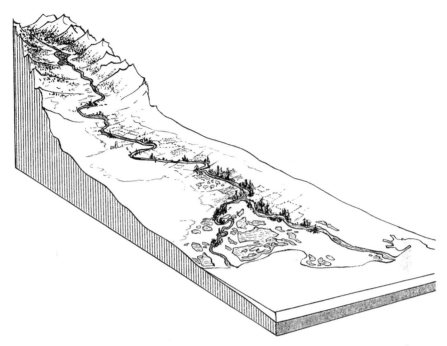

Fig. 5.3 A typical large Mediterranean river originating in nearby mountain ranges and emptying into the sea, forming deltas with a mosaic-like structure of habitats.

restricted to high altitudes in the mountain ranges that were affected by Pleistocene glaciations (Sierra Nevada, Pyrenees, Alps, Apennines, Dinaric Alps, and the Atlas). Most of them are nutrient-poor, steep-sloping, deep, and have little or no emergent vegetation. Some inland lakes in Italy, however, are of volcanic origin and occupy ancient calderas. A series of inland lakes in the Balkan peninsula are parts of karstic formations resulting from fracturing of limestone blocks. The most famous of these are the interconnected Greek lakes Megali Prespa and Mikri Prespa, and also Lake Vegoritis, which are among the richest and most productive aquatic ecosystems of the Mediterranean. They are rich in both emergent plants (*Phragmites australis*, *Typha latifolia*, *Scirpus lacustris*, *Iris pseudacorus*) and submerged vegetation (*Nymphaea alba*, *Ranunculus* spp., *Nymphoides peltata*, *Myriophyllum verticillatum*), and are biological hotspots for endemic fish species and wildlife. A large diversity of birds, including the white pelican (*Pelecanus onocrotalus*), the Dalmatian pelican (*P. crispus*), as well as several herons, spoonbills, cormorants, and ibises breed there in large colonies.

Deltas and coastal lagoons

The most extensive wetlands in the Mediterranean are alluvial flood plains, coastal lagoons, and deltas of the main rivers flowing down from nearby high mountain ranges: Guadalquivir and Ebro in Spain, Rhône in France, Pô in Italy, and Axios and Evros in Greece (Fig. 5.3). The most important lagoon systems stretch over more than 200 km from the mouth of the Rhône to the French–Spanish border, and from Venice to Trieste along northern Italy's Adriatic coast. Large deltaic systems do not occur in the Maghreb part of western North Africa, since the short, highly seasonal rivers found there do not provide enough sedimental material for delta formation. The only large delta in North Africa is that of the Nile, whose waters flow down from tropical Africa.

The formation of extensive deltaic systems is favoured in the northern Mediterranean by the small amplitude of tides, which allows the development of offshore sand banks inside which alluvial sediments deposit in shallow waters. Coastal lagoons are produced by the accumulation in coastal waters of sand and silt deposits that are brought by rivers and continuously reshaped by marine currents and wind. This results in the building of offshore bars that are more or less parallel to the coast and encircle inland lagoons. When powerful rivers empty into the Sea, they form deltas with numerous channels that continually change course. This gives rise to a maze of marshes and lagoons interspersed by sand dunes or mud flats, arising from the river's meanderings and oxbows over the course of centuries and millennia. The ongoing changes of river arm configuration within a delta result in a perpetual upheaval in habitats. Thus, a large Mediterranean delta is a moving mosaic of wetlands with contrasting salinity, and seasonal water levels usually no more than two metres deep. Salt concentration varies widely in space and time from freshwater to hypersaline waters (up to 40 g l^{-1}) in relation to rainfall and seasonal water levels. Coastal lagoons are typically isolated from the sea by sand dunes that are open here and there allowing connections with the sea. Coastal dune vegetation varies in importance from a narrow spit with *Ammophila* grasses to extensive woods of red juniper (*Juniperus phoenicea*) and stone pine (*Pinus pinea*).

Wetland plant assemblages are largely determined by fluctuating salinity and water levels. A recurrent feature in all coastal Mediterranean aquatic systems is the huge variations they undergo in biologically important factors such as flooding periodicity and water or soil salinity, all of which have profound effects upon the structure and dynamics of plant and animal communities. Many of these factors vary enormously during the course of the year, from year to year, or over even longer periods of time. As a result, plant and animal communities are highly dynamic and do not exhibit long-term predictable successional changes in species composition. Wherever water levels widely fluctuate current assemblages of plant species usually reflect recent past events.

In areas that are flooded for a few months, and where salt concentration remains high, vegetation is mostly composed of halophytes* in the Chenopodiaceae, especially several species of *Arthrocnemum*, and salt grasses such as *Aeluropus* and *Paspalum*. In these flatlands, which often cover large areas, the only tree species are tamarisks that thrive under a wide range of salinity and water levels (e.g. *Tamarix africana* and *T. canariensis* in the Iberian peninsula, *T. gallica* in France, and *T. tetranda* in the Balkan peninsula). They can survive flooding for up to six months, sometimes more, at a water depth of one metre. Where salinity in the soil is lower, wet grasslands including many Papilionoid chamaephytes

Box 5.1 Characeae as breeding and foraging algae for animals

In many bodies of water, including those that dry up for several months in summer, plant communities of shallow water include Characeae, an important group of freshwater green algae that resemble vascular plants. Characeae have the exceptional ability to fix lime in their tissues (up to 70% of the plant) so that they become brittle and coarse to the touch. These submerged plants are of primary importance as food for the hundreds of thousands of ducks that overwinter in the Camargue. Thick carpets of Characeae are also excellent breeding sites for many fishes, amphibians, and aquatic insects while oogons* of these algae constitute an important part of the diet of teal (*Anas crecca*) in winter (Tamisier 1971).

may extend over huge areas. These shrubs include *Dorycnium jordani* and, in the eastern Mediterranean, *Prosopis farcta* and wild licorice (*Glycyrrhiza glabra*).

Aquatic vegetation in permanent water bodies varies according to salinity. It includes *Ruppia* spp. and various algae (e.g. *Ulva* spp., *Chaetomorpha linum*) in saline waters, passing to freshwater plants (e.g. *Potamogeton pectinatus*, *Ranunculus baudotii*, *Myriophyllum*, *Zannichellia* spp.) and large reed beds as salinity decreases.

Mediterranean wetlands are rich in fish species that occupy the various habitats according to their tolerance of highly saline, brackish, or fresh water. Sea fishes include sea-bass (*Dicentrarchus labrax*), gilt-head (*Sparus auratus*), and sole (*Solea vulgaris*), while carp (*Cyprinus carpio*), pike (*Esox lucius*), and pike-perch (*Stizostedion lucioperca*) live in fresh water. Species that tolerate the large fluctuations in salinity levels found in brackish lagoons include eels (*Anguilla anguilla*), several species of mullets (Mugillidae), sand-smelt (*Atherina boyeri*), and a tiny Mediterranean seahorse (*Syngnathus abaster*). Many fish species, called diadromous* and euryhaline, depend completely on brackish lagoons for spawning so that large migrations of these fishes occur between the Sea and these inland bodies of water.

The periodic but unpredictable drying out of many Mediterranean wetlands has resulted in the evolution of several strategies that allow animals to escape the effects of drought and desiccation. For example, several bird species such as the flamingo (*Phoenicopter roseus*), marbled teal (*Anas angustirostris*), stilt (*Himantopus himantopus*), and ruddy shelduck (*Tadorna ferruginea*) are peripatetic and opportunistic, taking advantage of temporarily favourable conditions wherever they occur. These birds can breed in quite different areas from one year to the next depending on water levels. Similar flexibility in behaviour is found in other animals as well. In many invertebrates, especially small crustaceans such as amphipods, ostracods, and copepods, reproduction and growth occur in winter and early spring. These animals await the return of favourable conditions by spending the hot dry summer in quiescent stages such as eggs. Thus their life cycles are often seasonally reversed by comparison with similar organisms in central Europe.

Productivity of coastal lagoons is exceptionally high, having been estimated to be 8–10 times greater than that of the Sea. The economic value of lagoon fisheries along the French Mediterranean coast exceeds that of the Mediterranean trawling fleet (Britton and Crivelli 1993), and coastal lagoons yield 10–30% of the total Mediterranean production of fish. In the saline lagoons of the Camargue the fauna is often reduced to a few highly adapted species (Britton and Johnson 1987), notably the brine shrimp (*Artemia salina*), which can survive salinity levels up to 300 g l^{-1}, larvae of the Dipteran *Ephydra* spp., a copepod (*Cletocampus retrogresses*), a few Microtubellaria, and nematodes. J.-N. Tourenq has estimated that the biomass of invertebrates may reach 500–1000 g m^{-2} in late spring when water temperature is rising and most organisms are at their peak of annual growth. There may be as many as 30 000–50 000 brine shrimps per square metre at this time of year! Such high concentrations constitute the main food of flamingos, shelducks, avocets, Kentish plovers, and swarms of migrating waders.

Taking advantage of the exceptional diversity and productivity of Mediterranean wetlands, many species of birds breed together in mixed colonies. Extending over 145 000 ha in southern France, the Camargue is one of the most famous and best preserved wetland areas of the Mediterranean. More than ten species of terns, gulls, and waders breed together on small islets scattered in its large lagoons. Colonies may have several thousand breeding birds and attract secretive species, such as ducks (*Netta rufina*, *Tadorna tadorna*) and redshank (*Tringa totanus*), that benefit from the protection provided by other species. From the core breeding area, these birds then disperse to a variety of feeding grounds according to species-specific habitat preferences and foraging techniques (Fig. 5.4).

Some species, for example avocet, stilt (*Himantopus himantopus*), and Kentish plover (*Charadrius alexandrinus*), do not leave the lagoon and forage only in shallow water at a depth proportional to the length of their legs. They mostly feed on small crustaceans, including phyllopods (*Artemia*), amphipods (*Gammarus locusta*), and on a large variety of aquatic insects (Ephydridae, Syrphidae, Chironomidae, and Dolichopodidae). They also occasionally eat molluscs (*Hydrobia*) and marine worms (*Nereis diversicolor*). The slender-billed gull is highly specialized to Mediterranean lagoons where it catches flat fish using the technique, unusual for a gull, of plunging its long neck into shallow water. The oystercatcher (*Haematopus ostralegus*), which is not very common in the area, regularly forages along the coasts of lagoons and the Sea, searching for large molluscs (*Cardium* spp.) and beetles (Tenebrionids and Carabids). Several species of fish-eating terns regularly forage at sea but only one of them, the sandwich tern, is restricted to this feeding area. The common tern (*Sterna hirundo*) and the smaller little tern (*S. albifrons*) forage both at sea and in the lagoons and canals. The rare gull-billed tern (*S. nilotica*) is more terrestrial and mostly feeds in inland rice fields on large insects (mole crickets, grasshoppers, dragonflies),

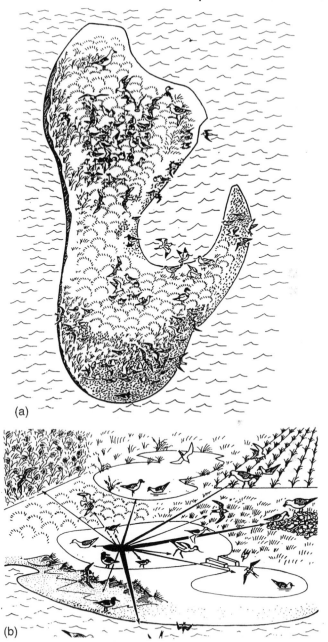

Fig. 5.4 Organization of a typical breeding colony of birds in the Camargue. (a) As many as 14 species breed together on an islet and benefit from mutual protection from predators. There are often several subcolonies, two of which are shown: in the upper part of drawing, a flock of black-headed gulls (*Larus ridibundus*) with several pairs of slender-billed gull (*Larus genei*), ducks, and redshank (*Tringa totanus*); in lower part, flocks of common tern (*Sterna hirundo*), sandwich tern, (*S. sandvicensis*) and other species. (b) Each species leaves the islet to forage in its species-specific habitat. (Reproduced with permission from Blondel and Isenmann 1981.)

amphibians, lizards, and crustaceans (*Triops, Branchipus*), sometimes far away from the breeding colony.

Finally, the two most common gulls, the black-headed gull (*Larus ridibundus*) and the yellow-legged gull (*L. cachinans*) are highly eclectic in their feeding habits and foraging habitats. As a consequence, they are also the species that have benefited most from human-induced changes of landscapes, as they make use of a large variety of food resources found in marshes, crop fields, rice fields, farmland, and garbage tips. Accordingly, their population sizes have increased tremendously during this century from some few hundred pairs of black-headed gulls in the thirties to nearly 10 000 breeding pairs today, and from a handful of yellow-legged gulls to more than 4000 pairs today (a 9% annual rate of increase). Lebreton and Isenmann (1976) demonstrated that population increase of the black-headed gull was mainly a result of a higher winter survival rate due to the use of predictable food resources provided by garbage tips and boat trawling. The yellow-legged gull became a threat for the breeding colonies of other gulls, terns, avocets, and even flamingos, because it steals their eggs and eats their young fledglings. Entire colonies would be completely destroyed by yellow-legged gulls if measures were not repeatedly taken to limit their numbers. From March to late July, the sky of the Camargue is the backdrop for a ballet of birds going to and from their breeding colonies, as they divide their time between foraging and nest caring.

Endoreic marshes*

Not all bodies of water are connected to the main rivers or to large permanent lagoons. One very interesting example is the typically Mediterranean endoreic marshes, which are completely dependent on rainfall and so dry up for several months each year. These temporary marshes are highly variable in size, duration of submersion, and salinity levels. They harbour rare and often threatened plant species such as *Marsilea strigosa*, *Pilularia*, *Lythrum*, *Juncus*, quillwort (*Isoetes setacea*), and *Damasonium stellatum* (Fig. 5.5), as well as loosestrife (*Lythrum tribracteatum*) and a buttercup (*Ranunculus laterifolius*). Some of these species are known from no more than a handful of locations. For example, the very rare *Teucrium aristatum* is only known from two sites, one in France (Crau), the other in Spain. As many as 4% of the plant species considered as being threatened in France occur in these temporary marshes. Because of the strong constraints linked to unpredictable water supply, plant and animal species specialized in this type of habitat have evolved remarkable growth forms and life-history traits especially with regards to the production and dispersion of seeds. Most plant species are annuals or biennials, with a life cycle lasting only a few weeks (Grillas and Roché 1997). They produce large quantities of seeds that may remain in the ground for a very long time, until favorable conditions for germination occur. This is an insurance against local extinction. As many as 320 000–530 000 seeds per square metre have been reported in the Doñana marshes in Spain (Grillas and Roché 1997).

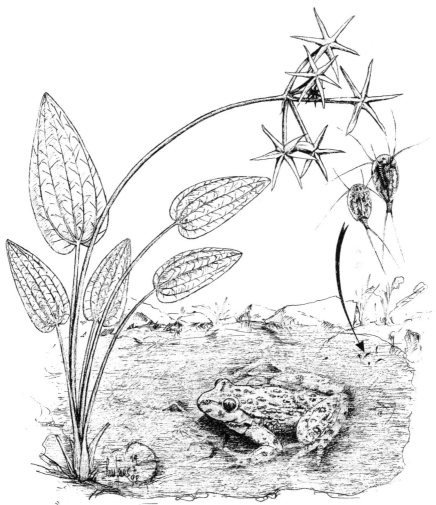

Fig. 5.5 Some inhabitants of temporary marshes: the thrumwort (*Damasonium stellatum*), the parsley frog (*Pelodytes punctatus*), and a micro-crustacean (*Triops cancriformis*).

This very distinctive habitat also harbours a large number of invertebrates that may occur in incredibly large numbers since fish predators are absent. Most of these invertebrates are adapted to prolonged droughts (copepods, phyllopods, ostracods). The unmistakable *Triops cancriformis*, which resembles a horseshoe crab in miniature (Fig. 5.5), lays eggs that may become enclosed in a cyst for several months after the marsh dries up, not hatching until it refills. Temporary endoreic marshes are also favourable breeding sites for numerous insects whose life cycle includes an aquatic stage, such as dragonflies and many beetles.

These marshes are also of paramount importance as breeding sites for amphibians. In March and April marbled newt (*Triturus marmoratus*), common newt (*Triturus vulgaris*), and several species of frogs and toads, especially the

Iberian spadefoot (*Pelobates cultripes*), the parsley frog (*Pelodytes punctatus*), and several others (*Alytes obstetricans*, *Bufo calamita*, and *Rana perezi*), lay their eggs here over a period of several weeks. As a response to the short period of flooding, which occurs mostly in winter, several species of amphibians have evolved a winter breeding season so that metamorphosis of the tadpoles occurs before the breeding site dries up. For example, the Iberian spadefoot spreads its breeding season from October to February in Andalusia, thus increasing its chances for successful reproduction. Of course, there are high extinction risks for these small populations tightly linked to fugitive habitats. Most of them presumably function as parts of a larger metapopulation* with exchanges of individuals among sub-populations breeding in discrete habitat patches scattered over short distances within a landscape. The more small ponds that occur in a landscape, the lower the extinction risks for populations of these amphibians. Unfortunately, as much as 30–50% of these habitats have been destroyed throughout the Basin, rendering many species characteristic of these habitats vulnerable to local extinction.

Chotts, sebkhas, dayas,* and gueltas*

In the most arid parts of the Basin, notably North Africa, central Turkey, and some parts of the Iberian plateau (Laguna de Gallocanta) where annual rainfall does not exceed 400 mm, there occur large endoreic temporary wetlands too dry and salty to be included in the previous section. The largest of these are called *chotts*. Many of the depressions where they occur in North Africa were once extensive freshwater lakes when the climate was more humid than today. Some of them, such as Chott Djerid in Tunisia, are among the largest wetlands in the Mediterranean. A large chain of endoreic drainage basins also occurs in Algeria at high altitudes (*c.* 1000 m) on the Plain of Chotts between the two main ranges of the Atlas Mountains. These temporary wetlands are usually devoid of aquatic vegetation and have a crust of rock salt or anhydrite covering the lake floor. Their margins are covered by a scattered vegetation mostly consisting of halophytic bushes (e.g. *Salicornia*, *Arthrocnemum*). Isolation from permanent water bodies, long periods of complete desiccation, and large seasonal variations in salinity make these habitats biologically very poor. Only a few species successfully colonize them, thanks to their having evolved resistance devices to desiccation (Cladocera), or else because they are long-distance colonists (Corixidae), which allows them to occupy these habitats for short periods.

In the arid zones of North Africa and the Near East, smaller depressions called *sebkhas* are occasionally filled following heavy rainfall. Since evaporation is about ten times higher than atmospheric precipitation in these areas, surface water remains for no longer than a few weeks. This may suffice, however, to provide breeding grounds for large colonies of nomadic birds such as flamingoes and stilts, as well as stopover sites for thousands of migrating birds. A large portion of central Turkey where rainfall is usually less than 400 mm yr^{-1} is drained by the Lake Tuz, a chott 90 km long and 32 km wide, but no more than 1.50 m deep. Although used as a salt pan yielding two-thirds of Turkey's industrial salt production, a colony of flamingos and several tens of thousands of wintering geese regularly visit this wildlife hotspot.

Some even smaller bodies of water in North Africa are also of great importance for wildlife in arid regions where water is scarce. Examples are *dayas* and *gueltas*. The former are small endoreic temporary ponds where water occurs for some weeks or months after large rainfall; the latter are deep holes in the bed of rocky wadis. Gueltas usually retain permanent water and are important spots for several species of plants, fishes, amphibians, and sometimes breeding birds, for example the ruddy shelduck (*Tadorna ferruginea*).

Intertidal mudflats

Since there are practically no tides in the Mediterranean Sea, there are also virtually no intertidal mudflats except in the Gulf of Gabès of the southern Tunisian coast, around the nearby Kneïs Islands, and to a much lesser extent at the head of the Adriatic Sea near Trieste. In the Gulf of Gabès a tidal amplitude of 3 m creates nearly 200 km^2 of mudflats, which support sea grasses such as *Zostera noltii* and *Z. nana*, and the salt grass *Spartina maritima*. These habitats are used as stopover places for thousands of migrating waders in both spring and autumn.

Summary

In this brief review of the main habitat types we have looked at the ecological diversity present in the Basin without going into detail about ecotones*, which constitute an entire array of habitats in themselves. In the next chapter we will use case studies to illustrate how organisms became adapted to, and thrive within, the mosaic structure of Mediterranean habitats that tend to present varying degrees of ecological and geographical isolation.

6 Populations, species, and community variations

As pointed out in several earlier chapters, levels of endemism vary widely among groups of organisms in different parts of the Basin and a clear relationship exists between endemism patterns and the particular geographical configuration of the spaces where species have evolved.

One Basin-wide 'constant' is the exceptional degree of environmental heterogeneity found at virtually any spatial scale, from regions and landscapes down to simple rectangles 0.1 ha in size, or even squares of 1 m^2. Mediterranean landscapes are 'patchy', and this multiscalar complexity has been propitious for the differentiation of populations and species over evolutionary time (Chapter 3), and of habitat-specific species assemblages over ecological time (Chapter 5). Resembling a fractal* mosaic in space, and a turning kaleidoscope in time, Mediterranean ecosystems and landscapes may be described as participating in a 'moving mosaic' best perceived when several different hierarchical levels and four dimensions are considered simultaneously. To study biota in such a patchy, and temporally changeable environment, populations and communities should not be approached as independent units, but rather as interactive ones, developing and interacting within a 'landscape' of other natural populations and communities.

The realization that regional dynamics can have a very marked influence on ecological and evolutionary processes at 'lower' hierarchical levels, those of species and populations, has been a major development in the field of ecology over the last 10 to 20 years. So much so that we tend to take it for granted nowadays, even if our working methods and conceptual tools have not evolved as quickly as one would like. Similarly, to understand how a Mediterranean landscape 'works', it is necessary to know something about how it fits into a larger picture as well, in space and in time.

In this chapter we examine some of the processes of differentiation that have occurred over evolutionary time in the Mediterranean area, at the levels of populations, species, and communities, within the prevailing context of a 'moving mosaic' of habitats. As emphasized in Chapters 1 and 2, the division of the Mediterranean into two well-defined halves (see Fig. 1.4), the many geographic discontinuities with islands, peninsulas, and mountains, as well as the highly dissected structure of landscapes and habitats, partly due to human interventions over past millennia, have all had many consequences for ecological and evolutionary processes among and within living systems. In this context, many questions are worth addressing, of which we will examine just a few. First, are there clear cases of west–east replacement of closely related species among species groups, that is, vicariant species, which would indicate that differentiation processes have operated on the same lineage in two or more different regions in the past? Second, what are the consequences of insular isolation on the organization of living systems and genetic diversity at the levels of communities and populations? Third, what is the importance of scale effects on the dynamics of populations, species, and communities in habitat mosaics within landscapes? Rosenzweig (1995) maintains that a great deal of the published data on species distribution and turnover needs to be re-examined with a closer look at area or scale effects. Finally, how do species and populations cope with habitat heterogeneity and isolation? What is the balance between local adaptation, that is, adaptive genetic differentiation, and phenotypic plasticity* of populations for adjustment to environmental variation?

West–east substitution patterns

A remarkable number of east–west species pairs, or vicariant species, have been identified among the dominant tree genera of the region, such as the two most widespread Mediterranean pines, Aleppo pine (*Pinus halepensis*) in the west, and Calabrian pine (*P. brutia*) in the east (Fig. 6.1). These species only co-exist (and form natural hybrids) in small districts of Greece, Turkey, and Lebanon (Barbéro *et al.* 1998). Similarly, *Juniperus thurifera* in the western Mediterranean is replaced by *J. excelsa* in the east (Barbéro *et al.* 1992).

Among the evergreen oaks, the western holm-oak is replaced by *Q. calliprinos* in the eastern half of the region (Fig. 6.1). Within the holm-oak complex itself there is a further differentiation, with *Q. ilex* in the eastern part of the range of the species and *Q. rotundifolia* widespread in south-western France west of the Rhône River, and especially in the Iberian peninsula. There is still uncertainty as to whether these two taxa should be recognized as 'full species'. Southern France, west of the Rhône, seems to be a zone of natural hybridization between these two forms (Michaud *et al.* 1995). From analyses using chloroplastic DNA, Lumaret *et al.* (1991) have shown that gene flow between the two forms is restricted to this zone. We do not know the history and evolutionary determinants of the differentiation of these two oak forms, but the holm-oak is clearly highly complex and genetically variable. The ongoing epigenetic role of humans cannot be neglected for these oaks (see Box 6.2), as is the case for many economically important plants throughout the Basin (see Chapter 8).

Among the deciduous Mediterranean oaks the situation is far more complex, with a large series of vicariant species that will be described later in the chapter. Similar patterns also occur in deciduous plant species, as for example in *Pistacia terebinthus* (west) and *P. palestina* (east). Examples of vicariant pairs among animals are common as well. Among birds they include the black-eared wheatear (*Oenanthe hispanica*) to the west and the pied wheatear (*O. pleschanka*) to the

(a)   EVERGREEN OAKS

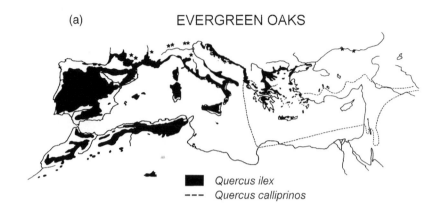

■ *Quercus ilex*
- - - *Quercus calliprinos*

(b)   PINES

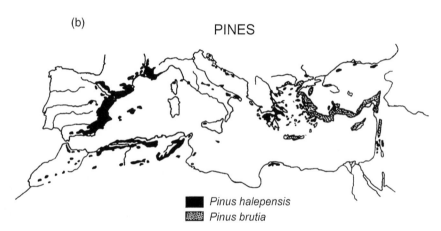

■ *Pinus halepensis*
▨ *Pinus brutia*

Fig. 6.1 Distribution of the most widespread oak and pine species in the Mediterranean Basin showing west–east disjunctions between (a) holm-oak (*Quercus ilex*) and *Q. calliprinos*, and (b) Aleppo pine (*Pinus halepensis*) and Calabrian pine (*P. brutia*). (After Quézel 1985; Barbéro *et al.* 1998.)

east, the Neumayer's rock nuthatch (*Sitta neumayer*) to the west and the eastern rock nuthatch (*S. tephronota*) to the east, the peregrine falcon (*Falco peregrinus*) to the north of the Mediterranean and the Barbary falcon (*F. pelegrinoides*) to the south. Numerous other examples discussed by Vuilleumier (1977) and Haffer (1977) support the view that many processes of allopatric speciation occurred through vicariance thanks to isolation of biota at different epochs in the past, especially in the eastern part of the Basin. In some of these examples, hybridization between species at the contact zone means that they are closely related and recently descended from a common ancestor.

An important framework for differentiation of organisms is provided by the extensive archipelago of small and large islands, which we will discuss next.

Life on islands

Since Darwin, and the seminal book by Wallace (1880), biologists have been fascinated with the evolutionary biology of island biota, particularly in relation to factors determining species diversity, adaptive radiation, and evolutionary changes within populations. An island is a self-contained region whose species originate entirely by immigration from outside. By self-contained one assumes that within the island each species has an average net growth rate sufficient to maintain a viable population size. Populations on islands cannot be 'sinks'!

Differentiation

When a propagule*, say a few individuals of a species or at least a seed or a fertilized female, succeeds in immigrating and then colonizing an island, it is confronted with new sets of environmental factors, both biotic and abiotic. Moreover, its genetic background is different from that of the mother population because of reduced diversity in the colonist population, which is a genetic bottleneck, and potential difficulties due to inbreeding effects. These new ecological and genetic conditions result in new selection pressures that inevitably lead the founding population to diverge from its mainland mother population. Divergence may eventually lead to speciation if reproductive isolation is attained before new individuals of the same species colonize the island. Two examples of such processes in Mediterranean islands are the co-occurrence of the blue chaffinch (*Fringilla teydea*) and the chaffinch (*F. coelebs*) in the Canary Islands, and that of the Corsican swallowtail butterfly (*Papilio hospiton*) and the European swallowtail (*P. machaon*) in Corsica. In both cases, the former species succeeded in colonizing the islands and differentiating from its mother mainland population into a full species before a second colonization by individuals of the same mother species occurred. Complete genetic isolation between species of the two pairs allowed them to co-exist in sympatry*. Because rates of evolutionary divergence are inversely proportional to the distance between the mainland and the island, endemism rates are higher on remote islands than on islands that are close to the mainland source. In addition, endemism rates are lower in groups with high dispersal aptitude than in those that disperse poorly. Indeed, endemism rates are higher in plants than in most vertebrates and much higher in less mobile animals, such as reptiles, amphibians or non-flying mammals, than in birds or bats.

In birds there are few endemic species in Mediterranean islands, except in the remote Canaries, because islands within the Basin are too close to the mainland for differentiation to have had a chance to occur between two colonization events (Table 6.1).

However, at the subspecies level, morphological changes on islands led taxonomists to recognize many subspecies of birds in Mediterranean islands. For instance, on Corsica more than half the species of birds are considered as subspecies. This does not necessarily mean, however, that morphological changes are associated with large genetic changes. For example, the citril finch (*Carduelis*

Table 6.1 Bird species endemic to Mediterranean islands.

Island(s)	Species
Canary Islands	*Columba bollii*
	Columba junionae
	Apus unicolor
	Anthus berthelotii
	Saxicola dacotiae
	Fringilla teydea
	Serinus canaria
Cyprus	*Oenanthe cypriaca*
	Sylvia melanothorax
Western islands (Balearics, Sardinia, Corsica, and nearby islets)	*Sylvia sarda*[1]
Corsica	*Sitta whiteheadi*

[1] Small populations of this species also breed near the coast in south-eastern Spain

citrinella) occurs on both the mainland and Corsica. Although this species is restricted to the Oro – Mediterranean life zone on the mainland, it is widespread from sea level up to high mountains in Corsica where its ecology and behaviour differ greatly from those on the mainland. However, there is a surprisingly low genetic divergence between the citril finch population of Corsica and those of the Alps and the Pyrenees despite the Corsican population having been isolated from the mainland populations a long time ago (Pasquet and Thibault 1997).

Depauperate biota

Evolution and genetic differentiation of endemic forms on islands is only one aspect of the story of island life. In the Mediterranean archipelago, differentiation of endemic species is, for most groups, just the visible tip of the iceberg. Comparing island patterns with mainland patterns may be conducive to inter-pretations of ecological and evolutionary processes that occur as a result of isolation. The most obvious and well known character of island communities is that they are impoverished in comparison with communities occupying areas of similar size on the nearby mainland. For example, 108 species of birds regularly breed on Corsica, an island of 8680 km^2, as compared to 170 to 173 species found breeding in three areas of similar size in continental France. Moreover, as expected from the classical species–area curves, the smaller the island the lower the number of species. One example for birds is given in Fig. 6.2. As a general rule, rates of species impoverishment are a function of the dispersal abilities of organisms. As compared to species richness on the mainland, impoverishment in

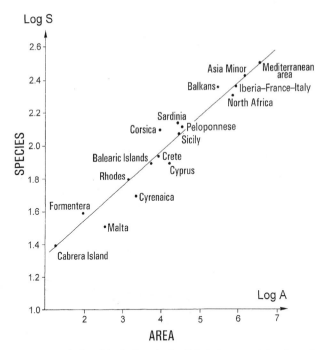

Fig. 6.2 Species–area relationship (in log–log) of birds in some islands and mainland regions (Reproduced with permission from Blondel 1986).

Corsica is 38% in birds, 43% in reptiles, and 68% in non-flying mammals.

Similarly, in butterflies, most of the variation in species richness on Mediterranean islands can be explained by differences in the islands' surface areas. For example there are only 24 species of butterflies on one of the smallest islands, Formentera (115 km^2) as opposed to a maximum of 89 species in Sicily, the largest (Hockin 1980). However, there are some exceptions to the regular trend of decreased species richness on islands. For example, insectivorous groups of vertebrates that feed upon small flying insects are much less impoverished than most other groups. There are as many swifts and swallows on Mediterranean islands as on the nearby mainland and 26 species of bats occur in Corsica, exactly the same number as on the mainland.

Area is an important determinant of the insular faunas for two reasons. First, as the size of the island decreases, so does the size of a species' population. Thus the probability of extinction increases as the island area decreases. Second, as the size of an island increases, so does its probability of encompassing a wider range of habitats that will be suitable for colonization by a broader range of species.

Changes in ecological processes

The general impoverishment in species numbers occurring on islands results in a cascade of changes in fundamental processes at the levels of communities, species, and populations. Studies conducted on Mediterranean islands provided good examples of these changes in birds. They may be summarized in five main points:

1. Successful colonization

First, not all the species of the source mainland have the same chances of becoming successful colonists on an island. There is some kind of selection among the candidates for island colonization. In birds the most successful colonists are species that tend to be small, widespread, and abundant on the mainland and rather flexible in their habitat selection and foraging habits (Blondel 1991). They are, so to speak, pre-adapted to cope with the different environmental conditions they will find in their new habitats. In other words, the small generalist species will be favoured. Similar strategies in plants are those of small, generalist, dioecious* flowers. In birds, the smaller the island, the lower the number of species and the higher the proportion of resident species that are widespread in the overall region. Thus, island communities are not a random subset of the communities on the mainland. For example, the breeding bird fauna of Malta includes only 28 species of land birds, examples of which are the house sparrow, the kestrel, the turtle dove, the barn swallow, the nightingale, and the goldfinch. All of them are extremely common and abundant throughout the entire Mediterranean basin. To take another example, the woodpecker family is represented by nine species in the Mediterranean but only the great spotted woodpecker (*Dendrocopos major*) and the wryneck (*Jynx torquilla*) regularly occur on large islands such as Corsica, Sicily, and Sardinia.

2. Increased population sizes

Population sizes of the species that occur on islands are often much larger than those of their mainland counterparts. This process, known as 'density inflation', has often been interpreted to be a result of the ecological space that would have been occupied by the missing species being actually filled by the smaller number of species present in insular communities (see Table 6.2). But this interpretation has been challenged because the mechanisms of population increases on islands are still largely unknown (Blondel *et al.* 1988).

3. Niche enlargement

Many species on islands often occupy more habitats and forage over a larger spectrum of microhabitats than on mainlands, and prey upon a larger range of food items. This is a process called 'niche enlargement'. As a consequence, species impoverishment on the scale of a whole island is not necessarily reflected on the scale of local habitats. For example, some habitat types on Corsica have actually more species than their mainland counterparts as shown by changes in species diversities along two habitat gradients that match each other on the basis of vegetation structure (Table 5.2). A higher number of bird species in Corsican matorrals is a result of niche enlarging whereby species 'spill over' from forests to

Table 6.2 Niche enlargement and density inflation in tit species (*Parus*) in matching habitat gradients on Corsica compared to the nearby mainland. Habitats range from 1 (low matorral) to 6 (old mature oak forest). Each species occupies on average more habitats and has higher population sizes (breeding pairs 10 ha^{-1}) on Corsica than on the nearby mainland.

Species	Population density in each habitat (breeding pairs 10 ha^{-1})					
	1	2	3	4	5	6
Mainland						
Great tit				2.2	3.1	3.2
Blue tit						11.5
Coal tit						0.2
Crested tit						1.8
Total mean				2.2	3.1	16.7
Corsica						
Great tit	1.6	1.7	2.5	3.6	2.6	4.7
Blue tit		0.5	0.2	3.3	7.9	14.2
Coal tit				1.2	2.1	4.1
Total mean	1.6	2.2	2.7	8.1	12.6	23.0

Source: Blondel 1985b

matorrals, hence increasing local species diversities (Blondel *et al*. 1988). All these processes result in larger population sizes since populations are more flexible in habitat selection, hence reducing extinction risks. The process of niche enlarging has been observed in many species of both animals and plants.

A classical explanation of higher population densities and niche enlarging on species-poor islands is that extra resources become available because of the lack of competitors. As a consequence it is argued that island habitats include similar numbers of individual birds to mainland habitats, but with fewer species. These responses of niche breadth and population density are often called 'competitive release' and 'density compensation', respectively (MacArthur *et al*. 1972).

4. Changes in behaviour

Changes in territorial behaviour and aggressiveness are often observed in insular populations of vertebrates. The social behaviour of many reptiles, mammals, and birds reveals remarkable shifts, such as reduced territory size, increased territory overlap, acceptance of subordinates, reduced situation-specific aggressiveness, and abandonment of territorial defence, as compared to mainland populations. These changes are often associated with unusually high densities, niche enlarging, low fecundity, and the production of a few, competitive offspring. One explanation of these shifts in behaviour is the 'defence hypothesis' (Stamps and Buechner 1985), which suggests that a release in aggressiveness and

Box 6.1 Black rats: friends or enemies?

On a very small islet (6.4 ha) off the coast of Corsica, as many as 141 black rats lived in close proximity and did not compete severely for resources. As a result of a 'stranger–neighbour effect' they knew each other and lived on good terms, so to speak. This did not mean, however, that they had lost their aptitude to be aggressive towards unknown intruders. Granjon and Cheylan (1989) introduced five rats (four males and one female) from Corsica onto this small islet after having equipped them with radio tags. Within a few hours, all five were recovered dead, killed by the resident rats. The introduced rats were unknown and therefore treated by the local territory owners as enemies to be exterminated.

territorial defence may occur as a result of trade-offs between defence costs of the territory in crowded populations and costs of reproduction. If defence costs become exaggerated, animals would benefit by expending less energy in territory defence and reallocating their resources into breeding activities to produce young that are more competitive. One example of such shifts is that of the black rat in small islands (see Box 6.1).

For birds too, on Mediterranean islands, there is evidence that territory owners are often much less aggressive than on the mainland, accept subordinates in their territories and allow 'intruders' to breed between adjacent territories. For instance, compared to their mainland conspecifics, Corsican blue tits are much less aggressive and readily share their territory with other pairs.

5. Changes in body size

Changes in body size are known as the 'van Valen rule', according to which large species tend to become smaller and small species tend to become larger; hence, body sizes on islands tend towards uniformity (van Valen 1973). However, there are so many exceptions to this trend that the rule has been questioned by many scientists. Amongst birds, for example, take the large raptors. Although it has often been claimed that there should be fewer on islands because their large size prevents them from constructing viable populations and because of a reduction in the diversity of prey, there are in fact many large raptors on most Mediterranean islands: vultures, eagles, and falcons. In Crete, for example, there are up to 12 species: Lammergeyer, Egyptian vulture, griffon vulture, black vulture, buzzard, rough-legged buzzard, golden eagle, lesser kestrel, kestrel, Eleonora's falcon, lanner, and peregrine falcon! However, the overrepresentation of large raptors on many Mediterranean islands may have more to do with the relationship between each of these islands and the species that inhabit them. For large birds at least, islands that are not far from the mainland are not necessarily *biological islands* if the species that breed in them are part of a single larger population extending over large mainland and island areas. In that case, repeated exchanges of individuals between the mainland and the island make the latter but a part of a larger range of distribution.

One famous example of changes in body size on islands is that of the dwarf hippos and elephants which inhabited Mediterranean islands before their extermination by humans (see Chapter 9 and Fig. 9.1). In such relatively small mountainous areas as Cyprus or Malta a small body size should be selected for if it allows these species to make a better use of the food supply and to construct larger populations that are less vulnerable to extinction. The dwarf hippo of Cyprus (*Phanourios minutus*) presumably reduced its size, after crossing the body of water separating this island from Turkey some 100 000 years ago, from that of a normal hippo to that of a pig! Besides size reduction, these animals were also characterized by large changes in their legs, which were relatively shorter, more robust, and without fingerwebs, thus allowing them to walk on the tips of their fingers. All these morphological changes apparently evolved as adaptations to enable them to walk and climb in the mountainous terrain of Mediterranean islands. There has been much debate on the selection pressures which led to dwarfism or gigantism. It has been proposed that a release in predation and/or interspecific competition favours gigantism in small species, especially rodents. In this context, one interesting observation reported by J.-D. Vigne (1987) is the reverse trend of decrease in the size of endemic rodents *Prolagus* and *Rhagamys* during the Holocene in Corsica, presumably as a response to increased predation pressures from predators introduced by humans. This 'response', however, did not prevent these species from going extinct.

Community dynamics in heterogeneous landscapes

Turning now to communities in complex habitat mosaics, we will examine how these communities are organized at smaller scales of space and how their dynamics are driven by factors that make the mosaic of habitats 'move' in space and time. These dynamics involve repeated colonization–extinction processes operating at the landscape scale. A landscape is made up of many local habitats that continuously change over time so that species and communities that inhabit them are also perpetually changing. We will illustrate these processes using the turnover of habitats and species assemblages as a response to recurrent fires.

Dynamics of bird communities and fire

The desolate appearance of burnt areas provokes fear and panic in most people because fires are directly associated with the destructive behaviour of humans. However, provided their 'return rate' is not too high, fires are a natural component of the dynamics of most ecosystems, and more so in the Mediterranean than in many other regions. They do not seem to alter the physical structure of soils, nor do they destroy the organic matter in the soil unless recently burnt areas are too rapidly and intensively grazed by sheep and goats.

Fire provides an excellent illustration of the role of natural disturbance in the functioning of ecosystems at the scale of landscapes, the so-called patch dynamics (Pickett and White 1985). Schematically, Mediterranean landscapes periodically and frequently exposed to fires are characterized by a turnover of four habitat types, which replace each other in space and time like a 'moving mosaic'. These are grasslands, low matorrals, high matorrals, and forests with, of course, a huge range of local variants depending on substrates, periodicity of fires, history of land use, dominant plant species, and other factors.

Using birds as a model, an elegant study of post-fire successional processes was carried out by Prodon *et al.* (1987). The experimental design involved 186 study plots evenly distributed among a series of 11 habitats ranging from grasslands to mature forests in the holm-oak series. On this landscape scale, a census was carried out on 51 species. They ranged from species of open vegetation (e.g. woodlark, corn bunting, linnet) to those of the forest (e.g. robin, blackcap, chaffinch) with species of matorrals (e.g. several species of warblers, the nightingale) between them. At each census spot, the dominant plant species were recorded as well as vegetation profiles. The data were used to model the relationships between vegetation structure and bird communities and illustrate the turnover of species in the 11 habitats (Prodon and Lebreton 1981).

At the scale of the whole range of habitats within the mosaic, all the 51 species found in the post-fire gradient were already present somewhere within the landscape. Accordingly, the whole gradient is a closed system within which processes of local extinction and recolonization operate. Species of open vegetation colonize habitats immediately after fire and then are replaced by matorral species such as warblers and the nightingale as the resprouting vegetation grows up. In turn, matorral species will be replaced by forest species as the recovery process moves towards completion (return to a forest stage).

The turnover of bird communities as a response to fire-induced habitat changes highlights the importance of scale effects in community investigation. Because environmental heterogeneity plays a prominent role in structuring communities, ecological processes operate on spatial and temporal scales that are far larger than those usually used in field studies. At the scale of a landscape and over long periods of time, the survival of all the species of a mosaic of habitats involves the existence of a disturbance regime that is unpredictable in time and space in the short term but predictable in the long term. This regime periodically moves its position up and down any given habitat patch within a landscape including patches of grasslands, matorrals, and forests. On a broader geographical scale, the combination of a disturbance regime and community dynamics, resilience,* and inertia* results in a dynamic equilibrium that may be fairly stable in the long term. Such a system is characterized by (1) a given pool of species that is a legacy of history, and (2) a regime of disturbance that is specific to each region. To be sustainable in the long term, such a system requires areas large enough for the spontaneous occurrence of stochastic and chance events. Blondel (1987) coined the term 'metaclimax' to define both the spatial scale required for maintaining a self-sustaining system and the disturbance regime that guarantees all the habitat patches required for the regional survival of all the species determined by history.

Small scale disturbance and plant species diversity

In the previous example, disturbance events such as fire determine the dynamics of species and communities, such as birds or woody plants, that disperse over relatively large scales of space and time. At much smaller scales, disturbances such as ant hills, the holes dug by small mammals, and the grazing of plants by snails and insects, are sources of spatial and temporal variability that contribute to maintain the species richness of plants and animals. The role of these small-scale disturbances in the dynamics of plant species diversity has been recently studied in abandoned vineyards near Montpellier, southern France. 'Old fields' abandoned by farmers are known to be rich in plant species (Escarré *et al.* 1983), as is the case for most open habitats in the Mediterranean Basin. Lavorel and colleagues (1994) created artificial disturbances in three old fields that had been abandoned for 1, 7, and 15 years, respectively. Before the disturbance experiments began the vegetation was dominated by annual species, including grasses, legumes, and forbs* arranged in a dense mosaic pattern. There were 118 species in the three fields combined, with 81, 57, and 79 species in the 1-year-, 7-year-, and 15-year-old fields, respectively. Experimental disturbances mimicking herbivory, pathogens (using herbicides), and small mammal disturbance of the superficial soil layer were applied at periods coinciding with climatic conditions favourable for germination in Mediterranean habitats, that is, October, December, and March.

When applied in autumn, disturbance prompted the germination of annual grasses, followed by legumes, and then a variety of annual forbs. This resulted in a peak of species richness in early winter, which further increased in early spring as several new annual and perennial species germinated as well. Then species richness sharply decreased from May to August because species that became established in autumn and winter completed their life cycle and died before the onset of summer drought. In all fields the winter (December) and spring (March) treatments increased the contribution of perennials but only a few of them survived in summer so that species richness was very low in this season.

Four points are worth mentioning:

1. Total species richness always increased after experimental disturbance, but the composition of species assemblages differed relative to undisturbed 'control' sites and among experimental sites.
2. Effects of disturbance treatments depended greatly on their timing, with the highest number of species germinating after application of the October disturbance events.
3. Temporal variation in species composition was higher in species-poor sites than in species-rich sites, and disturbances promoted species co-existence if they reduced biomass of dominant species.

4. In the oldest field (15 years), heterogeneity in the vegetation was higher than in younger sites because there were more biennials and long-lived perennials.

Immediately after a disturbance, space was non-limiting so that many species could establish, but as recovery proceeded surface space progressively became saturated so that the first established plants prevented further germinations from taking place. Composition differences among sites and dates of experimental perturbation provide evidence for the importance of small-scale disturbances for early establishment of seedlings and space pre-emption as soon as a 'window' is open for colonization by species. This is especially important in Mediterranean habitats that are characterized by severe climatic constraints, because these mechanisms allow the persistence, at the landscape scale, of competitively inferior species such as early-successional and understorey annuals.

Large differences in species assemblages among experimentally disturbed and control sites indicate that responses to disturbances involve species-specific strategies. Chance and spatial variability in micro-environmental factors favour different species locally. The so-called 'regeneration niche' (Grubb 1977) of the component species, that is, life cycles and differences in the timing of germination, explain the seasonal patterns of species establishment and replacements. Timing of germination is particularly crucial in determining the way species can take advantage of suddenly available gaps. Hence co-existence of species with different regeneration niches is favoured if disturbances occur asynchronously. On the other hand, species within a given regeneration-niche group appeared as rather interchangeable and whether or not they will successfully establish themselves after a disturbance is mostly a matter of chance.

This case study provides evidence that all species in a given pool cannot co-exist in the same area. A certain amount of recurrent disturbances is a prerequisite for the survival of all of them in a mosaic-like landscape that is designed by disturbance events.

The spatial dynamics of predatory ants

Ants contribute a major part of the overall insect biomass in most terrestrial ecosystems and may include more than half of the individual insects (Wilson 1992). Therefore, their role in ecosystems at both scales of species-specific interactions and functional groups* must be important. Some important functions of ants are seed dispersal (myrmecochory), soil bioturbation, and predation, together with some extraordinary species-specific interactions with butterflies (*Maculinea* spp.). The caterpillars of these butterflies are delicately transported by ants to their nests, where the latter carefully raise them for the sweet secretions they produce, of which the ants are fond (Thomas *et al.* 1989). Caterpillars will stay in ant nests until they pupate and become full-grown butterflies. In a typical Mediterranean landscape, including several life zones along an altitudinal gradient (420–1880 m) at Mt. Ventoux, southern France, 64 ant species were counted (du Merle *et al.* 1978). A careful analysis in 58 study sites of ant species richness and species-specific distributional patterns revealed a significant discrimination of community structure in relation to vegetation belts and vegetation units. The richest communities occurred in the thermo-, meso-, and supra-mediterranean life zones and then declined with increasing altitude. Several ant species are strongly associated with particular plant species, for instance *Aphaenogaster gibbosa* with thyme and *Formica gagates* with downy oak.

An experiment designed to assess one important function of ants, namely their predation pressure on eggs of other insects, revealed that many species move seasonally across different habitat patches, resulting in different ant assemblages over time. Predation has been studied in a small mosaic of three habitat patches, a clearing (942 m^2), a forest edge (478 m^2), and a forest patch (952 m^2). For the three habitats combined, 14 species have been recorded (13 in the clearing, 10 in the forest edge, and 6 in the forest) and 175 nests located. More than half of these species (9) are insect-egg predators. The egg eating activity of ants has been measured using traps supplied with eggs of the Mediterranean flour moth (*Anagasta kuehniella*). Results showed that a very high proportion of the moth's eggs were eaten by ants in the three habitats (up to 76%) and that important between-habitat exchanges of ants occurred. The relatively low number of predatory ants resident in the forest was compensated by high seasonal migration rates from the two other habitats, especially the clearing. In particular, there was an intense exploitation, especially in mid-summer, of the forest habitat by ants invading it from colonies established in the two nearby habitats. Species involved in this dispersal were *Leptothorax unifasciatus*, *Pheidole pallidula*, and *Myrmica specioides*. Such between-habitat migrations of predatory ants suggest that clearings contribute to controlling populations of insects, including foliage-eating insects harmful to tree foliage.

Thus, the patchy geographical configuration of the landscapes plays an important role in the dynamics of insect populations at the scale of the 'clearing–ecotone–forest' system. This is because many forest insects and their predators carry out parts of their life cycle in habitats other than forests. This mosaic structure allows some highly mobile species (e.g. *Leptothorax unifasciatus*) to find optimal environmental conditions for breeding at any time of the year. Predator species that exploit forest resources from neighbouring habitats may reach higher population sizes, and hence achieve higher controlling effects on insect populations, at the scale of a combination of several habitats than at the scale of only one habitat. This example illustrates the importance of neighbouring effects in some important ecosystem processes at the scale of a mosaic of habitat patches.

Adaptation, local differentiation, and polymorphism

One may expect that besides the turnover of species described from these examples, the great diversity of habitats is conducive to a wide range of selection regimes. In other words, selection pressures that shape the life-history traits of populations must differ depending on local conditions. Species and populations may respond to these pressures in a number of ways. They may evolve 'local specialization*' whereby local specific combinations of genes adapt organisms to local conditions. They may also evolve 'phenotypic plasticity', whereby the same genotypes may express in different phenotypes according to environmental conditions.

Many studies in recent years have demonstrated that life-history traits which have important fitness consequences may evolve quite rapidly, even within a few generations, provided they are submitted to strong directional selection pressures. Given the great diversity and heterogeneity of Mediterranean land-scapes, it should be advantageous for populations to have high genetic variability that would allow them to respond to unpredictable changes of the environment. There is a growing body of evidence showing that plants and animals exhibit a high degree of intraspecific variation at different spatial scales: among individuals within populations, among subpopulations within a landscape, and among populations across a species' range. Many field studies in the Mediterranean area have addressed the problem of how the degree of adaptation of organisms to their environment relates to habitat heterogeneity. We shall describe five studies that have demonstrated the functional significance in Mediterranean habitats of a high intraspecific variation of life-history traits, in a perennial grass, a geophyte, a dwarf shrub, several tree species, and a passerine bird.

Orchard grass

One interesting issue in biological diversity is the contribution of phenotypic plasticity and adaptive genetic differentiation, or local specialization, to the variation of ecological and geographical distribution patterns in closely related taxa. Several studies have shown that variation in geographical distribution of a species is often associated with genetic differentiation between populations that may result from natural selection in different habitats. In this context, levels of ploidy* in plants, that is, the number of chromosomes in the nucleus, is an interesting question to address. Stebbins (1971) noticed that many polyploid* plant species have spread over a wide range of environmental conditions whereas closely related diploids* have remained restricted to much smaller areas. Has the greater environmental heterogeneity of the Mediterranean Basin resulted in a higher ecotypic variation* of local populations than that found in non-Mediterranean parts of the species' range? Lumaret (1988) has shown that the perennial orchard grass (*Dactylis glomerata*, Poaceae) consists of a complex of no less than 9 distinct subspecies, with several insular endemics, and as many as 15 diploid types, 3 tetraploid, and 1 hexaploid, the latter being confined to North Africa. Several tetraploid forms (e.g. *Dactylis g. glomerata*, *D. g. hispanica*, and *D. g. marina*) have wide distribution ranges and exhibit high morphological and physiological variation (Fig. 6.3). Stebbins and Zohary (1959) pointed out that the differentiation of closely related tetraploid forms of *D. g. glomerata* resulted from autopolyploidy* of diploids derived from both temperate and

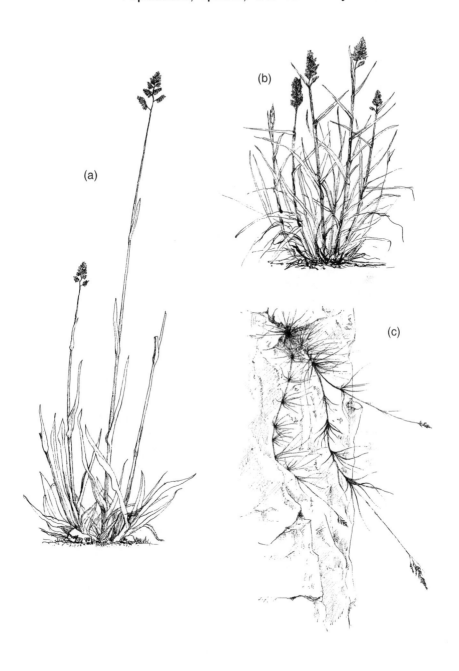

Fig. 6.3 Drawings of three taxa of the orchard grass (*Dactylis glomerata*) complex with differing levels of ploidy. (a) A typical diploid form; (b) a dwarf form of drier habitats (tetraploid); and (c) a cliff-dwelling sub-ligneous tetraploid. (Drawn from photos kindly provided by R. Lumaret.)

Mediterranean groups, thus providing the plants with ecological attributes adapted to both climatic regions. Polyploidy appears to be an evolutionary response to the wide range of ecological conditions within the Mediterranean region. In fact, almost all forms of *D. glomerata* in the Mediterranean group are to some extent adapted to drought conditions. Representatives of this complex of taxa range from sea level to well above the tree limit in high Mediterranean mountains with a predominance of tetraploids in more severe ecological conditions. Thus, in at least some cases, tetraploidy may widen the habitat range of this complex at both ends of ecological gradients. Hexaploidy in hyper-arid climates of Libya may be a further step of adaptation to extreme ecological conditions. In addition to providing a common means whereby plant taxa may diversify ecologically and genetically (Petit and Thompson 1997), polyploidy may confer enhanced resistance to attack by pathogens, insects, and nematodes.

The mechanisms allowing polyploids to adapt to a wider range of environments than their related diploids are associated with biochemical, physiological, and developmental changes (Bretagnolle 1993). For example cell size, which determines photosynthetic rates, as well as DNA content, is higher in tetraploids, so providing them with a greater photosynthetic capacity than diploids. More-over, autotetraploids* exhibit greater mean performances in fitness-related traits over a range of environments than their related diploids, as shown in two perennial grasses, *Dactylis glomerata* (Bretagnolle 1993), and *Arrhenatherum elatius* (Petit *et al.* 1996). In the latter species, the diploid subspecies *sardoum* is endemic to pine forests and open scree slopes in mountains whereas the tetra-ploid subspecies *elatius* occurs in a wide range of habitats such as road sides, open fields, waste ground, and woodlands all over Europe. Petit and colleagues experimentally demonstrated that the tetraploid *elatius* has better performances in all environments: it shows significantly higher values for stem height, leaf surface area, total seed number, and number of spikelets per flowering tiller than the diploid. Interestingly, these studies have demonstrated that the greater variation in the ecology and geographical distribution of tetraploids does not result from greater phenotypic plasticity but from between-population genetic differentiation. Thus, the greater vegetative stature and inflorescence size of the tetraploid suggest that their more widespread distribution is due to better performances rather than to a greater capacity to buffer environmental variation. Indeed, differences in vegetative stature were observed between open habitats and woodland in tetraploids, but not in diploids.

The highly dynamic evolution of polyploidy in many groups and greater genetic differentiation may be critical components of their success in spatially variable environments such as those found in the Mediterranean area. Adapta-tions to Mediterranean conditions in polyploid groups also include morpholo-gical traits allowing water-saving mechanisms and seed retention throughout the summer drought. For example, the seeds of the tetraploid *Dactylis g. hispanica* are not shed until autumn, when conditions become suitable for germination (Lumaret 1988).

Case studies on these perennial grasses support the contention of Stebbins

Fig. 6.4 *Cyclamen balearicum* in Mallorca. (Drawn from photos kindly provided by J. Thompson.)

(1971) that 'most of the variation within species with respect to visible characters is associated with adaptation to climatic and edaphic factors of the environment'.

Continental and island populations of *Cyclamen*

Many plant species occur as scattered and isolated populations both on islands and on the nearby mainland, which provides good opportunities to investigate evolutionary processes and the genetic structure of populations. *Cyclamen balearicum* (Primulaceae), one of the 20 or so species of this genus occurring in the Basin, is endemic to five geographically isolated sites in southern France, and to the Balearic Islands of Mallorca, Menorca, Ibiza, and Cabrera (Fig. 6.4).

On the mainland this geophyte lives in the understorey of Mediterranean evergreen oak and pine woodlands, usually on north-facing slopes or gorges in shady rocky forests or else on limestone outcrops (Debussche *et al.* 1995). It occurs over a much wider range of habitats on the Balearic Islands. The presence of this species, with its poor long-distance dispersal ability, in southern France and on the four Balearic Islands suggests that it was already present in all these regions by the time they became isolated from one another in the Pliocene (5 Myr BP). Therefore, they have been genetically isolated for several million years, with most of them having become further fragmented through climatic changes and human action.

The effects of habitat isolation on levels of genetic diversity and patterns of genetic differentiation in this species have been investigated by Affre *et al.* (1997) using biochemical techniques. Their study produced two interesting results. First, as island biogeography would predict, island populations contained less diversity (fewer alleles, less heterozygosity*) than continental populations, suggesting that the colonization of the Balearics caused a loss in genetic diversity. Second, and rather unexpectedly, the authors found that the genetic differentiation among continental 'islands' was greater than that among true islands (as measured by Wright's fixation indices *F*st). This greater differentiation among continental sites might be due to the fact that these populations may have undergone more severe isolation effects, because of either glaciation or human-induced habitat fragmentation. Within continental islands the differences among populations were even greater between individual valleys, further supporting the idea of greater isolation in mainland populations. Interestingly, the heterozygote deficiencies observed in all populations of this species are among the highest values so far reported in the literature. In summary, this endemic species of *Cyclamen* is characterized by a marked population structure, high effects of inbreeding on genetic diversity, and much lower levels of genetic diversity than most widespread species.

Stebbins (1942) stressed that endemic and rare taxa often contain significantly lower amounts of genetic variability as compared to related widespread abundant ones, and this has been confirmed by many researchers. A crucial point contributing to this trend is the small effective size of populations of endemic and rare species and the consequences of inbreeding on their genetic diversity. This is especially crucial in islands where genetic theory predicts that insular populations are genetically less diverse than the populations of their mainland sources. The case study with Balearic cyclamen provides a striking example of how a combination of historical, ecological, and evolutionary factors can influence patterns of genetic variation within and among populations of a Mediterranean endemic species. It also highlights the risks of extinction faced by endemic species when they are restricted to very small isolated populations.

Sexual and chemical polymorphism: the story of thyme

Some fascinating examples of intraspecific variation in reproductive systems and the contents of essential oils can be observed in Mediterranean plants. Thompson *et al.* (1998) recently synthesized an enormous body of work that has been conducted on such variation in an emblematic aromatic shrub, the wild thyme (*Thymus vulgaris*), which forms a dominant component of open vegetation in many matorrals of the western part of the Basin. In the last 35 years, more than 20 scientists have endeavoured to decipher the genetic and ecological factors that determine the evolution and maintenance of polymorphic variation in the reproductive system and chemical compounds present in this species.

The first part of the story concerns sexual polymorphism. Thyme is a gynodioecious* species, which means that plants bear flowers that are either hermaphrodite or exclusively female. Sex determination is controlled by the interaction of cytoplasmic genes (that produce female phenotypes) and nuclear genes that restore the hermaphrodite phenotype. When the inheritance of sex is cytoplasmic, females produce only females because the male function is sterile. The frequency of females in natural populations averages around 60%, varying from 5% to more than 90% depending on populations. What maintains gynodioecy in thyme and why is there such marked variation in female frequencies? Since female phenotypes reproduce only via ovules, they suffer a loss of genetic diversity relative to hermaphrodites. What advantage compensating for the lack of pollen production allows them to persist in populations? Research has shown that if sex inheritance is cytoplasmic, females need only a slight advantage over hermaphrodites, in terms of seed production, to be maintained and to invade the populations (see Gouyon and Couvet 1987; Couvet et al. 1990). In addition, the genetic cost of females may be offset if hermaphrodite progeny suffers inbreeding depression (thyme is self-pollinating) or else if plants reallocate to seed production the resources they save by not producing pollen. Indeed females produce 2–3 times more seeds than hermaphrodites and their offspring are more vigorous. However, the maintenance of the gynodioecious polymorphism requires that hermaphrodites also occur and provide advantages to the population, which implies that restoration of male function must sometimes occur. Sexual phenotypes are thus determined by the interplay of cytoplasmic male sterility genes and nuclear genes that restore male fertility. The explanation of high and variable female frequencies is that populations are not at equilibrium, that is, there is a high variation in nuclear and cytoplasmic gene frequencies. Such a variation arises from differences in local fitness of plants, depending on whether they are females or hermaphrodites. Female frequencies depend on the age of populations, with high female frequencies in very young populations that invade newly abandoned cultivated sites or sites that have recently been burnt. Founder effects during early colonization cause a lack of variation in cytoplasmic feminizing genes (Manicacci et al. 1996; Tarayre et al. 1997). In association with an absence of appropriate nuclear restorer alleles, this results in cytoplasmic sex determination and the production of only females, which rapidly invade the site. Later, the arrival of nuclear restorer alleles via seed or pollen dispersal causes female frequency to decline. Thus, there is in this species a 'moving mosaic' of mating systems that is driven by the occurrence of repeated extinction–recolonization events resulting from ecological disturbances such as fire that cause spatial variation in the genes which determine sexual phenotype. This is a fine example of metapopulation dynamics, with non-equilibrium prevailing in young populations where female frequencies may be as high as 90%, and equilibrium being maintained across suites of interacting populations (Olivieri et al. 1990).

The second part of the thyme story concerns polymorphism in the composition of essential oils in plant cells. A comparative study of plant species in the Mediterranean Basin has shown that as many as 49% of evergreen, xeromorphic* woody shrubs produce aromatic volatile oils (Ross and Sombrero 1991). Indeed, many Mediterranean plants are renowned for their fragrances, which give an exquisite smell to matorrals, and a large number are widely utilized as dried or fresh herbs for cooking.

In thyme, six genetically determined forms of biochemical compounds (monoterpene) have been recognized on the basis of the dominant forms produced in glands on the surface of the leaves (Vernet et al. 1986). We might expect that such polymorphic variation is maintained because the various forms differ in fitness according to spatial or temporal variation of ecological factors, whether biotic or abiotic. It appears that phenolic compounds dominate in dry hot sites whereas non-phenolic chemotypes dominate in inland and hilly sites that are often colder and moister. However, spatial differentiation in the distribution of chemotypes also occurs at much smaller spatial scales (Gouyon et al. 1986), sometimes even at the scale of a few metres. Such patterns suggest that chemotype variation has something to do with local adaptation. Potential functions of biochemical compounds in aromatic plants include, among others, resistance to drought, a cooling function during their vaporization, initiation of stomate closure, resistance against herbivores, and inhibition of the germination and growth of other plants.

There is evidence that at least the two latter functions work in thyme. For example, there is a marked variation between chemotypes in their palatability to slugs, and the snail *Helix aspera* definitely prefers non-phenolic phenotypes. Interestingly, phenolic compounds may differ within the same genotype depending on the age of the plant. Seedlings, which are expected to be highly vulnerable to herbivory, have a phenolic phenotype that protects them, but this phenotype will be replaced by a non-phenolic one as the plant gets older and becomes less vulnerable to herbivores. As explained by Linhart and Thompson (1995), the phenotype most preferred by snails may 'hide' behind a less palatable phenotype during early seedling development. These authors have subsequently found that different herbivores prefer different chemotypes. Hence variation in herbivore preference and abundance may well influence the maintenance of polymorphism in essential oils.

Inhibition of the germination of other plants may also contribute to spatial variation in chemotype abundance. Tarayre et al. (1995) demonstrated that phenolic chemotypes have a greater inhibitory effect than non-phenolic chemotypes on the germination of an associated grass species (*Brachypodium pinnatum*). Thus, local variation in the composition of plant communities may determine and maintain the variation in chemical compounds. This story provides a nice opportunity to examine how secondary compounds confer adaptive mechanisms in Mediterranean plant species, with environmental heterogeneity playing a key role in the determination and maintenance of diversity at the level of local populations.

Oaks, pines, and pistachio trees

Mediterranean oaks, pines, and pistachio trees play important roles in almost all ecosystems of the Basin, and are all rich in systematic complexity and ecological

Box 6.2 Consequences for genetic variability of human selection on oaks

By comparing many genetic loci of individual trees of holm- and cork oaks in a mixed population, Lumaret and co-workers showed that genotypes of the cork oak are consistently clustered while those of holm-oak are scattered and show little similarity (Fig. 6.5). What emerges most clearly from these patterns is not habitat selection or adaptation on the part of ecotypes within species, but rather effects clearly related to human selection processes. Whereas the holm-oak has been managed by people throughout its distribution range for many millennia, it has never undergone specific selection for any trait (except for isolated cases in central Spain where it was considered as a fruit tree, and selection was carried out for sweet acorn production to feed pigs). In contrast, cork oak has been subjected to ongoing selection to improve its highly useful outer bark everywhere it occurs in the western

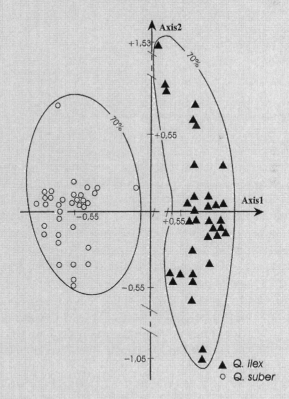

Fig. 6.5 Position of the multilocus genotypes of 24 holm-oak (*Q. ilex*) and 24 cork oak (*Q. suber*) trees sampled at Mauguio, southern France, according to polymorphism ascertained at 11 different loci from 7 enzyme systems. Genotypes were clustered at the 70% level by a hierarchical cluster analysis. (R. Lumaret and L. Toumi, unpublished data; see Toumi and Lumaret (1998) for description of methods.)

Table 6.3 Vicariant species among Mediterranean oaks (*Quercus* spp.)

Leaf form	Group[1]	North Africa	Western Mediterranean	Eastern Mediterranean	Zagros[2]
Evergreen					
Q. coccifera	B	*Q. coccifera*	*Q. coccifera*	*Q. calliprinos*	
Q. ilex	A	*Q. ilex*[3]	*Q. ilex*	*Q. aucheri*[4]	*Q. baloot*
Q. suber	B	*Q. suber*	*Q. suber*	*Q. alnifolia*[5]	
Deciduous					
Q. ithaburensis	B			*Q. ithaburensis*	*Q. brantii*
Q. cerris	B		*Q. cerris*	*Q. hartwissiana*	
Q. infectoria[6]	A	*Q. faginea*	*Q. canariensis*[7]	*Q. boissieri*	*Q. boissieri*
Q. libani	B	*Q. afares*	*Q. pyrenaica*	*Q. libani*	*Q. libani*
Q. robur	A		*Q. humilis*[8]	*Q. anatolica*	

1. A: acorns fall six months after ripening; B: acorns usually fall 18 months after ripening.
2. Zagros, Caucasus, or Himalaya.
3. Includes *Q. rotundifolia* of Spain.
4. Endemic in Greece and Aegean Islands.
5. Endemic in Cyprus.
6. Special group of semi-deciduous species that lose only some of their leaves in autumn.
7. Occurs only in a restricted area of southern Spain, near Gibraltar.
8. Extends into parts of central Europe.
Sources: (A. Shmida, unpublished and M. Barbéro, P. Quézel and F. Romane, personal communication)

diversity. Much information is available on the genetic structure and breeding systems of these species, which are dioecious or monoecious and highly outbred. In species that have been of great importance to humans for centuries, the genetic structure of populations may have been modified to some extent through human selection pressures, thus obliterating or at least superseding the expression of natural processes (see Box 6.2). Yet it is worthwhile considering whether any of the trends we have examined thus far in this chapter can be discerned operating at the species level among these long-lived species.

Variable life-history traits of clear adaptive significance in oaks, pines, and pistachios include growth form, leaf and fruit longevity, sexual reproductive and seed dispersal systems. Among Mediterranean oaks and pistachio trees, for example, we find species contrasting in one important life-history trait, evergreenness versus summergreenness. Six species of Mediterranean oaks are evergreen but the other 35 to 40 species are all summergreen or 'deciduous'. Among *Pistacia* species, the ratio is steeper still: one evergreen species to about ten deciduous. Almost all the evergreen species of both genera are widespread while only among the deciduous ones do we find endemism. In both groups, vicariant species and inter-fertility among sympatric species are the norm (see Table 6.3).

In fact, a continuum exists in the Mediterranean between evergreen and deciduous oaks and pistachios. At one extreme, some deciduous species have no overlap between leaf generations punctuated by cold winter seasons, while at the other extreme, leaves of evergreen species of the previous season fall in late spring, after the current year's new leaves have emerged. So in this situation there is a nearly total overlap in annual leaf 'crops' on a given tree. An intermediate 'semi-deciduous' group also occurs among oaks, as shown in Table 6.3. One similar case among pistachios (*P. saportae*) appears to be a natural hybrid between a deciduous and an evergreen species. Finally, a great deal of leaf longevity variation is observed both within, and especially among, populations of a given species.

In general terms, evergreen oaks and pistachios can be considered old 'primitive' taxa, with tropical or subtropical affinities, while deciduous ones are presumed to be more 'advanced', having resulted from adaptive radiation into colder, extra-tropical latitudes in more recent times (see Chapter 3). Notably, the evergreen species of both groups are restricted to warmer life zones of a given region and, within landscapes, to the warmer drier habitats.

These observations have prompted several researchers to ask whether a clear difference between related deciduous and evergreen species could be determined in terms of ecophysiological water use efficiency or, conversely, water loss prevention under conditions of water stress. In a detailed comparison of several evergreen and deciduous oaks of the same age growing in a controlled uniform environment, Achérar and Rambal (1992) found no clear difference in water use strategy, except among very young seedlings (see also Chapter 7, p. 171).

Contrary to the distribution pattern described above for the oaks, a very different pattern occurs in the Mediterranean maples: the deciduous Montpellier maple (*Acer monspessulanum*) is widespread and occurs in many life zones, while the two evergreen ones are more narrowly distributed (*A. sempervirens* in Crete and *A. obtusifolium* subsp. *syriaca* in the eastern Mediterranean), and both are restricted to cool, moist mountain zones. A third group includes a number of deciduous species, for example *A. campestre*, *A. hyrcanum*, *A. platanus*, and *A. tataricum*, all in the north-east quadrant. This group is primarily restricted to cool, mixed deciduous forests of premontane formations (meso-, supra-, and oro-Mediterranean life zones).

The 16 species and subspecies of Mediterranean pines (Table 6.4) are also quite varied in their life-history traits, climatic and soil requirements, and biochemical composition (Barbéro *et al.* 1998). Considerable variation occurs in their reproductive biology. Generally, when moving from subtropical zones towards the temperate zones, there is a change in both pines and oaks from biennial to annual fruit maturation, which is presumably a response to increasingly cold winters. By reducing fruit maturation time, these trees no longer need to maintain viable fruits on the tree in a period when they would be subject to freezing, but rather are able to 'release' them for germination in the fall of the same year of their development.

Some Mediterranean pines are widespread and may often be invasive (e.g. *P. brutia* and *P. halepensis*), while others are highly sensitive to fire, long-lived, and also limited to particular soil types. In terms of area of distribution, the same

Table 6.4 Distribution of the Mediterranean pines with respect to substrates, climatic zones, and life zones. Number of + is proportional to abundance of the species in each category.

Species	Substrates[1]						Climatic zones[2]				Life zones[3]				
	Ma	Ca	D	Si	Ub	Ss	H	SH	SA	A	TM	MM	SM	MtM	OM
Pinus halepensis	+++	++	++	+		+	++	+++	+++	++	++	+++	++		
P. brutia	+++	++	++	+	+++	+	+	+++	++	+	+	+++	++		
P. p. pinaster			+	+++		+++	++	+++				+++	+		
P. p. hamiltonnii			+	+++		+++	+++	+++			+	+++	+		
P. p. maghrebiana	+++	+	++	++	+	++	+	+++	+		+	+++	++		
P. pinea			+			+++	+	+			+	+++	+		
P. canariensis				+++			+	+++			+	+++	+		
P. nigra salzmannii		++	+++	++			+++	++				+	++	+++	
P. n. laricio				+++			+++	++					++	+++	
P. n. nigra		+++	+++	+++	++		+++	++					++	+++	
P. n. mauretanica			+++				+++						++	++	
P. n. dalmatica			+++				++	+++				+	+		
P. n. pallasiana	+++	++	++	++			+++	++					++	+++	
P. sylvestris	+++	++	++	++			+++	++					++	+++	
P. heldreichii	+++		+++		+++		+++						++	+++	++
P. uncinata		++	++	++			+++							+++	+

[1] Ma: marls; Ca: limestone; D: dolomite; Si: acidic sands; Ub: ultrabasic volcanic rocks; Ss: sandstone
[2] H: humid; SH: sub-humid; SA: semi-arid; A: arid
[3] TM: thermo-Mediterranean; MM: meso-Mediterranean; SM: supra-Mediterranean; MtM: montane-Mediterranean; OM: oro-Mediterranean
Source: Barbéro *et al.* 1998

pattern prevails as among oaks and pistachios, namely that the widespread generalists are in lower life zones (most affected by humans) while restricted taxa and endemics are most common either in mountainous areas (e.g. *P. heldreichii, P. uncinata*) or else on islands (e.g. *P. canariensis*). A good example of this group is provided by *P. heldreichii* in Greece and the closely related *P. leucodermis* in the Calabrian mountains in southernmost Italy. These are ancient (Miocene) species restricted to dolomite substrates, which do not survive fire and do not invade or regenerate readily, but can live for very long periods, up to 3000 years (M. Barbéro, personal communication). In the highly variable black pine (*P. nigra*), however, and in maritime pine (*P. pinaster*) biochemical studies indicate that more successful colonizers among subspecies (e.g. *P. nigra cluziana*) are characterized by greater intrapopulation variability of certain chemical markers (e.g. flavanoids and protoanthocyanids) than edaphically restricted or otherwise narrowly distributed subspecies, such as *Pinus n. salzmannii* in southern France or *P. n. mauretanicus* in North Africa, both of which are limited to dolomite substrates (Barbéro *et al.* 1998). In pines, as for the evergreen oaks, the long history of human manipulation of genotypes must be taken into account when attempting to interpret the intricate tapestry of their taxonomy and distribution within the Mediterranean space.

It was suggested in Chapter 5 that former Mediterranean forests showed a much denser, more intricate mixture of oaks, pines, junipers, and deciduous trees than today. Only fossil pollen or charcoal data allow attempts at forest 'reconstruction', or 'landscape archaeology', but one essential theme stands out clearly: it is usually possible to associate at least one pine to each of the broad-leaved and sclerophyllous oak formations, especially at lower altitudes. For example, *Pinus halepensis* occurs with *Quercus ilex*, while *P. brutia* occurs with *Q. calliprinos*. *P. pinaster* subsp. *hamiltonii* occurs with *Q. suber* and *P. sylvestris* with *Q. humilis* (Barbéro *et al.* 1998). Several junipers that were formerly part of the conifer canopy layer have been removed or drastically reduced by humans. Pines recover better, and spread faster than most junipers, and they are also far more frequently planted than any conifer or other kind of Mediterranean tree.

Blue tits in habitat mosaics

The blue tit (*Parus caeruleus*) is a small passerine bird common in broad-leaved and mixed forests at low and mid-altitudes. It prefers oak forests, both summergreen and evergreen wherever they occur, but also occupies a wide range of deciduous and mixed deciduous–coniferous forest types. Its distributional range covers the whole Mediterranean region, including the Canary Islands where several subspecies have been described. Important life-history traits of the blue tit, such as the time of breeding and clutch size, vary enormously depending on whether habitat patches are dominated by deciduous downy oaks (*Quercus humilis*) or evergreen holm-oaks (*Q. ilex*). In the vicinity of Montpellier, most forested landscapes are mosaics of habitats with zones where evergreen holm-oaks are dominant and patches where deciduous oaks are dominant. The entire

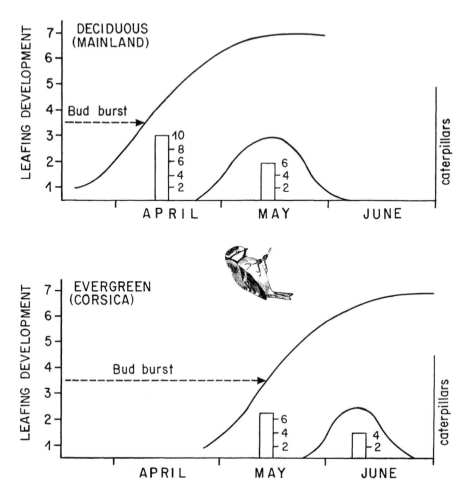

Fig. 6.6 The breeding schedule of blue tits (*Parus caeruleus*) in Mediterranean habitats dominated by the deciduous downy oak on the mainland near Montpellier (above), and in habitats dominated by the evergreen holm-oak in Corsica (below). On each graph the curve on the left indicates the process of leafing development from stage 1 (dormant bud) to stage 7 (young leaf fully developed); the two columns indicate clutch size and fledging success, respectively, and the bell-shaped curve illustrates the timing and abundance of caterpillars, the preferred diet of the birds. Note the delay of four weeks in Corsica in the timing of all events; that is, the leafing process, laying date, and the period of caterpillar availability. Differences in the timing of reproduction of tits is genetically determined.

area is relatively close to the large deciduous forest blocks of central France. By contrast, in many parts of Corsica most of the habitats suitable for tits are evergreen oak forests and matorrals, interspersed with very small patches of deciduous trees. The spring development of new leaves occurs four weeks later in holm-oak than in the deciduous downy oak, which results in a corresponding four-week interval in the development of the leaf-eating caterpillars the tits feed upon. Birds which have the best breeding success are those that synchronize the

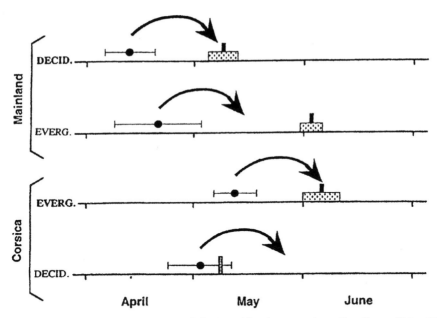

Fig. 6.7 Schematic representation of the matching between breeding time of blue tits (*Parus caeruleus*) (black dots = laying date $\pm$ SD) and the time of maximum abundance of caterpillars in the commoner habitats (deciduous on the mainland, evergreen on Corsica), and the mismatching in the less common habitat. The black arrows indicate the mean date of maximal food demand, when the young birds in the nest are approximately 10 days old. Closeness between arrow and black bar indicates good match (Reproduced with permission from Dias and Blondel 1996).

period when they raise their nestlings with the brief two- to three-week period of caterpillar availability in spring. Indeed, both in deciduous oak habitats on the mainland and evergreen oak ones on Corsica, blue tits start to breed at a time that allows fairly good synchronicity of nestling period and the period of caterpillar availability. This means that they start to breed four weeks later in the evergreen habitats of Corsica than in the deciduous habitats of the mainland (Fig. 6.6). These two populations, which are isolated by the sea and thus deprived of all contact between them, are assumed to be adapted to their local habitats (Blondel *et al.* 1993). Indeed, experiments in aviaries have demonstrated that the difference in laying date between these two populations of tits has a genetic basis. Hence they are locally genetically specialized to their respective habitats.

From these observations, and assuming that tits evolved similar adaptations to local food resources *within* each of the two habitat mosaics (mainland and Corsica), as they did *between* them, one may predict that there would be a similar difference of four weeks in the start of breeding depending on whether they breed in deciduous or in evergreen oak habitat patches. Observations do not support these predictions. Tits in the patches that are the less common in each landscape (evergreen on the mainland, deciduous on the island) start to breed approximately at the same date as those in the most common patches. They breed too early in evergreen mainland patches and too late in deciduous Corsican patches in relation to the peak of food abundance (Fig. 6.7). This mismatching between the time of breeding of the birds and food availability results in a lower breeding success than in the commoner habitats. The most likely explanation for the weak inter-habitat differentiation of blue tit life-history traits in each region is that gene flow across the different habitat patches within each landscape prevents birds from evolving adaptation to the mosaic character of their local environment (Lambrechts *et al.* 1997). This hypothesis is further supported by genetic analyses (Dias *et al.* 1996).

However, in the topographically highly dissected Mediterranean landscapes, there may be local adaptation at still finer scales. For instance, in a small isolated catchment in Corsica cut off by mountain ranges, the dominant tree species is the deciduous downy oak. Comparing breeding patterns of birds in this habitat with those of birds in a large forest block of the evergreen holm-oak only 25 km away, Lambrechts and his colleagues found striking differences in the main breeding traits of the two populations. Birds from the small catchment with deciduous oaks started egg laying on average one month earlier, laid on average about 1.5 more eggs, and fledged about 35% more young than those from the large evergreen holm-oak forest (Lambrechts *et al.* 1997). In both study sites birds nicely synchronize their breeding time to the local patterns of food abundance. This case study shows that the degree of isolation of local populations and landscape structure have profound effects on local differentiation in life-history traits that are crucial for the survival of populations.

This study of Mediterranean blue tits shows that population processes operate at the scale of a landscape, not just at that of habitats. In habitats that are clearly isolated, selection regimes may result in a fine local specialization of life-history traits. But wherever the dispersal range of a species by far exceeds the size of local habitat patches, differences in selection regimes may result in local maladaptation of populations due to gene flow. Thus, most life-history traits in Mediterranean species confronted with high environmental heterogeneity must result from a fine balance between local specialization and phenotypic plasticity. These scaling effects must have consequences at the community level as well: if the dispersal range of a species is larger than the size of habitat patches, the persistence of local populations in patchy environments and, thus, related components of community diversity, may depend on habitat-specific demographic rates in other habitats.

Species turnover in time: migrating birds

In his seminal book, *The Palaearctic-African bird migration system*, published in 1972, Reginald Moreau described and analysed the magnitude of the task that faces the estimated five billion birds leaving the Palaearctic region each autumn to spend the winter in more favourable areas further south. If we are to dare a comparison, the gigantic seasonal shift of birds back and forth between Eurasia and Africa each year is similar to the series of contraction–expansion cycles of

Table 6.5 Examples of bird species of the four main categories that occur in Mediterranean habitats according to seasons.

Residents	Summer visitors	Long-distance migrants	Winter visitors
Green woodpecker	Scops owl	Tree pipit	Wren
Crested lark	Bee-eater	Garden warbler	Meadow pipit
Fan-tailed warbler	Roller	Willow warbler	Water pipit
Cetti's warbler	Tawny pipit	Wood warbler	Song thrush
Moustached warbler	Melodious warbler	Pied flycatcher	Dunnock
Serin	Subalpine warbler	Wheatear	Chaffinch
Goldfinch	Nightingale	Rock thrush	Robin
Mallard	Turtle dove	Redstart	Chiffchaff
Kestrel	Lesser kestrel	Icterine warbler	Teal

biota across Europe as a result of glacial episodes during the Pleistocene. The only difference is in time scales, one year in the first case, tens of thousands years in the second. Given its strategic location at the border between the two major continental land masses of the Old World, the Mediterranean region, with a moist and mild climate from autumn to late spring, plays a key role in this system. Thus, onto the kaleidoscopic picture of plant and animal communities in space is superimposed a kaleidoscope of bird communities that seasonally change in time. As soon as the breeding season is completed and sometimes even before, many Mediterranean habitats are invaded by migrating birds that will either stay in the region for some time before going further south in tropical Africa, or will stay for the whole winter. Surviving individuals of long-distance migratory species will pass again through the Mediterranean in spring on the way to their northern breeding grounds. Thus the Mediterranean is crossed twice a year by myriads of migrants; the autumn migrants far outnumber the spring migrants because many die during their journey south or in their winter quarters. Moreau calculated that in order for the surviving individuals to make it back to their breeding grounds the next spring, some 1200 birds must cross the Sea each day per linear kilometre for a full three months!

Seasonal aspects

Thus, there is in the Mediterranean a permanent turnover of birds in time, with four main categories involved (Blondel 1969): residents that remain in the region all year round; long-distance trans-Saharan migrants that live in the region only for brief periods in autumn and spring; summer visitors that invade the region in spring from their African winter quarters for breeding; and winter visitors that do not breed in the region but invade to spend the whole season there (Table 6.5). Blondel (1969) has shown, from bird censuses carried out all year round for four years in a sample of habitats in the Camargue, that from the total census of several

Box 6.3 Mediterranean wetlands as winter quarters for wildfowl

Because they rarely freeze Mediterranean wetlands are of prime importance for migratory birds fleeing cold weather further north. They are
wintering grounds for a large variety of ducks, as well as coots, geese,
cranes, and sometimes swans that leave their breeding grounds in central,
eastern, and northern Eurasia in late summer. Some wintering species
such as the wigeon (*Anas penelope*) breed as far east as eastern Siberia. In
some large wetlands such as the Camargue, the Guadalquivir delta or Lake
Ichkeul (Tunisia), more than 150 000 individuals may overwinter. But
there are regional differences in the composition of local assemblages. For
example, Lake Ichkeul is very important for greylag geese (*Anser anser*)
(8% of the overwintering waterfowl), wigeon (*Anas penelope*) (60%), and
pochard (*Aythia ferina*) (30%), whilst the Camargue is a winter ground of
primary importance for shoveler (*Anas clypeata*), teal (*Anas crecca*), red-
crested pochard (*Netta rufina*), and gadwall (*Anas strepera*).

Wintering ducks belong to two main categories. The first consists of
granivorous species (e.g. mallard and teal) that feed on the protein-rich
seeds of a variety of submerged macrophytes such as pondweeds (Po-
tamogetonaceae) and the cosmopolitan algae Characeae. The second
category includes herbivorous species that exploit vegetative tissues
such as leaves and stems (e.g. gadwall, wigeon). Ducks spend the
daytime in large concentrations in large bodies of water without vegetation. These are used as day roosts where they rest, groom and, at the end
of the wintering season, start to display before leaving the site. At dusk
they leave the pond and disperse to a variety of habitats where they
forage in farmland and shallow waters of seasonal marshes for most of
the night. Since the main diet of the second group is less rich than that of
the first, herbivorous species are obliged to extend their feeding time
into the daylight hours. Large concentrations of wintering ducks inevitably draw in their wake large raptors such as spotted eagles (*Aquila
clanga*) and white-tailed eagles (*Haliaeetus albicilla*), which survey the
ducks' wintering places from nearby trees and take every opportunity to
feed upon them for the greater pleasure of bird-watchers!

thousand individual birds, 13% were long-distance migrants, 24% were breeding
birds, either resident or summer visitors, and 63% were winter visitors. This latter
category also includes the swarms of ducks and waders that overwinter here (Box
6.3). A schematic picture of this seasonal ballet is given in Fig. 6.8. Of course
there is much overlap between these categories. Many resident species start to
breed in the heart of winter, in December–January, for example the griffon vulture
or the eagle owl, while the first autumn migrants are already appearing in
Mediterranean wetlands in late June. Among the earliest autumn migrants are

Fig. 6.8 The Mediterranean is a key place for migratory and wintering birds. Most migrating species of the Palaearctic region either migrate through to reach their African winter quarters or overwinter in the Basin. Few of them overwinter in tropical regions of southern Asia. The figure also shows (bottom right) the turnover through the seasons of breeding, migrating, and wintering birds. Note the overwhelming importance of autumn migrants and wintering birds, and the length of the breeding season, which largely overlaps with the spring migration.

wood sandpipers (*Tringa glareola*), which leave their breeding grounds soon after their young are emancipated and arrive in Mediterranean wetlands at the very beginning of summer. At a time when many species in the Mediterranean, bee-eaters, rollers, and herons, are still rearing their young. In fact, migrating or wandering birds may be observed all year round in any part of the Mediterranean. There are no 'holidays' for bird-watchers in the Mediterranean! In any place, at any time of the year, and in any habitat, there is always a rare bird to see.

Nearly 230–50 species of birds leave their Eurasian breeding grounds in autumn and winter in the Mediterranean or further south. Migrants passing through the western half of the Mediterranean are mainly drawn from the western Palaearctic, Scandinavia, central Europe, and western Russia, while those passing through the eastern half come from further east. Some amazing travellers such as the wheatear (*Oenanthe oenanthe*) and the willow warbler (*Phylloscopus trochilus*) come from as far east as eastern Siberia while others, for example the barn swallow (*Hirundo rustica*), travel as far south as South Africa. About one-third of the migrants do not go further south than the Mediterranean and stay in the Basin for the whole winter. The moist and frost-free winter climate is particularly favourable for birds that forage on the ground and in shallow water. Large numbers of wintering passerines such as thrushes, dunnocks, wrens, and kinglets benefit from the diversity of insects, snails, earthworms, and other invertebrates they find on the ground, as well as from the plentiful fruits produced by a wide range of bushes and trees (see Chapter 7). Contrary to the picture in central and northern Europe, where birds are much more abundant in summertime than in wintertime, the opposite is true in the Mediterranean (see Fig. 6.8).

Long distance strategies

The second category, about two-thirds of the Palaearctic migrants, is that of the many birds which depend for their survival on small flying insects. From late October until the end of March, they are unable to sustain themselves in Europe so that they must escape to the tropics. Over 130 species of land and freshwater birds winter wholly or mainly south of the Sahara. In autumn they have to negotiate the crossing of the Mediterranean Sea and the Sahara Desert. Only soaring raptors and storks avoid crossing the Sea, using flyways at the two extremities of the Basin, the Bosphorus–Dardanelles to the east and Gibraltar to the west. Thousands of birds may be observed in a single day, an unforgettable sight. But nearly all small species cross the sea on a broad front. As Admiral Lynes wrote in 1910, 'there is not a single hectare of land or sea that is not over-flied by migrant birds in the Mediterranean region'. This was confirmed by radar observations in the 1960s. In springtime, when weather is favourable, it is an amazing experience to sit on a beach and search through a telescope for tiny black spots a few metres above the sea, approaching the coastline. When they get close enough it is always a surprise to see that these spots are pied flycatchers, nightingales, rollers, rock thrushes, or any other of the dozens of species that had left the coast of Algeria the previous evening. All of them seem exhausted when they land. Quite often, alas, they end their journey in the beak of a wandering yellow-legged gull that has purposely lingered along the coast in wait of its prey. Sometimes, when a small tailwind and a clear sky make flying conditions especially good, an avalanche of birds descends upon the first bushes at the coastline. Then, driven by the necessity to reach their breeding grounds without delay, they continue their journey, 'gliding' from bush to bush in a north-easterly direction while feeding as much as possible. It is only at dusk that they take off again for a non-stop nocturnal flight.

Refuelling

In autumn especially, Mediterranean habitats are of paramount importance for all these migrants because they must provide stopover sites and the large amount of food necessary for accumulating the stock of energy that they need to cross the Sea, and then the Sahara Desert. The genetics, ecology, morphology, and physiological adaptations of small long-distance migrants are amazing. They have been beautifully studied in recent years by a handful of enthusiastic scientists, and the findings summarized in a special issue of the ornithological journal *Ibis* (Crick and Jones 1992, see also Berthold 1993). Moreau (1972) thought that migrants crossed the Sahara in a single non-stop flight of at least 40–60 hours because conditions there were too difficult to allow small migrants to find resting and feeding places. However, Franz Bairlein (1992) showed that at least some trans-Saharan migrants regularly stopover in suitable desert habitats where they rest and 'refuel'. This suggests that the fat load accumulated in the Mediterranean prior to migration is not sufficient for a non-stop crossing of the large Sahara Desert.

During their stopovers in the Mediterranean, of some days to several weeks, long-distance migrants prefer open or semi-open habitats such as matorrals and various types of ecotones where they find a large food supply. They accumulate a considerable amount of fat, which will be used as fuel for their journey. Birds stuff themselves with food, mainly fruits which are rich in sugar and carbohydrates that can be readily and quickly transformed into glycogen and fat. Nearly all species, including those that are mainly insectivorous during the breeding season, such as warblers and flycatchers, eat fruits for fattening themselves. It has been calculated that premigratory fat deposition can be very rapid. Within one week, a 20 g bird may increase its body mass by 40–50%. When flying, small birds expend about four times more energy than when they are resting. Body weight decrease as reserves are consumed during migratory flight is of the order of 0.8% per hour in a bird weighing 10–20 g. Thus, since birds can accumulate up to 50% of their fat-free weight in fat reserves, they can make a non-stop flight of 30–50 hours, allowing them to cross the Mediterranean Sea and Sahara Desert without refuelling. They leave the northern shores of the Sea in autumn when they are fat enough to risk the journey. By contrast, the birds are particularly lean when they return in spring, having expended great amounts of energy in recrossing these same large, inhospitable barriers. For example, in southern France, redstarts weighed 13.8 g in spring, on average, but as much as 17.5 g in autumn (Blondel 1969).

Summary

In this chapter we have surveyed some components of the evolutionary and ecological variation of living systems as a response to environmental heterogeneity in the Mediterranean Basin. From vicariance effects at large scales of space and time to the evolution of life-history traits at the scale of local habitats, patterns and processes result in a huge genetic and ecological variation of species and populations. On islands, communities seem to be highly integrated constructions composed of relatively small numbers of species that have evolved adaptations to cope with the particular conditions of life they encounter there. In complex habitat mosaics, species and populations must cope with a high diversity of habitats and associated selection regimes. They may evolve local specialization, whereby local specific combinations of genes adapt organisms to local conditions. They may also evolve 'phenotypic plasticity' whereby the same genotypes may express in different phenotypes according to environmental conditions. One important conclusion is that processes operate at the scale of landscapes, not just at that of habitats. Such scaling and harlequin effects have consequences at the community level as well: if the dispersal range of a species is larger than the size of habitat patches, the persistence of local populations in patchy environments and, thus, related components of community diversity, may depend on habitat-specific demographic rates in other habitats.

7 Life history traits and ecosystem functioning

It would need an entire book to describe in detail what is known of the evolutionary consequences of 'mediterraneity' on living systems. We will limit ourselves to looking at some recent developments and landmarks in the immense field of evolutionary ecology in Mediterranean biota. These examples are just the tip of a largely unknown iceberg being examined by that fascinating field of study that explores the extent to which the biological characters (or traits) and suites of traits we observe are evolutionary responses to mediterranean-specific selection pressures, or else result from other factors such as phylogenetic constraints or historical effects not related to the mediterranean bioclimate specifically.

Evergreenness* and sclerophylly

Evergreenness is a recurrent and striking feature in plant assemblages of the Mediterranean area. This trait of plant foliage is in fact one of the most ancient features of vascular plants since it characterized the primitive gymnosperms and primitive angiosperms in the Cretaceous. It is from evergreen ancestors that most modern floras have evolved (Raven 1973; de Lillis 1991), and evergreenness may have been an adaptation to the uniform and warm climate of the Cretaceous and a large part of the Tertiary. In contrast, summergreenness, or the deciduous leaf habit, apparently evolved later as a response to climate changes, notably the appearance of climatic seasonality in the lower Oligocene, about 35 My BP. Seasonality may have been logically conducive to the evolution of changes in the phenological cycles of leaves, leaf longevity, and changes in the energetic balance and maintenance costs of photosynthetic tissues, which all differ sharply between the various kinds of deciduous and evergreen leaves. Evergreen broad-leaved plants survive and even dominate today in the largely discontinuous areas mostly located between the subtropical deserts and the temperate zone that are distinctly characterized by summer aridity, that is, the five mediterranean-type regions of the world. They also dominate in the majority of vegetation types in the tropics. What, if anything, is special about evergreen leaves of mediterranean trees and shrubs? For one thing, they are mostly sclerophyllous.

Is sclerophylly a Mediterranean adaptation?

The term sclerophylly designates rigid, more or less leathery leaves that present a low surface-to-volume ratio, a large development of nervature per unit of leaf area, a thick cuticle, and a tendency to brittleness when hit or folded. Some are relatively large, as in laurel, Lauristinus (*Viburnum tinus*), or ivy (*Hedera helix*), or else pinnately compound (carob and lentisk). But most sclerophyllous leaves found in the Mediterranean area are small and undivided, as in the holm-oak, boxwood, myrtle (*Myrtus communis*), and smilax (*Smilax aspera*) (Fig. 7.1). All are relatively long-lived, and for this reason the ancient Greeks and Romans considered such evergreen plants as myrtle and laurel to be symbols of love and eternity.

Five lines of evidence attest to the adaptive value of sclerophylly in mediterranean ecosystems.

1. A classical argument for explaining evergreenness and sclerophylly is to describe them as straightforward adaptations to environmental conditions such as the prolonged drought in the hottest part of the year which is typical of all mediterranean-climate areas. It is commonly argued that sclerophyllous leaves allow plants to control transpiration by closing stomata under water stress, which in turn enables them to survive periods of summer drought. In all five mediterranean-type ecosystems of the globe, the large number of plant species that exhibit sclerophylly are thus considered as having independently evolved a 'water-saving' strategy (Mooney and Dunn 1970). This feature has often been seen as signalling an evolutionary convergence among the five mediterranean-type regions (see Chapter 3). The rationale is that if the survival of perennial species such as shrubs and trees depends on their ability to maintain a positive carbon balance over the year, the evergreen habit gives the advantage, as compared to the deciduous habit, of a longer photosynthetically active period that, in addition, coincides with the summer drought.

2. Sclerophylly, it has also been suggested, is associated with the chronic state of nutrient deficiencies found in the soils of most mediterranean-type ecosystems (Specht and Rundel 1990). Indeed, evergreen species tend to predominate in nutrient-poor habitats, such as in dry and/or sterile environments (Monk 1966). This argument is supported by the large variation in the percentage of sclerophyllous species within floras, depending on the soils, nutrient loads, and microclimates. Within any given Mediterranean region, sites with very poor soils, such as gypsum or bauxite badlands, or dolomitic sands, show a preponderance of evergreen sclerophyll shrubs in families and genera as varied as *Achillea, Daphne, Erica, Genista, Globularia, Helianthemum, Helichrysum, Juniperus, Lavendula, Rosmarinus, Thymelaea,* and *Thymus*. The rationale is that a balance between photosynthetic gain and maintenance costs of non-photosynthesizing tissues can only be achieved, under limiting environmental conditions, by reducing the allocation of photosynthates to leaf growth, and by prolonging leaf longevity. Thus, it is

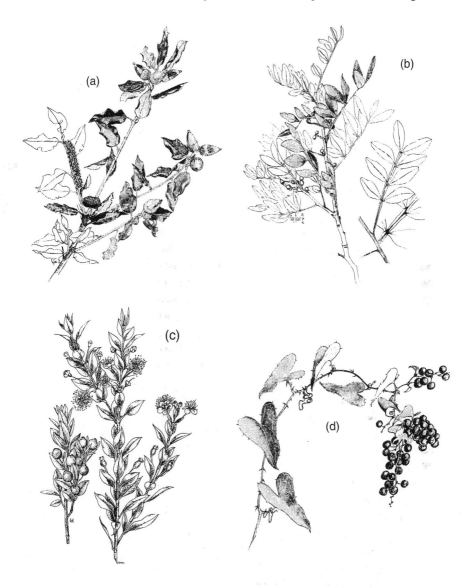

Fig. 7.1 Four evergreen sclerophyllous plants of differing life forms and families common in the Mediterranean: (a) holm-oak (*Quercus ilex*); (b) false olive (*Phillyrea media*); (c) myrtle (*Myrtus communis*) (reproduced with permission from Zohary and Feinbrun 1966–86); (d) smilax (*Smilax aspera*).

not surprising that evergreen sclerophyllous leaves also figure prominently in the floras of upper montane forests in temperate zones and in nutrient-poor, tropical lowland forests (Beadle 1966). Moreover, the lifespan of most sclerophyllous leaves exceeds one or even two years, averaging three years in holm-oak and the olive, and five to six years in kermes oak.

3. Another apparent adaptation of hard sclerophyllous leaves is that they offer little reward to herbivorous animals, and thus appear to suffer less phytophagy* than softer, broader, and relatively more sweet-leaved trees and shrubs. This 'anti-herbivory' effect is achieved partly by virtue of unpalatable and bad-tasting compounds that deter many herbivores.

4. One additional explanation for sclerophylly being so widespread in contemporary mediterranean ecosystems may be the overall resilience of shrubs and trees sharing this trait in the face of perturbations. Resilience in mediterranean plants allows them to respond successfully to events that affect plant growth and survival, such as summer drought and repeated fires or cutting. For example, holm- and kermes oak, boxwood, and cade juniper (*Juniperus oxycedrus*) all resprout vigorously from the stump when the tree is cut or burned down. Deciduous oaks, in contrast, generally die following such treatment. A high carbon:nitrogen ratio in the leaves also acts to inhibit or retard decomposition once leaves finally do fall off the plant. This in turn may have significance in setting up feedback cycles wherein low nitrogen soils and leaves of low nitrogen content are favoured at the ecosystem level. Certainly the carbon-rich and nitrogen-poor leaf form influences many biotic and abiotic processes.

5. Finally, additional support for an adaptationist view of sclerophylly comes from quantitative studies showing that plant associations in progressively higher life zones, that is, those with milder, wetter climates, show a declining percentage of sclerophyllous species in six different Mediterranean mountain ranges studied along altitudinal transects (Barbéro *et al.* 1991). The fact that so many groups of different phytogeographical origins and affinities, from boxwood to heathers, carob, oleander (*Nerium*), and olive, have sclerophyllous leaves gives additional credence to adaptationist arguments.

However, despite many efforts to understand its ecophysiological function, sclerophylly is still a rather empirical term and the function of sclerophylly presumably differs considerably between mediterranean-type plant species and tropical plants. The precise selective forces contributing to the evolution, geographical distribution, and functional role in ecosystem dynamics of sclerophylly and some of its correlates (e.g. leaf longevity and evergreenness) are still poorly understood (Aerts 1995).

Is sclerophylly an artefact of the past?

There are also arguments against an adaptive value of sclerophylly. Recent studies (e.g. De Lillis 1991; Damesin *et al.* 1998) focusing on the water and gas exchange responses to summer drought by sclerophyllous and deciduous species suggest that evergreen species are in fact not better adapted to water stress than deciduous species. An insight into the differences in ecophysiological performance of the two leaf morphotypes may be provided by comparing them where they co-occur in the same environment. Comparing the carbon balance and water and nutrient relationships of deciduous and evergreens, Schulze (1982) showed that the two groups operate with similar success in the same habitat with different adaptive responses to environmental constraints. Detailed physiological studies at the leaf scale revealed that, despite differences in biochemical composition, size, and mass per unit area, the leaves of the two species respond similarly to water limitation (Damesin *et al.* 1998). Similar results were obtained in a comparison of evergreen and deciduous oaks of the same age growing in a controlled environment (Achérar and Rambal 1992). It may be, however, that differences in function between the two groups become more apparent at the ecosystem level than at the leaf scale because of differences in leaf area indices between the two leaf types. A high total leaf area per unit of soil area could allow the evergreen species to maximize their winter and spring carbon gain at a time when the water supply is not limiting. In addition, differences between the two morphotypes in the timing of leaf fall and leaf litter decomposition may result in functional differences in the economy of nutrients, favouring evergreens at the ecosystem level.

Going even further, Schulze (1982) suggested that sclerophylly should be considered as an 'epiphenomenon' of the water-stress adaptation, that is, a by-product rather than a response to water stress. The same could be said of the argument that sclerophylly is an adaptation to nutrient-poor conditions. Some support for this view may be seen in the fact that many, possibly even most, bath'a and phrygana shrubs produce seasonally dimorphic leaves (Margaris 1981). In other words, they bear relatively thin, large leaves with high photosynthetic rates in winter and spring, as opposed to reduced, thickened ones of the typical sclerophyllous type in summer. This seasonal dimorphism is attained by alternating shedding and new growth of different types of branches and leaves in response to changing levels of water availability in the different seasons. Furthermore, experimental additions of phosphorus and nitrogen have been shown to alter the structure of leaves on adult sclerophyll shrubs (Beadle 1966).

An important additional perspective on sclerophylly can be gained by considering phylogenetic constraints. The Mediterranean vegetation originated from tropical and temperate sclerophyllous ancestors of the Upper Cretaceous that were uniformly distributed over the Laurasian continent (Raven 1973; Axelrod 1975). The Eurasian flora originally evolved from that flora, which was rich in Ginkoaceae and other hard-leaved families during the Mesozoic. Later, during the Tertiary, so-called lauriphyllous forests spread along the continental borders of the Tethys Sea and gave rise to the extant evergreen sclerophyllous flora (Axelrod 1975). Thus, phylogenetic relationships may explain better than any ecophysiological adaptations the widespread occurrence of this life form in the Mediterranean flora today.

Other leaf types

In addition to sclerophyllous leaves, many other leaf types occur in the Mediterranean flora that also provide protection against desiccation while still allowing photosynthesis. These include woolly leaves (e.g. *Phlomis*), succulent leaves (*Euphorbia*, *Sedum*) and tiny, short-lived leaves such as are found on the numerous brooms and broom-relatives (e.g. *Genista*, *Lygos*, *Retama*, *Spartium*, and *Teline*). Several groups bear evergreen stems that are photosynthetically active all year round. These species, called retamoid, have small deciduous leaves that fall during drought periods. In the drier areas particularly, many species are aphyllous*, or nearly so, and confine their photosynthetic activity to their stiff, evergreen stems (one example is *Aphyllanthes monspeliensis*). Finally, a few species, such as the widespread butcher's broom (*Ruscus aculeatus*), a prickly perennial herb of forest and matorral understoreys, produce long-lived sclerophyllous phyllodes* as well as evergreen photosynthesizing stems. Thus, Mediterranean plants show a remarkably wide range of leaf types.

Precocious-flowering geophytes

In addition to sclerophylly and evergreenness, one distinctive trait of the Mediterranean flora is the bulbous life form. These so-called geophytes have a fleshy, subterranean storage organ which is usually the only portion of the plant that survives the extended period of summer dormancy. A great number of unrelated species in many different families have evolved this growth form throughout the Basin. For example, there are 217 species of geophytes in the Israeli flora, which represent about 10% of the total (Fragman and Shmida 1998). Well known geophytes are found in monocot groups such as the lily, crocus, tulip, iris, amaryllis, and orchid, and a few in dicot groups, for example cyclamen. With about 250 species, the Euro-Mediterranean orchid flora is particularly rich although it includes only terrestrial species. The largest genera in the regional flora are *Ophrys* (20–50 species), *Orchis* (25–40 species), *Dactylorhiza* (5–28 species), and *Cephalanthera* (2–10 species) (Dafni and Bernhardt 1990). While some orchid species occur as far north as Scandinavia (e.g. 20 species in Norway), at least two subfamilies are mainly east Asian (Raven and Axelrod 1974) and differentiated extensively in the Mediterranean area upon arrival there. Species richness may be very high in some parts of the Basin, for instance, there are 100 species in Greece, but declines to the north and towards the arid zones, with 86 species occurring in Turkey, 40 in Syria, 29 in Israel, and only one in Egypt (Dafni and Bernhardt 1990). Orchid species richness also declines with increasing altitude in mountainous areas.

In addition to avoiding drought by shedding their above-ground parts when water is limiting, geophytes gain a head-start over the annual plants with which they compete for resources. In early springtime, the subterranean bulbs, corms, or rhizomes of geophytes allow them to begin growth as soon as temperatures rise above a certain threshold. Annuals, by contrast, spend a number of weeks germinating and gaining initial establishment for their roots, during which time they are especially vulnerable to grazing and trampling by animals.

One typical feature of Mediterranean geophytes is the wide range of their

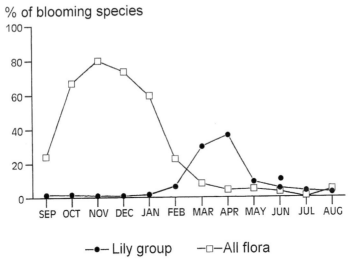

Fig. 7.2 The blooming pattern of the lily group (25 species in 10 families) as compared to that of the whole flora of Israel (2241 species in 130 families). (After Shmida and Dafni 1990.)

flowering time compared to that of the overall Mediterranean flora. Whereas most Mediterranean plants flower in spring, many of the geophytes are hysteranthous*, that is, they flower in a leafless state during autumn. In Israel, 10% of the native flora flowers in autumn, and this group consists mostly of geophytes in the 'lily group' (Fig. 7.2).

Two different types of hysteranthous geophytes have been identified in the Israeli flora, an 'Urginea-type' and a 'Crocus-type'. In the former, flowering is progressively delayed in the course of evolution until an autumn-flowering strategy is established. In the latter, the opposite trend occurs (Fig. 7.3). The hysteranthous habit appears to confer several adaptive advantages, in terms of optimizing both water and nutrient uptake and in terms of attracting potential pollinators. Thus, flower size, plant height, and seed dispersal mechanisms have all been found to have significant correlations with the atypical flowering time of these 'precocious' geophytes.

Annuals in highly seasonal environments

Even more abundant than geophytes, in terms of species richness and biomass, are the annuals or ephemerals*, which figure so prominently in almost all the Basin's regional floras and plant communities, especially in the eastern Mediterranean. In the hills near Jerusalem, an average of 143 (and a maximum of 189) plant species, mostly small annuals, were found in 1000 m^2 quadrats in a particularly rainy springtime (Shmida 1981). This is a world record in alpha level

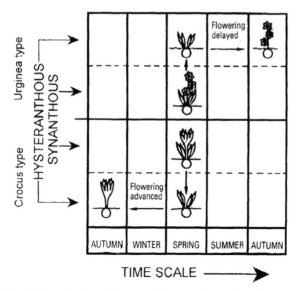

Fig. 7.3 Probable evolution of hysteranthous or 'precocious flowering' among Mediterranean geophytes. The initial condition where flowers and leaves appear together is termed 'synanthous* flowering'. (After Dafni *et al.* 1981.)

diversity for vascular plants *at this particular spatial scale*. Indeed, annual plants may well be considered one of the 'specialties' of the Mediterranean Basin, often constituting half of the dominant vegetation present whereas they rarely amount to more than one-tenth in other parts of the world (Raven 1973).

To cope with strongly seasonal environments, being an annual is a highly successful life-form strategy among plants (Grime 1979). Many annuals have a relatively small number of large seeds, which are highly sophisticated storage organs that allow survival during the unfavourable dry season. Seeds of such annuals, or therophytes* (see Table 5.2), are relatively short-lived, lasting only a single drought season. Many other annuals produce long-lived diaspores* that protect seeds over several years until an appropriate season for germination presents itself. Annuals of both types are found in large numbers in a variety of Mediterranean habitats. Following agricultural abandonment they are the first colonizers, only to be replaced by perennial grasses, shrubs, and trees as successions progress. In many heavily grazed areas ecosystems are artificially maintained in what amounts to an early successional stage, and this has possibly resulted in the recent evolution of many grazing-tolerant annuals from ancestral herbaceous perennials (Pignatti 1978). As mentioned in Chapter 3, there are over 1500 species of ruderals, segetals, and also desert ephemerals, that are subtly adapted to one or more situations in the Mediterranean and have latterly become highly successful 'weeds' in many other parts of the world.

Adaptive shifts in the timing of phenological phases ('phenophases') such as germination and the onset of flowering and seed production are common in Mediterranean annuals, as is morphological variability in height, shape, and overall biomass. To a large extent, they share these features with annuals of deserts but differences do occur between the two groups at both species and population levels (Aronson *et al.* 1992, 1993b). A common habitat feature that annuals often share in Mediterranean environments is the unpredictability of resources in space and in time.

Common adaptations of annuals to disturbance in the Mediterranean include delayed germination, via hard-seededness and other mechanisms (e.g. in *Medicago*, see section on biological invasions in Chapter 9), and both genotypic and phenotypic responses to interactions between temperature and moisture (Groves 1986). Another common adaptation is the possession of morphological features predisposing the fruits to dispersal by grazing animals and to burial in the soil. Genera such as *Bromus*, *Erodium*, *Emex*, *Hordeum*, *Medicago*, *Stipa*, and *Trifolium* all provide examples of such adaptations (Shmida and Ellner 1983). These traits are thought to be an adaptation to spatial heterogeneity and unpredictability, that is, a means of spreading the risk of offspring mortality in space, just as delayed germination is a means of spreading the risk in time. Given their panoply of adaptations for dispersal and for enduring disturbed sites, it is not surprising that Mediterranean annuals are among the most widely dispersed plants in the world today.

Coping with fire

The consequences of wildfires in the reproductive biology of plants is a major factor in Mediterranean ecosystems because of their importance in determining vegetation dynamics and dispersal patterns of many species. In particular, many fire-resistant or even 'fire-stimulated' plant species occur here, particularly in the low altitude life zones. In the Mediterranean area many plant species were preadapted—'abadapted' according to Harper (1982)—to human disturbances as a result of them having evolved life-history traits under the pressure of a high regime of natural disturbances, especially fire. These so-called 'pyrophytes' (Kuhnholtz-Lordat 1938) include two main types. The first are 'resprouters' that have the ability to do so quickly after fire, thanks to a strong and diversified root system that often constitutes most of the living biomass of the plant. Examples are the holm-, kermes-, and other evergreen oaks, as well as lentisk, many heath family species, and the various strawberry trees. In some resprouter species, fire defence mechanisms also include a thick and corky bark, as in cork oak (*Quercus suber*), low inflammability due to high mineral content of wood (e.g. *Tamarix* spp. and *Atriplex* spp.) (Le Houérou 1981), or underground regenerative organs that resprout after fire. These enlarged root crowns, or 'burls', are found in Barbary thuja (*Tetraclinis articulata*) and false olive (*Phillyrea media*), for example, and show a highly unusual and ornamental structure in cross-section.

The second type of active pyrophytes include the 'seeders', like rock-roses (*Cistus* spp. and related taxa), whose seeds require a thermal shock to germinate (Trabaud and Oustric 1989; Roy and Sonié 1992). For example, some *Cistus* spp.

that normally show no more than 30 seedlings m^{-2} can produce as many as 5000 seedling m^{-2} shortly after a fire (M. Etienne, personal communication). In contrast to resprouters, seeders must recolonize the recently burnt area, and seedling recruitment is restricted to a narrow 'window of time' during the first growing season after fire. Thus, there is a clear advantage for the so-called 'temporal dispersers' among 'fire recruiters' (Keeley 1991), which spread the risk of seedling mortality in time rather than space. For such 'temporal dispersers' the bulk of the seed pool is deposited near the parent plant. Seedling recruitment after fire is often superabundant, but since germination events are rare, an equilibrium of sorts is established. Temporal fire dispersers also tend to be highly drought-tolerant and physiologically able to withstand extreme water stress. As a consequence, they have rapid growth rates on open sites but are shade-intolerant. Granivorous animals (insects, vertebrates) may affect this narrow 'window' of reproductive opportunity, through direct consumption that reduces the seed bank or by redistribution of seeds through dispersal, a process called zoochory*.

With one or two exceptions, such as the Canary Island pine which, like the cork oak, has evolved a thick, partly fire-resistant bark that protects the tree from mortality during fires, most pines do not resprout after fire and are readily killed. The relatively fast-growing but short-lived pines rely on their small wind-dispersed seeds to colonize new sites. The most common strategy of pines is to occupy recently perturbed areas such as those that are available just after fire. With the many seeders that colonize recently burnt areas, they are often the pioneering species that will constitute the first stages of post-fire succession.

* * *

Perhaps the most fascinating adaptations to Mediterranean ecosystems are those that associate plants and animals in close interactions. The study of plant–animal interactions began in the Mediterranean area as early as 370 BC with the Greek philosopher Theophrastus, who was a fine scientist closely associated with Aristotle (Thanos 1992). He wrote two books, *De causis Plantarum* and *Historia Plantarum* (both recently translated into English), that have led him to be considered as the founder of the science of Botany. Theophrastus paid considerable attention to seeds that are consumed by the larvae of beetles (thought by him to be produced by the seed itself) and he was intrigued by the galls as a tannin source, especially the galls produced by certain oaks and pistachio trees (for example *Quercus infectoria* and *Pistacia terebinthus*) (Fig. 7.4).

Theophrastus also recognized the repellent function of plants producing toxic compounds, which may cause poisoning or death to animals that eat them. As examples he cited the black hellebore (*Helleborus cyclophyllus*), lethal to horses and cattle, the deadly root of nightshade (*Aconitum* spp.) which is avoided by sheep, and spindle bush (*Euonymus* spp.), whose fruits and leaves are lethal to sheep and goats if eaten in large quantities. Theophrastus was the first to observe the role of animals in seed dispersal, citing the cormlets of the corn-flag (*Gladiolus segetum*) as being dispersed by moles, the caching of acorns by jays, and the mechanisms of mistletoe fruit dispersal. He identified the two species of

Fig. 7.4 Stem of terebinth (*Pistacia terebinthus*) with galls of an aphid (*Pemphigella cornicularia*).

mistletoe occurring in Greece (*Loranthus europaeus* and *Viscum album*), and wondered how it was that these curious plants grow only on specific host trees. He came to the right conclusion: birds consume mistletoe berries, the seeds of which pass unharmed through their digestive tract. The seedlings then establish themselves from bird droppings that happen to fall in a good place on a suitable host tree.

Herbivory and plant defences

In its simplest form herbivory is the removal of plant parts, especially young leaves, by animal consumption. Although all plant species are to some degree at risk from attack by plant predators, herbivory usually concerns only a small proportion of the net primary production in most terrestrial ecosystems. However, outbreaks of chewing insects may result in the complete defoliation of trees, as is sometimes the case with the 'bag worm' larvae of the processionary moth (*Thaumetopoea pinivora*) in plantations and natural stands of various pine species, or that of the related moth *T. pytiocampa* in both oak and pine woodlands. Sometimes insects may cause the death of trees. One example is the bug *Matsucoccus feytaudi*, which caused the near total eradication of the maritime pine (*Pinus pinaster*) in huge plantations of this tree in southern Europe in the 1960s.

The formation of galls

A special case of 'herbivory' is that of the production of galls. The relationship between a gall-making insect and its host plant is considered parasitic. The insect induces growth deformities in the plant that are then used as food resources and sites for larval development. In *Phillyrea angustifolia* (Oleaceae), the gall-inducing insect is a cecidomyiid fly (*Schizoma phillyreae*). Adult flies emerge from the galls during the flowering period of the shrub and seek suitable breeding sites. Females oviposit one egg in the ovary of open flowers. The gall begins to develop 6–8 weeks after flower fertilization. Then the larvae grow inside the ovary and remain in a larval stage for at least three years (Traveset 1992). The cost of this to the plant is a drastic reduction in the number of viable seeds since up to 97% of the initiated fruits of a plant may become galls.

Herbivory may have important effects on a series of processes such as plant dispersal, growth patterns, reproductive success, and plant forms (Ginocchio and Montenegro 1992). By removing only parts of the plant, herbivores may leave other parts capable of regeneration through the iteration of new modules. Hence, galls may have important effects on the structure and shape of vegetation.

Herbivory and the structure of plant communities

Since grazing is usually selective, floristic composition and the relative abundance of species change as a result, which thus may be a powerful force in shaping the structure and species composition of plant communities. In particular, seedlings of many species may be vulnerable because they lack physical and chemical defences that develop only later. During the first growing season after fire, when seedlings of many species develop together, preferential consumption of one species over others by herbivores may shift the outcome of interspecific competition. Annual composites, crucifers, umbellifers, grasses, and legumes all tend to increase under moderate grazing, while perennial forbs, grasses, and tall annual grasses tend to decrease (Noy-Meir *et al.* 1989). If shifts occur among shrub species that may become dominant, they can change the species composition of communities. There have not been many case studies of these processes in the Mediterranean, but Mills (1983) has shown in the chaparral of California that herbivory on seedlings affects the status of the perennial shrub California-lilac (*Ceanothus greggii*) in the community. In mixed seedling populations of California-lilac and chamise (*Adenostoma fasciculatum*), small mammals eat a much larger proportion of the former, thus favouring the establishment of the latter. Exclosure experiments showed that when all species are protected from herbivory, the balance between the two species is reversed. In the Mediterranean area, the long-standing practice of overgrazing and over-browsing in most areas has resulted in plant communities that are heavily unbalanced in favour of plant species which are avoided by livestock, such as asphodels, brooms, and wormwood (*Artemisia* spp.).

Noy-Meir (1988) showed that normally dominant grasses can be massively replaced by ruderal forbs in a 'vole year' in Mediterranean grasslands. The unusual outbreak of large populations of the small rodent *Microtus socialis* wreaked such havoc in the grass swards that opportunistic ruderals were able to invade. Interestingly, Noy-Meir and colleagues came to the conclusion that *undergrazed* grasslands and shrublands in the Mediterranean are particularly vulnerable to serious perturbation. Similar events have been observed in North African steppes where 'eruptions' and abrupt crashes of populations of rodents such as *Psammomys obesum* and *P. meriones* can have tremendous impact on grassland and shrubland ecosystems (E. Le Floc'h, personal communication).

In an evolutionary perspective, such 'vole years', and similarly 'catastrophic' years due to other causes, can explain in part how annual forb species that now occur in grazed grasslands could have existed 'before the coming of domestic grazers, when pressure from large wild grazers was probably too light to open gaps in the dense tall grassland' (Noy-Meir 1988).

Defences against herbivory

Plants have evolved several types of defence, both physical and chemical, against herbivory. The most common 'anti-herbivore' defence mechanisms include the production of secondary metabolites, especially tannins and terpenes, and physical defence mechanisms such as long and sharp thorns that limit accessibility or availability to foliage. As part of their protection 'package' against herbivory, many Mediterranean bulbous plants are rich in bitter or toxic 'secondary' compounds that deter grazing animals and insects. These alkaloids and other compounds accumulate not only in leaves but also in underground storage organs that are normally attractive to rodents, boars, and other burrowing animals since they contain moisture as well as carbohydrates, even in summer. One example is the squill, or sea onion (*Urginea maritima*, Liliaceae), which is found in sandy coastal areas of North Africa and the eastern Mediterranean. Its bulbs contain a series of glycosides, some of which are used as cardiac stimulants or diuretics, while others are well known as rodent poisons (Fig. 7.5).

Fig. 7.5 Squill (*Urginea maritima*), showing large flowers and leaves, and enormous storage bulb packed with rodent-deterring alkaloids (reproduced with permission from Zohary and Feinbrun 1966–86). Specimens have been found with bulbs weighing more than 2 kg.

Comparable to the well-defended bulbs of squill, many evergreen Mediterranean shrubs contain volatile essential oils in their leaves that appear to play a role in deterring herbivores. Examples include myrtle, thyme, mint, sage, basil, lavender, coriander, dill, oregano, rue, laurel, rosemary, and fennel (Fig. 7.6). The great majority of Mediterranean taxa producing these aromatic compounds are found primarily in the mint, parsley, and, to a lesser extent, sunflower families.

The ecological role of volatile aromatic compounds in leaves (and, to a much lesser extent, seeds and roots) of Mediterranean plants is complex. These oils are highly flammable, which has led to the suggestion that they are involved in fire-return feedback dynamics. It has also been suggested that they may inhibit germination of the seedlings of competitor species (a process called 'allelopathy'), mimic insect pheromones as a means of attracting pollinators, and even reduce

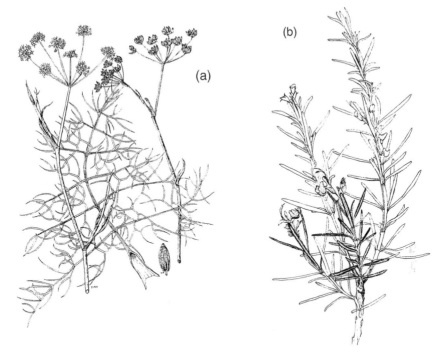

Fig. 7.6 Two examples of aromatic Mediterranean plants. (a) Fennel (*Foeniculum vulgare*), a plant with chemical-rich leaves and bulbs. Several varieties have been selected and bred for large, toxin-free bulbs (reproduced with permission from Zohary and Feinbrun 1966–86); (b) rosemary (*Rosmarinus officinalis*).

water stress by providing anti-transpirant action (Margaris and Vokou 1982).

For the best studied aromatic species groups, at least, there is both significant among-species variation and genetically controlled within-species variation in the oil content (Thompson *et al.* 1998). Building on a large body of previous work, Gouyon *et al.* (1986) concluded that the distribution of intraspecific variability (chemotypes) in thyme (*Thymus vulgaris*) appears to be heavily influenced by the environment (see Chapter 6, p. 152).

Pollination

Many plant species need a vector to transport pollen from one individual to another. The two most important vectors are wind and insects. In the Mediterranean other animals such as mammals, including bats, and birds, are not involved in the pollination process, in contrast to what happens in tropical regions. The mutualistic plant–animal systems involving insect pollinators imply that both partners gain some reward from their association. Insects transport the pollen for the host plant, usually organized as compact pollinia (pollen sacks), and in return are offered nectar, pollen, stigmatic secretions and, sometimes, other resources such as edible oils and floral fragrances (Simpson and Neff 1981). Indeed, it is selectively advantageous for the plant to tempt insects with a suite of attractants such as floral shape, pigmentation, scent, and edible rewards (nectar, pollen, oils) that are generally concealed in more or less long floral spurs.

Generalist insect-pollinated plants

A prominent feature in the Mediterranean flora is that most entomogamous* (i.e. insect-pollinated) plant species are generalist, that is, they can be pollinated by many insect species over a wide range of families. Exceptions to the widespread generalist habit include the fig tree (see Box 7.1) and many orchid species, which will be discussed below.

However, although ants are among the most abundant insects on Earth and are well known as dispersing agents for seeds, in their relationships with flowers they seem to be mostly nectar robbers. This is because most ants are thought to secrete toxic liquids that render pollen grains unviable. However, Gomez et al. (1996) have shown that several plant species in the high mountains of southern Spain (e.g. *Alyssum purpureum*, *Arenaria tetraqueta*, and *Sedum anglicum*) and arid lands (e.g. *Lepidium subulatum*, *Gypsophila struthium*, and *Retama sphaerocarpa*) are mostly pollinated by ants. This is presumably because in the extreme environmental conditions that prevail in these habitats, ants by far outnumber other potential pollinators.

In Mediterranean habitats, most pollinating insects are flying insects, for example flies, hoverflies, bee-flies, butterflies, wasps, beetles, and especially bees of two different categories (O'Toole and Raw 1991). The first category includes small (3–7 mm) slow-flying bees with a short tongue and relatively low energetic demands. The second is dominated by medium to large (10–25 mm) fast-flying bees with long tongues and much higher energetic demands. In fact, a loose correlation between flower traits (e.g. size, tube length, and nectar production) and pollinator traits (e.g. size and tongue length of pollinators) indicates that insects of various sizes visit flowers of a large range of sizes. However, in a detailed study of the relationships between flower colours and the natural colour vision of insect pollinators in the flora of Israel, Menzel and Shmida (1993) found a general trend towards higher frequencies of ultraviolet blue and blue colours in flowers predominantly visited by bees, as compared to higher frequencies of blue-green and ultraviolet green colours in those predominantly visited by flies and beetles. In Mediterranean ecosystems, these authors argue, a highly competitive pollination 'market' exists, which is dominated by hymenopteran insects characterized by (1) a high colour detection capacity with ultraviolet, blue and green receptors, and (2) a strong learning capacity at the level of individual bees for floral features that guide the pollinators according to reward experience rather than by innate search images. Thus, floral colours and shape are adaptive 'advertising' signals recognized by fast-learning pollinators.

Competition among plants

The spring peak of flowering in Mediterranean landscapes produces a surplus of flowers relative to the number of potential pollinators. This may result in sharp competition among plants for pollinators, especially because most plants are generalist vis-à-vis their insect pollinators. To simplify greatly, early-flowering plants invest largely in rewards (nectar and pollen) and 'advertisements' (large colourful flowers) (Cohen and Shmida 1993). Later in the year there is a surplus of insects over flowers, and by flowering in mid-season some plants try to 'sell' their rewards 'cheaply' rather than competing for the services of pollinating insects earlier in the year (Dafni and O'Toole 1994). The fact that summer-flowering plants tend to reduce their investment in flower size and rewards, in comparison with the spring-flowering ones, could be a response to a large number of non-competing potential pollinators.

One interesting example of competition among plants for pollination in the Mediterranean area is that of the almond tree (*Amygdalus dulcis*). The almond tree blooms in January–February, the rainiest and coldest months of the year and the time of lowest activity for pollinating bees. At that time, when almond trees are in full bloom, the potential pollinating honeybees are mostly inactive, remaining in their hives even on sunny days, unless ambient temperature rises above 15 °C. Why did the almond tree not evolve a more 'appropriate' blooming time? Eisikowitch *et al.* (1992) gave an answer: studying the foraging behaviour of honeybees, both on almond trees and on competing flowering plants, they showed that almond trees have a very low competitive ability. Trees would not achieve pollination unless they shifted their blooming time toward the lowest period of competition with other flowering plants. This resulted in a potential pollination period lasting only 18 days, but which is nevertheless sufficient for survival of the species. This is a typical case of avoidance of competition whereby a specific blooming period results from a trade-off between two selection pressures, one towards an early flowering to avoid competition among plants for pollinators and the other towards more favourable climatic conditions later in spring.

A plethora of strategies

Sophisticated pollination systems are also found in the orchids, as first described and interpreted in an evolutionary context by Darwin in his seminal book *The various contrivances by which orchids are fertilised* (published in 1897). Although several orchid species are autogamous* (e.g. *Ophrys apifera*), a great many are insect-pollinated by one or several species. Some groups (e.g. *Platanthera* and *Anacamptis*) are pollinated by hawkmoths or butterflies, and others (*Herminium*) by parasitic dipterans. But the great majority (e.g. *Epipactis* and *Himantoglossum*) are pollinated by a large number of different species, including sawflies, carpenter bees, flies, syrphids, or honey bees. However, solitary bees are by far the dominant pollinators of most orchid flowers. Usually a small number of pollinating species

Box 7.1 The sophisticated fig–wasp system

Some sophisticated mutualistic associations between insect-pollinated plants and pollinating insects justify using the term coevolution as defined by Janzen (1980): 'an evolutionary change in a trait of the individuals in one population in response to a trait of the individuals of a second population, followed by an evolutionary response by the second population to the change in the first'. Coevolution implies ecological relationships that are beneficial for both species. One example is that of the fig tree (*Ficus carica*), which has been a close companion of humans in the Mediterranean area since the Bronze Age. Flowers of the fig are allogamous*, and hence must be pollinated by an insect. All species of fig (over 800, all but one in the tropics) are pollinated by species-specific chalcid wasps of the family Agaonideae. The species that pollinates the Mediterranean fig is *Blastophaga psenes* and the pollination story of this fig–wasp mutualism has been nicely described by Kjellberg *et al.* (1987) and Anstett *et al.* (1995). It is a story of great importance for humans since unpollinated figs abort, except in parthenocarpic cultivars. Dried figs, especially, which constitute the great bulk of the world market, must be pollinated by hand to insure adequate sugar content and good flavour.

Over a thousand tiny flowers are organized like a carpet inside the inflorescence of a fig, which is a compact infolded structure called 'syconium' (Fig. 7.7). A female wasp loaded with pollen enters the syconium. While penetrating the opening at the top of the fig she may lose part of her antennae and wings. She then visits one flower after another and pollinates in passing the stigmas of several flowers. In the course of probing different styles, she oviposits in some of them and then promptly dies. After the eggs hatch, the young larvae feed inside the developing ovules of the fig. Several weeks later, when the fruit is almost ripe, the young wasps emerge and the wingless males fertilize the females. This allows the young winged female, loaded with pollen, to leave the fig and fly to another fig host plant. A new cycle will thus begin. This system is a strict mutualism because each species benefits from the other, and indeed is necessary for its survival: the plant for pollination and the insect for food, oviposition sites, and shelter. In some countries, cultivated figs are 'caprified' (the process of fig pollination) by hanging male figs on cultivated fig trees or by planting male trees in elevated sites upwind of the fig orchard in such a way as to ensure the pollinating insects an easy flight downwind. This enables the farmer to plant only one male for every 20 female (fruit-producing) trees. Elsewhere, there is a very active market for male figs for use in caprification.

Box 7.1 continued

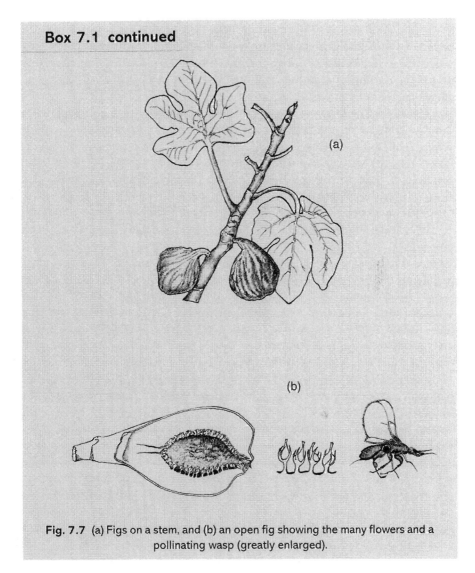

Fig. 7.7 (a) Figs on a stem, and (b) an open fig showing the many flowers and a pollinating wasp (greatly enlarged).

are involved. For example, although *Platanthera chlorantha* is visited by 28 species of insects, only 6 of them are responsible for 97% of the total pollination events (Nilsson 1978).

It is well known that in many more or less species-specific systems, flowers resemble the target insect species that is supposed to pollinate them. The lower petal (labellum) of the flower has a certain shape and bears certain spots or ornaments that are 'designed' to attract insects. In these cases one can speak of 'legitimate pollinators' and the patterns and shape of the flower probably result from an adaptation to mimic those of the insect. Well known examples are found among the *Ophrys*, some species of which have been given the name of an invertebrate that resembles the design of the flower (e.g. *Ophrys fusca*, *O. tenthredinifera*, and *O. araneifera*, Fig. 7.8). Some Mediterranean orchids may flower as early as late November, but the peak flowering season is March–April, a period that coincides with the highest density and diversity of actively pollinating solitary bees and wasps. The degree of match between the activity of pollinating insects and the flowering period of orchids may differ among regions of the Basin, depending on latitude and altitude.

In many cases, plant–insect interactions may be much more complicated than a simple symmetric reward for the two partners. In some cases, orchid flowers (e.g. *Serapias*) offer insects a shelter in unfavourable weather conditions, wherein the flower acts as a 'greenhouse' whose temperature can be up to 3 °C higher than that outside. This provides an energy gain for the insect (Dafni *et al.* 1981) and also a safe nocturnal lodging. In such cases, the flowers mimic a sleeping hole for the solitary bees that are the main pollinators of the orchid. It has also been noted that insects sleep in the tunnel-shaped flowers of *Cephalantera*, which results in pollination.

Some other orchids have evolved a 'trickster' behaviour for deceiving their pollinators by exploiting naive insects that are conditioned to forage among certain floral models. These 'fraudulent' orchids imitate the suite of cues that normally attract pollinators, but without giving them any reward (Bernhardt and Burns-Balogh 1986). This trickery may be particularly rewarding because nectar and carbohydrate resources are costly for the plant to produce. This strategy is selectively advantageous in perennial plants that must withstand seasonal aridity.

Another plant strategy is to mimic other species that attract pollinators. Biologists recognize two kinds of mimic systems. The most widespread is Batesian mimicry*, in which a low-density non-rewarding species mimics the flowers of a model species that offers rewards at a much higher level (Schemske 1981). Examples in the Mediterranean area are the orchid *Cephalantera longifolia*, which mimics the rock-rose (*Cistus salviifolius*), and the related *C. rubra* that mimics several species of *Campanula*. Sometimes the mimic species lacks a specific model and flowers early in the season when there are large numbers of inexperienced generalist pollinators on the wing. This mimicry-based-on-naivete, as Little (1983) called it, is widespread in *Orchis*, *Ophrys*, and *Dactylorhiza*, three genera common in the Mediterranean Basin. Of course, such systems can only have evolved and persisted where such plants share habitats with a number of other species that bloom at the same time, bear similar attractive cues, and are truly rewarding.

The second kind of mimic system is Müllerian mimicry, which implies that the set of partners includes at least three species blooming at the same time and showing a convergence in floral attractants. At least two species mimic each other but both offer some version of the standard edible reward. In the Mediterranean area, *Ophrys caspia* offers no reward, yet attracts four species of solitary bees. These bees actually pollinate the orchid but other species as well, for example *Asphodelus aestivus* (Liliaceae) and *Salvia fruticosa* (Lamiaceae), for which they are 'legitimate' pollinators because these plants offer rewards. *Ophrys caspia* takes advantage of bees that are unable to discriminate between similarly attractive models which offer a reward, and the non-rewarding orchid.

There are still other 'odd' systems among Mediterranean orchids, for example the 'brood site deception' whereby flowers simulate the oviposition site of a pollinator species and thereby attract a female insect ready to lay eggs. For example, *Epipactis consimilis* attracts female hoverflies (Syrphidae) to lay their eggs on the labellum of the flower by mimicking an aphid, which is the normal oviposition target for these flies. Finally, there are examples of 'sexual deceit'. This system involves 'pseudo-copulation' since it relies on inducing male pollinators to attempt copulation with parts of the flower that resemble the female insect. Similarly, an alternative to an edible reward can be an attractive olfactory stimulus for insects. For example, flowers of *Orchis galilea* emit a strong, musky smell that acts as a species-specific sexual attractant for the male of the pollinating bee, *Lasioglossum marginatum* (Bino *et al.* 1982). A study of the chemical composition of the odour of *Ophrys* species has shown a strong chemical similarity with that of

(a)   (b)

Fig. 7.8 Two common orchid species of the genus *Ophrys*, showing the spectacular flowers that attract insect pollinators, and their typical 'pseudo-bulbs'. (a) *O. tenthredinifera*, and (b) *O. fusca*.

the male insect pollinators (Borg-Karlson and Tengoe 1986). These males attempt to copulate with the flower but only receive depositions of pollinia for their efforts. Orchids with a pseudo-copulatory syndrome have a labellum that resembles in shape and colour the dark body of female insects (Dafni and Bernhardt 1990). The shiny spots that decorate the labellum of several *Ophrys* species (e.g. *O. specula*) are similar to those that appear on the back of a female wasp when she crosses her wings. The Mediterranean Basin is especially rich in such 'deceptive' orchid species.

Fruit dispersal by birds

If the animals most frequently responsible for plant pollination are insects, it is vertebrates, primarily birds, that are responsible for the seed dispersal of plants producing large edible fruits. Plants producing bird-dispersed fleshy fruits are a prominent component of most Mediterranean woodlands and shrublands (Herrera 1995). Birds eat whole fruits and then regurgitate or defecate intact seeds suitable for germination. Rewards provided by plants to the birds are, so to speak, the price they must pay for having their seeds dispersed far from the mother plant. Bird-dispersed plants are taxonomically very diverse in the Mediterranean area and include dozens of species in the families of pistachio, olive, myrtle, laurel, honey-suckle (Caprifoliaceae), and others. More than half of them rely entirely on birds for their seed dispersal.

The rewards of fleshy fruits

Happily for the birds, fruit production by these woody plants is usually high in terms of both number and biomass. For example, annual fruit production in some Spanish habitats can attain as much as 1 400 000 ripe fruits ha^{-1}, corresponding to a range of 6–100 kg dry mass ha^{-1} (Herrera 1995). Nearly all ripe fruits in Spanish matorrals (90.2%) are consumed by avian dispersers (Herrera 1995) (Table 7.1). These fruits are characterized by 'flag' features such as bright red, black, or blue colours, and their conspicuous location at the end of vertical stems makes them readily visible to birds. They are nutritionally rich, containing high levels of fat and protein, and also have secondary metabolites in their pulp which may confer an enhanced detoxification ability against toxic compounds for birds that eat them (Herrera 1984). Moreover, there is a fairly close match between the ripening season of fruits and the seasonal patterns of occurrence and migration of avian dispersers in Mediterranean shrublands.

The functional significance of the links between seed-dispersing birds and bird-dispersed assemblages of plants in Mediterranean woodlands and matorrals must be approached at the community level because there are hardly any species-specific mutualistic systems involved. Instead, several species of birds eat and disperse the seeds of many species of plants, such that it is unlikely that the local extinction of one partner would seriously affect the local survival of the other.

Table 7.1 Proportion of ripe fruit crops taken by avian seed dispersers in various woody Mediterranean plant species in a wide range of families.

Plant species	Fruits eaten by birds (%)	Plant species	Fruits eaten by birds (%)
Asparagus aphyllus	100	*Rhamnus alaternus*	61–93
Pistacia lentiscus	91–100	*Lonicera etrusca*	80–90
Smilax aspera	86–100	*Viburnum tinus*	51–75
Phillyrea angustifolia	72–99	*Prunus mahaleb*	50–68
Osyris alba	76–98	*Cornus sanguinea*	36–49
Daphne gnidium	92–97	*Pistacia terebinthus*	25–30
Myrtus communis	89–95		

Sources: various in Herrera 1995

The only known example of a species-specific interaction of this kind occurs in the southern Alps, and involves the nutcracker (*Nucifraga caryocatactes*) and the rock pine (*Pinus cembro*) (Crocq 1990). As much as 95% of the nutcracker's food is provided by the large oily seeds of the rock pine. In turn, the tree relies almost completely on the bird for dispersal of its seeds. Nutcrackers harvest the entire production of seeds of this pine in autumn and then bury small clusters of 7–10 seeds in small holes scattered throughout their territory. The birds return in winter and spring to eat the seeds, but some caches are forgotten, or abandoned. Seeds from these caches will insure the regeneration of the pine.

Two categories of fruit-dispersers

In terms of vegetation dynamics, and of the survival of the billions of birds that have to leave their northern breeding grounds in winter, the process of fruit consumption by birds in the Mediterranean area is of paramount importance. As in the case of plant–insect interactions leading to flower pollination, this mutualistic system implies that both partners gain some reward from their association. Birds transport the seeds of a plant that in exchange offers them a rich food supply. There is much evidence that Mediterranean avian seed dispersers have evolved morphological, physiological, digestive, and behavioural adaptations to take advantage of the abundant and profitable fruit supply provided by many woody plants. Morphological and physiological pre-adaptations, such as flat broad bills and various digestive adaptations, allow these birds to handle and swallow the fruits efficiently. Most of them are only seasonal frugivores, shifting from an insect-dominated diet in their spring–summer breeding grounds to a fruit-dominated one in autumn-winter. Plant–bird associations primarily involve a large number of passerine birds of small to moderate size (10–110 g) in the families of thrushes, warblers, and flycatchers. Members of the tropical bulbul family (Pycnonotidae) are also important seed dispersers in the eastern

Mediterranean and North Africa (Izhaki *et al.* 1991). These species are active during the period of fruit availability in the same habitats as the plants whose seeds they disperse. Seed dissemination is basically a within-habitat or landscape-scale process since most seeds only remain in a bird's digestive tract for a short time (20–30 min). Thus, they travel only small distances from the mother plant to where the birds eventually drop or eliminate them.

There are two main categories of avian fruit-dispersers in the Mediterranean area, corresponding to two distinct strategies for overwintering. The first category involves the millions of long-distance trans-Saharan migrants that cross the Mediterranean between August and late October en route to their tropical winter grounds (see Chapter 6, p. 162). Examples are the whitethroat, garden warbler, and pied flycatcher. Extensive consumption of sugar- and carbohydrate-rich fruits by these birds at their Mediterranean stopover sites may play an important role in determining their migration schedules (Bairlein 1991). In fact, as pre-migratory fat deposition is a prerequisite for successful migration across the Sahara Desert, the consumption of large quantities of fruits in the Mediterranean area appears to be a *sine qua non* in the Palaearctic–African migration system of these birds, as explained in Chapter 6.

The second category is that of bird species which overwinter in the Mediterranean from October to early March. As many as 12–16 species, reaching densities of up 15 individuals ha^{-1}, belong to this group with the robin, blackcap, song thrush, and blackbird as dominant species (Herrera 1995; Debussche and Isenmann 1992). Fruits of 20–30 plant species contribute at least 75%, and often more than 90%, of these birds' food supply. Most birds in this category usually eat the fruits of several plant species, each for a short time. This presumably helps to compensate for the unbalanced nutritional composition of each species' fruit pulp. However, lipid-rich fruits such as those of pistachios, olive tree, and Lauristinus (*Viburnum tinus*) invariably predominate in the diet of overwintering birds. Water and lipid content increases in the fruits of winter-ripening plant species as compared to summer- or even autumn-ripening ones. Fruits with a high content in energy-rich lipids are thus most abundant at a time of year when the energy demand of avian seed dispersers is highest. In a Spanish lowland shrubland, the three most common plant species (*Pistacia lentiscus*, *Myrtus communis*, and *Smilax aspera*), which together make up 63% of the fruit-producing cover, all have a remarkably long fruiting period of 2.2–3.5 months, which corresponds well with the overwintering period of birds (Herrera 1984). These same winter-fruiting species also have the highest levels of digestible lipids (up to 58% of pulp dry mass) and, not surprisingly, represent the most important food plants for this second category of overwintering birds.

Coevolution or serendipity?

Woody plants in the Mediterranean tend to ripen their fruits much later in the year than those in temperate regions. This has sometimes been interpreted as resulting from a coevolutionary process. However, the late-fruiting season of

Mediterranean woody plants could simply be the result of physiological factors related to evergreenness and the mild, rainy winters that occur in many parts of the Basin (Debussche and Isenmann 1992). Moreover, as seen in Chapter 3, the extant flora of the Basin includes many species of fleshy-fruit bearing plants of tropical origin that appeared long before the differentiation of most of the contemporary bird species that disseminate their seeds today. The fossil record reveals that at least 83% of genera of bird-dispersed woody plants in southern Spain were already present in western Europe before the onset of the Mediterranean climate in the Pliocene, 3.2 Myr BP (Herrera 1995). This does not mean, however, that the system did not originally result from coevolutionary processes. The more or less tight linkages existing between fleshy fruit-bearing plants and birds presumably evolved as far back as the lower Tertiary, when both the plant and bird species present were different. Many of the lineages that initiated the process were presumably the ancestors of birds that only occur today in tropical Africa, having definitively left the Palaearctic region during the Miocene climatic deterioration. Subsequent extinction and speciation events occurring in both plant and bird groups did not necessarily cause any breakdown of the mutualistic association. The present-day system may therefore be the result of an opportunistic reshuffling over time of different sets of species.

However, coevolutionary processes may be occurring currently as well. For example, the rather close fit between fruit size and disperser gape width may result from bird selection against large fruits in the plant species they visit regularly. The gape width of bird dispersers sets an upper size limit to the fruits they can ingest because birds usually do not nibble at fruits. Hence, the fraction of the fruit crop not removed by birds significantly increases with increasing fruit diameter, both among species and among individuals within species. Similarly, migratory habits, increased food passage rates, conferred detoxification ability, and seasonality in food preferences and size of digestive organs are all likely to reflect current adaptations to exploit the superabundant food supply provided by Mediterranean plants.

Decomposition and decomposers

Earthworms

Since Darwin (1859), the contribution of earthworms to physical, chemical, and biological soil processes has received much attention. These animals may constitute a dominant part of the life of soils, with biomass amounting to 1000 kg ha^{-1} in rich soils of temperate Europe and up to 4000 kg ha^{-1} in permanently grazed pastures (Lavelle 1988). The amount of soil they process may reach 230 tons ha^{-1} yr^{-1}. As in many other groups of animals, species diversity of earthworms is much higher in the Mediterranean Basin (150 species in southern France alone) than in all of northern Europe (30 species). This difference can be explained by Pleistocene glaciations that destroyed most species occurring north of the Mediterranean area (Bouché 1972).

Earthworms are generally assigned to four main functional groups on the basis of their location and activity in the soil:

(1) *epigeic* (acid tolerant species living in the leaf litter they feed upon);
(2) *anecic* (burrowers that also feed upon leaf litter but are sensitive to acidity);
(3) *endogeic* (deep-dwelling species that actively process and mix soil particles through continuous intestinal transit);
(4) *epianecic* (deep burrowers active in rich forest soils, being more acid-tolerant than the *anecics*).

Each of these functional groups is presumed to contribute differently to such major ecological processes as organic matter recycling, water drainage, and bioturbation.

The main aspects of earthworm soil activity are (Granval and Muys 1992):

(1) physical effects whereby earthworms promote soil porosity, aeration, water drainage, and bioturbation;
(2) biogeochemical effects such as litter decomposition, organo-mineral migration, P and N recycling, and shifts in pH and C:N equilibria;
(3) indirect contributions to maintaining high plant diversity by providing better germination and early growth conditions for trees and herbaceous plants;
(4) earthworms are important source of proteins for many predators. Up to 200 species of birds and mammals, that is, around one-third of all European taxa, regularly feed on them in the Mediterranean area (Granval 1988).

Dung beetles

With herbivory by insects, grazing and browsing by wild ungulates and livestock are the main types of predation of green plants by animals. Recycling the non-digested part of this food is a major process of ecosystem function. It allows the return to the soil of organic matter and minerals, especially nitrogen and phosphorus. It also contributes to the proliferation of micro-organisms, detritivore communities, and earthworms. In addition to the enrichment of soil in organic matter, a rapid decomposition of faeces decreases the risk of disease transmission among animals, because vertebrate parasite populations that reside in dung are killed in the process, and this may in turn locally play a role in increasing areas available for plants.

Each day an individual sheep or goat drops *c.* 2 kg of faecal matter while a cow can deposit as much as 28 kg (Lumaret 1995). Decaying faeces are recycled by large and complex assemblages with a subtle turnover through time of various micro-organisms, flies, and dung beetles. Animal dung represents both habitat and resources for at least 135 species of dung beetles (*Aphodius, Geotrupes, Bubas, Scarabaeus,* etc.) in the Mediterranean Basin. All these insects use this highly concentrated resource as food, as a substrate for laying eggs, and as a food source for their larvae. Many of them complete their life cycle in or near dung before it disappears completely. A key factor in the colonization of newly deposited dung

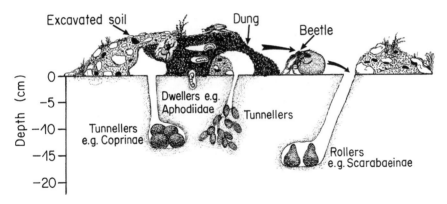

Fig. 7.9 The three main categories of dung beetles: 'dwellers', 'tunnellers', and 'rollers'. (After Lumaret 1995.)

is its odour, to which insects are highly sensitive. Dung must be moist in order to be colonized by insects, and this represents a constraint in Mediterranean habitats where high temperature and low precipitation cause dung to desiccate quickly. The summer drought period also slows down insect activity so that the recycling of dung is much slower in Mediterranean habitats than further north. Lumaret (1995) showed that the complete drying out of dung dropped in summer is achieved within two or three weeks in Mediterranean habitats as compared to two or three months in winter. Indeed, dung beetles are mostly active in spring and autumn and then decrease in numbers in summer. High desiccation rates associated with reduced beetle activity in summer are the reason why 30–50% of total summer droppings still remain unrecycled after eight months (Lumaret and Kirk 1987). These climatic constraints explain why only a few species of dung beetles have successfully colonized the most arid parts of the Mediterranean.

Dung beetles have evolved different strategies to exploit their unpredictable and temporary food resource. Schematically, three main groups are recognized (Fig. 7.9), of which the first seems to be the least adapted to Mediterranean bioclimates. These are the 'dweller' species (e.g. *Aphodius*, *Ontophagus*) that live inside the dung they use as food for themselves and for their larvae. Dwellers colonize dung as soon as it has been dropped and their eggs are laid directly in it. The larvae pupate in the ground under the dung and emerge as fully developed insects through a tunnel they dig themselves. It is crucial that the dung remains moist for the entire life cycle of the insects because they are unable to dig through the hard desiccated crust of dry dung. Females of some Mediterranean species have evolved an adaptation to cope with drought. They dig tunnels under the main dung pad and lay eggs in small pieces of dung that they put in small lodges, a behaviour very similar to that of the second category, the 'tunnellers'.

The 'tunnellers' are represented by species of *Ontophagus*, *Bubas*, *Copris*, and many species of *Geotrupes*. They avoid desiccation during hot Mediterranean summers by burying small pieces of dung in the ground down to a depth of 1.5 m. For example, *Bubas bubalus* individuals bury 200 g dry faecal matter for their breeding activities. At the bottom of the tunnel, they prepare a 'breeding chamber' filled with pellets of dung, each of which will receive an egg.

The third category is that of 'rollers' such as *Scarabaeus* and *Sisyphus* spp., and is another Mediterranean specialty. Like the tunnellers they also bury dung with eggs inside, but initially they prepare a large ball of dung and roll it for up to several metres until they find a suitable place for digging a tunnel. The famous nineteenth-century entomologist of southern France, Jean-Henri Fabre, was the first to describe this behaviour for the sacred scarab beetle, *Scarabaeus sacer*. The size of the ball is closely adapted to the size of the hind legs of the insect. A related species, *Scarabaeus semipunctatus*, lives in sandy regions of the Mediterranean. Lumaret has shown that females of this species will stop rolling their balls of dung and bury them as soon as they find a suitable place. It is possible to stimulate the burying behaviour by wetting the soil just in front of an actively dung-rolling insect.

Because of the many constraints in the Mediterranean caused by drought, which makes dung available for only a short period, tunnellers and rollers are more abundant than dwellers. The proportion of the first two categories combined is 46% as compared to only 28% in central Europe (Lumaret 1995). Moreover, several dwellers in the Mediterranean have completely reversed their life cycle as compared to that of tunnellers, concentrating their activity in the coolest and wettest months of the year. One threat to the biological diversity and abundance of dung beetles is the generalized use of veterinary drugs which makes domestic animal dung poisonous for these insects (Wall and Strong 1987).

Summary

We have looked at examples of traits and features that some species have retained as ancient historical constraints or else evolved as a response to the present-day Mediterranean bioclimate. Our goal was not to make an exhaustive review but rather, using case studies, to draw attention to the huge number of traits, or suites of traits, that allow Mediterranean species to survive and interact with other species. Biological diversity is also the diversity of life styles and functions within ecosystems. We also showed, from the example of fruit-eating birds, that ecosystem function in the Mediterranean can have profound influences well beyond the region's limits. The abundant fruit production of many Mediterranean plant species contributes greatly to the wealth of the bird fauna of the whole Palaearctic region.

8 Humans as sculptors of Mediterranean landscapes

Human pressures on Mediterranean ecosystems have existed for so long that di Castri (1981) did not hesitate to argue that a complex 'coevolution' has shaped the interactions between them and humans through long-lasting but constantly evolving land use practices. Studies of intraspecific variation within plant and animal species, such as those reported in Chapter 6, show that organisms may evolve life-history traits as a response to human-induced habitat changes. The transformation of landscapes and habitats has also had profound consequences on the distribution and dynamics of species and communities. One cannot understand the components and dynamics of current biodiversity in the Mediterranean without taking into account the history of human-induced changes. So much of the territory has been so profoundly transformed by more than 300 generations of human occupation that some process of 'landscape archaeology' is necessary in order to understand what we see day.

A 10 000-year 'love story' between humans and nature

The Mediterranean and Middle East were the cradles of some of the world's most ancient civilizations, each of which has left its stamp on parts or all of the Basin. From the dawn of human history there were a great many populations of *Homo erectus* all around the Mediterranean. At Atles del Tell, Algeria, a jawbone of this primitive man was recently found that appears to be around 700 000 years old. The Tautavel man (450 000 yr BP) found in the eastern Pyrenees in 1971 is considered to be one of the most ancient *Homo* fossils in Europe. In the Levant, Turkey, and Mesopotamia, records of permanent human settlements, including fair-sized cities, go back to the inter-glacial periods of the upper Pleistocene, when humans lived as hunter-gatherers in caves. Mediterranean islands, too, have been colonized and human-transformed for nearly as long as the continental lands. Fossil remains of humans have recently been found in Sardinia, dating back to 20 000 yr BP, and recent evidence from archaeological sites in Cyprus shows that human colonization of that island began as early as 10 500 yr BP, or very soon after the end of the last glacial period (Simmons 1988). Cyprus, Malta, and Crete in particular were home to some of the most brilliant ancient civilizations in the Old World. Many islands endured repeated invasions and incessant wars because of their much-coveted strategic position for defence (e.g.

Box 8.1 Corsica, a many-fold invaded island

According to the historian Michel Vergé-Franceschi, Corsica has been invaded no less than twenty times over the last 2500 years, and each time by different peoples. The first invaders were the Phoenicians (565 yr BC), soon to be followed by the Etruscans (540 yr BC). Since 270 yr BC, the Carthaginians (270 BC), Romans (259 BC), Vandals (AD 455), Byzantines (534), Goths (549), Sarrasins (704), Lombards (725), Pisanos (1015), Genoans (1195), Aragons (1297), the Genoans again (1358), 'Milanians' (1468), Franco-Ottomans (1553), French (1768), British (1794), and finally the German–Italian Axis during the Second World War, have all invaded the island. This does not mean, however, that all these peoples successfully established a foothold. Only the Romans (seven centuries) and Genoans (four centuries) stayed for long. Instead, most invaders established posts along the coast but did not venture far inland, and at such times the inland mountains served as refuges. (Traditional villages on islands and other fought-over ground were always situated on high, impregnable sites if at all possible, as demonstrated by the many 'oppida' on hilltops throughout southern Europe.) In these inaccessible regions it was easier to resist invaders (and wandering bands of robbers), and from them the natives could, when forced, fight back. Which explains the outburst of Sir Gilbert Elliot, English viceroy of Corsica, in 1794: 'Corsica is an ungovernable rock'!

More generally, however, trouble came from the sea and nearby plains. It is this repeated pattern of invasion, followed by retreat to the mountainous inland, that has resulted in a strange paradox: while a remarkable mixture of peoples inhabit coastal areas throughout the Basin, in the interior mountains of each country human populations are still extraordinarily distinct linguistically and behaviourally (McNeil 1992).

Malta, Cyprus), trade, or natural resources (see Box 8.1). Analysis of charcoal remains found in Crete suggests that during the Middle and late Minoan culture (c. 4000–3000 yr BP) the landscapes around the city of Kommos consisted of intricate mosaics of cultivated fields and orchards alternating with semi-natural woodlands exploited for wood and other products (Shay et al. 1992).

The Basin is also a crossroads and meeting point for humans, just as it is for plants and animals. Since the palaeolithic many waves of human migration have been absorbed, most from the east and the south, their biological and cultural characteristics being added to and overlaid on the previous ones to produce mosaic-like assemblages of peoples.

A succession of civilizations

The Mediterranean has been the theatre for the birth, blooming, and then collapse, of some of the most prestigious and powerful civilizations on record, and their successive turns of fortune have had great impact on biota and ecosystems everywhere in the region. Building on a thousand years of Sumerian, Akkadian, and other Semitic and non-Semitic civilizations in the lands between the Persian Gulf and the Mediterranean, the Phoenicians and Greeks extended their spheres of influence from the eastern Mediterranean to Iberia in the first millennium BC. Subsequently, Greek, Roman, and Ottoman empires vied for influence and domination (Fig. 8.1). During the first century BC, however, the Mediterranean became, for a period unique in its history, one single political entity and benefited for several centuries from the 'Pax Romana' (Fig. 8.1).

Then, in the fourth century, the Roman Empire split into several entities, one of which, the Roman Empire of the Orient, lasted until 1453, when it was replaced by the Ottoman Empire, which in turn collapsed in 1923. Yet until well into the seventeenth century, as a legacy of the Pax Romana, the Mediterranean remained the centre of gravity of world economic activity (Braudel 1949). The Romans brought the art of land management to its highest known degree with a series of techniques particularly suited to conditions in the Mediterranean area.

The collapse of the Roman Empire in the western Mediterranean, in the early fifth century, presumably had both positive and negative effects on biodiversity. While the Vandals and other invaders spreading out in search of new lands tended to sweep away the works-in-progress and the prosperity gained under the Pax Romana, the Visigoths, Sarrasins, and Francs, by contrast, appear to have sought to build upon Greco-Roman cultural foundations. They retained and improved the land use practices and resource management schemes developed and disseminated by them. Moreover, a general return of ancestral cultures and regional specificities, mainly Slav, Greek, and Turkish in the east, Arab in the south, and Latin and Celtic in the north-west, resulted in many changes in landscape and habitat design as well as in human pressures. Thus, it would be a mistake to think that the history of the varying landscapes and regional biota we perceive today has been linear and without oscillations, or that ill-considered resource depletion has been its only theme.

Intensive redesign

Apart from sheer, vertical cliffs and some remote mountainous areas, there is probably no square metre of the Mediterranean that has not been directly and repeatedly manipulated and, one might say, 'redesigned' by humans. An undeniable part of the 'love story' is one of forest destruction and replacement by simpler systems and a more convivial 'space' for people. The forest historian Thirgood (1981) has written of ten or more millennia of resource depletion to describe the interaction of humans and the Mediterranean forest. However, from Phoenician, Greek, and Roman times, a number of land use models have existed

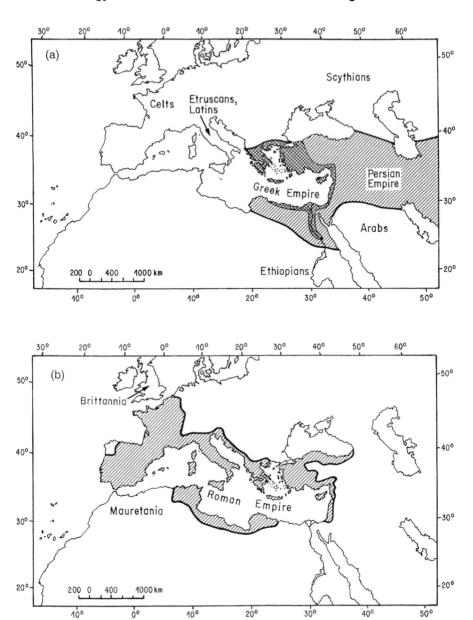

Fig. 8.1 Greatest extension of (a) the Greek and Persian empires, about 500 yr BC, and (b) the Roman Empire in AD 44, at the death of Julius Caesar. The most powerful neighbouring kingdoms or tribes are indicated on each map.

whose aim was to achieve what today is called sustainable long-term development of resources and ecosystem management.

According to Naveh and Dan (1973) human impact has had direct and sustained effects on Mediterranean living systems for at least 50 000 years, even though the first 'possession' of fire apparently dates back to at least 300 000 yr BP

(Trabaud *et al.* 1993) or even earlier. A veritable 'revolution' occurred about 10 000 years ago, when hunters in the Near or Middle East began to create and maintain their own food supply and thus to lay foundations for the domestication of plants and animals. Up until the industrial revolution in the nineteenth century, all Mediterranean civilizations were basically founded upon a fine balance between agricultural and pastoral activities, with forests and woodlands being cleared and exploited as needed in the near vicinity of farms and villages.

No less than farming, livestock husbandry has been the cause of a vast array of ecological changes. Over the millennia a series of pastoral and agricultural practices evolved from the earlier patterns of nomadism and semi-nomadism that characterized hunters and gatherers. These included burning, clearing, cutting, coppicing, terracing, cultivating, animal breeding, irrigating, and all the other innumerable activities related to a sedentary or semi-nomadic life in the Mediterranean area. There gradually evolved a remarkable structural and habitat diversity, at the landscape and regional scales, with approximately half of the Mediterranean space being occupied by agriculture, and the rest by forests, matorrals, and range-lands. Of course the precise situation varied considerably from one region to the next, and from one historical period to the next, as a result of changing demographic, sanitary, and socio-economic conditions. It also varied from one year to the next and this yielded the typical 'moving mosaic' effect of Mediterranean landscapes. Rural peoples were self-sufficient from their own crops, wheat, olive oil, milk, cheese, wine, cooked pork meats, figs, nuts, and the innumerable natural products they could find in matorrals and woodlands, being neither rich nor really poor. They succeeded in establishing an 'oecumene', that is a convivial, sustainable system of close and 'friendly' interactions with their near environment.

Forest destruction and transformation

Ups and downs

It is hard to imagine that the elephants which were captured by Hannibal for his armies (218–220 BC) belonged to a forest-dwelling subspecies that lived in the huge forested areas of southern Tunisia. Mesopotamian cuneiform texts reveal that 3000 years ago the Assyrian king Assurbanipal passed through large forests to reach Damascus in his conquest of Syria. In Roman times, more than half of northern Africa was still covered with dense forest. Even as recently as the sixteenth century, the armies of Charles-Quint travelled across Spain and France without ever leaving the shadow of tree canopy!

As explained in Chapter 3, a great many palaeobotanical, archaeological, and historical records demonstrate that the Mediterranean area was much more forested in the distant past than today. Beginning as early as the neolithic, the history of Mediterranean forests is one of terrible cycles of destruction. No more than 9–10% of the Mediterranean area is forested today (Marchand 1990). The first significant deforestation began as early as 8000 yr BP (Thirgood 1981) and dramatically increased by the end of the neolithic.

Particularly well marked in archaeological and fossil pollen records are the periods of expansion of the Persian, Hellenistic, and Egyptian (7000–4000 yr BP) civilizations, which were all eras of intensive land clearing and concomitant increase in the number and size of flocks and herds. This led to an expansion of cultivated fields at the expense of range-lands and an expansion of range-lands at the expense of forests (Trabaud et al. 1993).

In the north-west quadrant of the Basin, pollen diagrams show that large-scale neolithic deforestation in the Alps and the Pyrenees coincided with a warmer, moister climate, and the steady expansion of cereal culture as a result of human demographic expansion, first at low altitudes in southern France, throughout the Atlantic period (7500–4500 yr BP), and at middle altitudes towards 5000 yr BP (Triat-Laval 1979). Heavy human pressures occurred again as a result of large population increases during the chalcolithic period, especially the Sub-boreal period (c. 4500–2500 yr BP), and continued right up to the decline of the western Roman Empire in the fifth century AD. Based on historical records, fossils, pollen, and charcoal remains at archaeological sites, it is clear that there have been many historical 'ups and downs' corresponding to the waxing and waning of human activities in the various regions. The decline of each major civilization was almost always followed by a wide-scale recovery of forested areas.

A spontaneous recovery of western Mediterranean forests occurred after the decline of the western Roman Empire. It lasted until a resurgence of human activities in the Middle Ages. In Corsica, for example, palaeobotanical data show a huge expansion of agriculture during the early Middle Ages accompanied by rapid deforestation after the colonization of the island by the Republic of Genova (Pons and Quézel 1985).

Enormous amounts of wood have been consumed in the course of history for such varied purposes as domestic firewood, furniture, charcoal, shipbuilding, other forms of construction, and clearing of land for agriculture and livestock husbandry. From as early as the seventh century AD, the entire Mediterranean Basin was subjected to timber-based industries (Thirgood 1981). With the development of powerful empires, all easily accessible forests were heavily damaged and sometimes completely destroyed. For example, when Spain and Portugal were major naval powers, in the fifteenth to seventeenth centuries, much of the Iberian forests were cleared to allow shipbuilding, especially near the coasts and along the major rivers. When industrial activities began enormous quantities of wood became necessary for glass manufacturing, mine shafts, and fuel for metallurgy and other modern industries.

Following the catastrophic second half of the fourteenth century when the Black Plague struck southern Europe, a demographic renaissance took place, accompanied by renewed clearing and cultivation of lands which had been abandoned during an entire century of famine and plague. Finally, the nineteenth and, especially, the twentieth centuries have brought increasingly severe destruction of vegetation in many parts of the Basin. In most Euro-Mediterranean countries, however, the massive substitution of fossil fuels for wood after the First, and especially the Second World War, resulted in a generalized recovery of forests whereas forest destruction is continuing at alarming rates in all the Afro-Mediterranean countries (see Chapter 10).

Regional histories

Each individual region, however, experienced a different history of forest destruction. For example, the beginning of the nineteenth century was a very rich, prosperous period for south-western Europe, with high population density, and widespread destruction of woodlands. That same period was a desolate, depauperate time for peoples in the eastern Mediterranean and North Africa, which allowed a certain amount of forest and woodland recovery to take place. Naveh and Dan (1973) relate that 200 years ago Palestine had only 200 000–300 000 inhabitants, as compared to 5 million from the second to the fifth centuries AD, and again today. Similarly, North Africa had only half as many inhabitants in the eighteenth century (6 million) as it did during the seventh century, at the beginning of the Arab conquest (Le Houérou 1981).

In southern Europe, a spectacular demographic rise has taken place over the last 2000 years. In the eastern Mediterranean, a similar process took place, but two or three millennia earlier. The Lebanese and Palestinian mountain forests were heavily exploited by Egyptian pharaohs, King Solomon, and other ancient near-eastern rulers, starting at least 3000 years ago. In North Africa the Roman legions also took a heavy toll on forests. When Julius Caesar's fleet was destroyed by a storm during the war against Pompeius (46 BC), legionnaires were sent to rebuild it from trees harvested in the Sousse region of Tunisia, which was therefore still heavily forested at that time. No trees at all, except cultivated ones, can be found in this region today. A similar near-total removal of trees occurred in the Sharon plain between Tel Aviv and Haifa (the Hebrew word *Sharon* means *forest*) (Reifenberg 1955). Today, no vestiges whatsoever remain of these historic forests.

Mediterranean forests extend today over 85 million hectares, 9.4% of land area of the Basin (Marchand 1990). Quézel (1976a) estimated that no more than 15% of the 'potential' Mediterranean forest vegetation remains today, with the rest in more or less advanced stages of deforestation and soil degradation. Table 8.1 provides estimates of the area occupied today by forests and high matorrals in the various countries of the region. A high proportion of forested areas are actually plantations of pines and eucalyptus.

Successional dynamics

Given sufficient time, something resembling the primeval forest in a given site is the theoretically expected end result of secondary succession following destruction of the original formations. From grasslands, shrublands are expected to emerge, and then secondary forest of one sort or another. Quite often this

Table 8.1 Forested areas in Mediterranean countries.

Country	Area with a mediterranean-type climate (ha $\times$ 10^3)	Area covered by forests and matorrals (ha $\times$ 10^3)	%
Spain	40000	9200	23
France	5000	2186	43
Italy	10000	1570	16
Malta	32	0	0
Ex-Yugoslavia	4000	960	24
Albania	2000	248	12
Greece	10000	1568	16
Turkey	48000	6051	13
Cyprus	925	171	18
Syria	5000	440	9
Lebanon	1040	95	9
Israel	1000	116	12
Egypt	5000	2	< 1
Libya	10000	501	5
Tunisia	10000	840	8
Algeria	30000	2424	8
Morocco	30000	5190	17
Total	211996	31562	15

Sources: after Quézel 1985; Le Houérou 1990

dynamic is blocked at some stage of shrubland, or even degraded grassland. However, Mediterranean ecosystems can follow many different trajectories in the course of degradation, transformation, and secondary recovery, depending on land use practices, as well as regional or even local variations in substrates and locally available seeds of colonizing plants. To give just two examples, Fig. 8.2 shows the differences in typical secondary succession on calcareous vs. acidic rock substrates in Languedoc, southern France, and on the island of Corsica. These models are overly simplistic but do give an idea of the cumulative effect that human actions have on ecosystem trajectories and how easily positive feedback systems* tending to block secondary succession can be set off by these actions.

Consequences of deforestation

Large-scale deforestation has had two main consequences. The first was the progressive replacement of deciduous broad-leaved forests by evergreen sclerophyllous forests and matorrals of different physiognomy and composition. In the course of the last two to three millennia holm-oak and kermes oak have progressively replaced downy oak over wide areas in southern France, Spain,

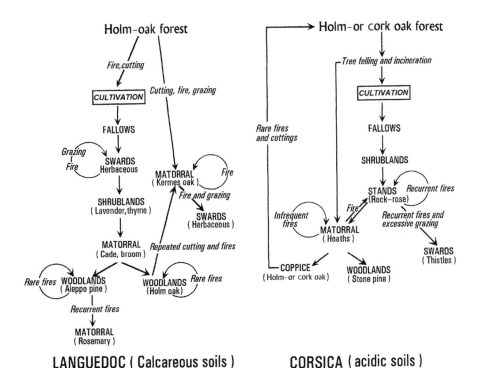

Fig. 8.2 A comparison of two successional trajectories. (After Trabaud 1981.)

Morocco, Corsica, and elsewhere (Reille *et al.* 1980). The substitution of deciduous trees by sclerophyllous evergreen ones, as well as the increase of habitat patchiness over time, has had many consequences on the distribution of populations and species as well as on their genetic diversity.

Changes in the genetic make-up of populations also arise from changes in selection regimes across habitat mosaics. For example, in the blue tit (see Chapter 6, p. 159), shifts in life-history traits of the Corsican population have presumably occurred as an indirect result of human action 2000 to 3000 years ago. Before human impact on the island's vegetation became strong, most of the habitats suitable for blue tits on this island were deciduous, not evergreen as they are at the present time.

In some instances, invasive pines have accompanied evergreen oaks in the process of secondary succession; examples are the black, Aleppo and Calabrian pines, all prolific producers of long-distance wind-dispersed seeds that readily colonize recently perturbed areas. Fire-sensitive species were also reduced or eliminated, and those not adapted to grazing and browsing were also increasingly confined to the few habitats where large ruminants could not reach them. In the eastern Mediterranean, a large number of Sudanian (Afro-tropical) species were probably eliminated or much reduced in distribution as a result of human action during the Holocene (Zohary 1983).

The second consequence of forest destruction was a generalized desiccation of the Mediterranean as a whole because of the rupture in water balance in many areas where forest cover was destroyed. Water-flow changes are among the most obvious consequences of deforestation. As plant cover decreases, surface runoff and stream flow increase. A dramatic increase in soil erosion occurred in many parts of the Basin in conjunction with deforestation, especially in North Africa. Exposure of the soil surface and breakdown of soil structure increase the amount of soil material that is carried away through runoff. Soil loss through gully erosion is several times higher when vegetation has been destroyed than on soils with a forest cover. For example, soil loss of a small area in North Africa that was ploughed after its woody plant cover had been removed amounted to 50 t ha^{-1} yr^{-1} as compared to 0.4 t ha^{-1} yr^{-1} on average in nearby forested areas (Dufaure 1984). In order to justify the huge reafforestation programmes that started in southern France at the end of the nineteenth century, foresters claimed that 600 000 m^3 of soil were washed away from Mt. Aigoual each year as a result of forest clearing. These enormous amounts of soil were said to be partly responsible for the obstruction of the port of Bordeaux (Nègre 1931). In a region of southern France where vegetation spontaneously recovered from covering 7% of the surface area in 1946 to 49% in 1979 as a result of rural depopulation, Rambal (1987) calculated that the stream flow decreased by 11%. Deforestation and its consequences explain the high vulnerability of many Mediterranean hilly landscapes where steep slopes make the effects of erosion even more pronounced and often irreversible.

* * *

Despite its long-term destructive consequences, forest clearing was a prerequisite for the creation of a lived-in convivial space for humans. Deforestation allowed them to expand areas for their experiments in domesticating and 'taming' selected plants and animals. Domestication of animals and plants began about 10 000 yr BP in the Mediterranean Basin, especially in its eastern part where modern humans evolved in an area of intense development among various populations of prehistoric man during the Mousterian period (100 000–33 000 yr BP) of the late palaeolithic. Permanent settlements in the Middle East became the foundations of the 'urban revolution' when humans built cities in the chalcolithic (8500–4000 yr BP) and then succeeded in domesticating animals and plants.

Animal domestication

The Mediterranean was home to the first successful experiments in the domestication of mammal species. This area is thus exceptionally rich in livestock genotypes selected by people over generations in each microregion of their centre of origin in the Near East and south-western Asia (Georgoudis 1995). The diversity of livestock breeds in the Mediterranean reflects the diversity of environments where humans selected their animals. Once achieved there, the practice of domestication and the various domesticated species spread through-out the world in a remarkably short period of time. One cannot imagine Mediterranean peasants without their donkeys, dogs, pigs, sheep, and goats. Domesticated mammals provide meat, milk, wool, and skins for the manufacture of tools, clothing, and tents, as well as an additional workforce that multiplied the possibilities for working the land and the woodlands, for transport, and for trade.

The first indication of domestication dates back more than 10 000 yr BP, to the time when climate was beginning to improve following the last ice age. The first domesticated animals were either predators, such as dogs or wolves, or generalist animals such as the wild boar that followed human groups to take advantage of their edible refuse. It is very easy to tame a young wild boar and then breed it, which explains why domesticated pigs appeared as early as 7000 yr BP (Pfeffer 1973). In fact, dogs were probably domesticated even before boars. They were important because they allowed the development of new hunting techniques and the protection of all other domesticated animals from wolves and other predators. The most ancient remains of domesticated dogs were found in German archaeological sites dating back 9500–10 000 yr BP. Independently or not, domestication of dogs also took place very early in the eastern Mediterranean, as testified by archaeological remains dating back to c. 9000 yr BP. Although bones of these early dogs resemble those of the dingo, a wild Australian dog, there is evidence that post-glacial human tribes established a hunting relationship with the wolf, and gradually succeeded in taming this species by capturing and raising young pups.

Horses were domesticated from a wild species (*Equus caballus*) that was widespread in European forests and steppes throughout the Quaternary. The horse was apparently first domesticated some 5000 yr BP in the steppes of Turkestan. Horses allowed the development of rapid transportation systems across dry steppes where cattle would have been too slow and food-demanding. The horse has been a decisive companion of humans for territorial conquest and waging wars, as well as for many daily tasks in rural life.

Perhaps even more important has been the donkey, first domesticated from a wild ass (*Equus asinus*) in Libya, 5000 yr BP by contemporaries of the ancient Egyptians. The donkey quickly spread and reached the Near East by 3400 yr BP and then rapidly extended to all the Mediterranean countries. The exceptional qualities of the donkey, including its physical endurance, resistance to fatigue, and proverbial frugality, made this animal an invaluable daily companion in the harsh, dry, and mountainous Mediterranean environments. The donkey's services to humans include transportation of heavy loads on large pack saddles, pulling of ploughs, turning of mill stones, raising of well water, threshing of wheat, and innumerable others. One of the most familiar pictures in all Mediterranean countries until recently was that of people moving along narrow streets of small villages on the backs of overloaded donkeys with their legs hanging down almost to the dusty ground. Additionally, the virtues of donkey's milk, very similar to that of human milk, are proverbial (the wife of Nero is said to have used the milk of five hundred she-asses for her daily baths!) When finally dead, the donkey provides excellent meat as well as tough hides used in the manufacture of parchment, clothing, and drum skins.

Domesticated races of wild cattle, deriving from the aurochs (*Bos primigenius*), appeared more than 6000 years ago, perhaps from forms that were domesticated in Mesopotamia (Pfeffer 1973). Cattle are quite often illustrated on ceramics from early Mediterranean civilizations and bull worship was long celebrated in many regions. This is best exemplified by the Egyptian Bull-God, Apis, and the famous Minotaur of Crete. Another Bull-God was celebrated in the Bronze Age in the western Mediterranean, at Mt. Bégo near Nice, where thousands of delicate carvings from this period are still visible in the rocks of the mountain.

Two types of water buffalo (*Bubalis bubalis*), first domesticated in India 5000 yr BP, still occur in the Mediterranean: the 'riverine' and, to a lesser extent, the 'Mediterranean'. These animals, which can forage in much deeper waters and on softer bottoms than traditional cattle, are still bred in Albania, Bulgaria, Egypt, Greece, Italy, Romania, Syria, Turkey, Tunisia, and ex-Yugoslavia. Their numbers are decreasing in all countries, however, except in Egypt, presumably as a result of decreasing areas of suitable range-land. Buffaloes are mainly bred for milk production, and only secondarily to provide meat and serve as draught animals. They are often closely associated with humans for the cultivation of rice.

The domesticated animals of most importance for Mediterranean peoples and which have had the most widespread impact on Mediterranean ecosystems through grazing and browsing are sheep and goats. The goat is a highly adaptable browser to mountainous terrain and semi-desert, capable of surviving on very sparse fodder. One shared feature that interests people is the fast herd build-up of sheep and goats. Therefore, traditional sheep and goat production has always been an important part of rural economy in the Mediterranean Basin. Local traditional systems of dairy sheep and goat production and the region's topography favoured the selection of genetically isolated populations which progressively evolved fixed characters. Local varieties of sheep and goat occur in almost all the large Mediterranean islands as well as in the oases on the borders of Morocco, Algeria, and Tunisia, and the oasis of El Fayum in Egypt (Georgoudis 1995).

Sheep were domesticated by the mesolithic, more than 7000 yr BP, among hunter-gatherers who succeeded in capturing young animals and raising them in captivity. The first unequivocal indication of sheep domestication appears to have involved two races of mouflon, an eastern one originating in Afghanistan and a western one that was widespread in south-western Asia. Today there are dozens of domesticated races in the Basin that differ in size, shape, coat, and horn patterns. Several varieties of the western race are still widely used in Turkey, Iran, and many Mediterranean islands.

The most ancient domesticated goat remains known were found in archae-ological sites near Jericho, in the Jordan valley, and are dated at *c.* 8700 yr BP. However, most domesticated goats (i.e. *Capra hircus*) appear to derive from the wild goat (*Capra aegagrus*), which is probably extinct, even on Crete where the distinctive goat population of *agrimi* probably descends from prehistoric domestic goats. The 'goat complex' clearly includes a large number of forms and hybrids, the systematic status of which is still unclear. It seems clear that much hybridization was carried out by early goat breeders with access to the many Palaearctic forms of ibex in the mountain ranges scattered from the Mediterranean Basin to eastern Asia (Fig. 8.3).

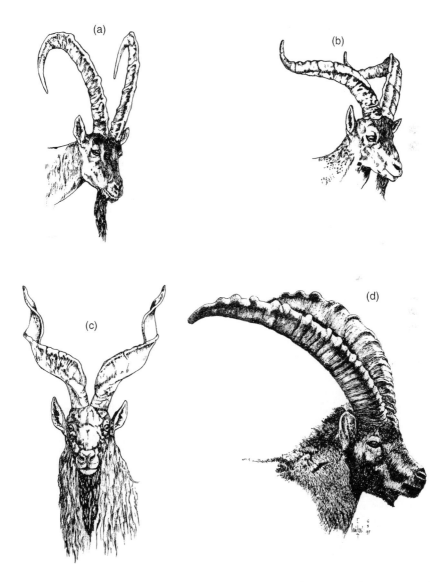

Fig. 8.3 Three examples (a–c) of the myriad goat races (*Capra hircus*) present in the Mediterranean today, and (d) the closely related ibex (*Capra ibex*). Note the variety of horn patterns.

Among smaller mammals, the cat was domesticated much later than the above-mentioned species, probably not from the European wild cat (*Catus sylvestris*), which proved to be impossible to tame, but rather from an African cat (*Felis cattus libyca*) imported from Egypt to the eastern Mediterranean three or four millennia ago. The history of its domestication is somewhat ambiguous. Even when partially domesticated, cats apparently long remained a commensal* companion of humans rather than being raised for a specific purpose. The carnivorous species that was used before cats to control mice populations in human settlements was the genet (*Genetta genetta*), which was raised in captivity from young cubs captured in the wild. Ancient Egyptians practised a religious cult in honour of the cat. Millions of cats were carefully mummified and preserved in tombs in the Nile Valley. Alas, literally tens of tons of cat mummies were exhumed and transformed into fertilizer at the beginning of this century (Pfeffer 1973), so that no genetic or phylogenetic information is available from this squandered primary source in the history of small mammal domestication.

Before closing this section brief mention should also be made of the elephant, which was first used for purposes of war by Alexander the Great, the Egyptian pharaohs, the Carthaginians, and the Romans but was never really fully domesticated. At best they have been tamed into compliance, especially in east Asia.

Plant domestication

The Mediterranean Basin, especially its eastern part, is one of the most important centres of origin for crop plants of worldwide importance (Hawkes 1995) (Table 8.2). In the 1920s the Russian plant explorer N. I. Vavilov noticed that many ancient cultures and cultivated plant prototypes come from the Fertile Crescent, the area that stretches from the Jordan valley through Syria, Turkey, and the mountains of Iraq and Iran. In 1927 he was impressed by the thousands of hectares planted with hundreds of wheat varieties in Cyprus, as well as by 'forested areas composed almost entirely of carob trees'. He advanced the opinion that Cyprus was the original homeland of the cultivated carob.

Though an exact figure for crop diversity is not available, estimates approach 522 cultivated crop species in the Mediterranean Basin, including both indigenous and exotic species first cultivated here. The precise origins of agriculture remain elusive, probably having occurred between 9000 yr BP and 11 000 yr BP. In any case, it was clearly concurrent and correlated with the domestication of animals. Fodder and pasture plants were given an evolutionary 'push' by humans at about the same time as the first temperate zone Old World food plants. In the Near East mortars and grinding tools, as well as sickles, have been found that date much further back than these dates, but they do not necessarily correspond to sowing of crops around villages. Whatever its

Table 8.2 The principal cultivated plants originating in the Mediterranean area.

Grain crops

Wheat (*Triticum*), 5 species
Oat (*Avena* spp.), 3 species
Barley (*Hordeum sativum*)
Canary grass (*Phalaris canariensis*)
Lentils, vetch, faba, erse (*Lens, Vicia, Lathyrus, Ervum*)
Peas, chickpeas (*Pisum, Cicer*)
Lupines (*Lupinus* spp.), 4 species

Forage plants

Cock's head (*Hedysarum coronarium*)
Clover (*Trifolium*), 3 species
Gorse (*Ulex europaeus*)
Fodder peas (*Lathyrus*), 3 species
Serradela (*Ornithopus sativus*)
Corn spurrey (*Spergula arvensis*)

Oil-producing plants

Flax (*Linum*), 2 species
Safflower (*Carthamus tinctoria*)
White mustard (*Sinapis alba*)
Rape seed, colza (*Brassica*), 3 species
Garden rocket (*Eruca sativa*)

Fruit crops

Olive tree (*Olea europaea*)
Carob tree (*Ceratonia siliqua*)
Almond (*Prunus amygdalus*)
Fig tree (*Ficus carica*)
Pomegranate (*Punica granatum*)

Vegetables

Beets (*Beta*), 2 species
Cabbage (*Brassica*), 4 species
Parsley (*Petroselinum crispum*)
Artichoke, Cardoon (*Cynara*), 2 species
Turnip, Swedes (*Brassica*), 2 species
Purslane (*Portulaca oleracea*)
Onion, garlic, leek, (*Allium*), 4 species
Salsify (*Scorzonera*), 2 species

Asparagus (*Asparagus officinalis*)
Sea-kale (*Crambe maritima*)
Celery (*Apium graveolens*)
Endive, chicory (*Cichorium*), 2 species
Garden chervil (*Anthriscus cereifolium*)
Cress (*Lepidium sativum*)
Parsnip (*Pastinaca sativa*)
Oyster plant (*Tragopogon porrifolius*)
Spanish oyster plant (*Scolymus hispanicus*)
Horse parsley (*Smyrnium olusatrum*)
Dill (*Anethus graveolens*)
Common rue (*Ruta graveolus*)
Sorrel (*Rumex acetosa*)
Blites (*Blitum*), 3 species

Condiments, dyes, and tanning agents

Black cumin (*Nigella sativa*)
Cumin (*Cuminum cyminum*)
Anise (*Pimpinella anisum*)
Fennel (*Foeniculum vulgare*)
Thyme (*Thymus vulgaris*)
Hyssop (*Hyssopus officinalis*)
Lavender (*Lavandula vera*)
Peppermint (*Mentha piperita*)
Rosemary (*Rosmarinus officinalis*)
Sage (*Salvia officinalis*)
Iris (*Iris pallida*)
Damascene rose (*Rosa damascena*)
Laurel (*Laurus nobilis*)
Hops (*Humulus lupulus*)
Madder (*Rubia tinctorum*)
Sumac (*Rhus coriaria*)

Source: after Vavilov 1935, reviewed by Hawkes 1995

precise age, the invention of agriculture divided the plant world linked to agricultural practices into two parts, the segetal and the non-segetal (Zohary 1973), and created an irreversible barrier between what might be called 'anthropogenic' and 'primary' plants. It also separated modern history into a 'segetal era', which began with the neolithic domestication of plants, and the earlier, 'pre-segetal era'.

Cereals

Many fascinating details have recently emerged concerning the origins of the four Old World cereals first found and domesticated in the Near East, that is, wheat, barley, oats, and rye. Cultivated wheats derive from a number of wild progenitors (Harlan and Zohary 1966). Diploid enkorn wheat (*Triticum monococcum*) and the tetraploid *T. timopheevii*, which are still cultivated on a small scale in the Balkans and Anatolia, apparently derive from the species *T. boeticum*, widespread throughout south-west Asia and the eastern Mediterranean. Tetraploid emmer wheat (*T. dicoccon*) and hard wheat (*T. turgidum*) both derive from the eastern Mediterranean wild emmer *T. diococcum*, first discovered by T. Kotschy in the anti-Lebanon in 1855.

The bread wheat we eat today (*T. aestivum*) is a hexaploid, and derives from hybridization between hard wheat and its diploid relative, *Aegilops tauschii* (Zohary 1969, 1983). The great advantage of hard wheat is the fact that the 'ears' of grain are not brittle like those of emmer wheat, and thus remain intact until harvest. In addition, once harvested at full maturity, the spikelets of durum wheat separate readily from their hulls and thus can be easily threshed.

Cultivated barley (*Hordeum vulgare*) was probably domesticated at around the same time as wheat, and served as the 'poor man's wheat' in areas of limited rainfall and poor soils. It ripens a full month before wheat but the quality of the grain is inferior and, since the Middle Ages at least, has primarily been grown for forage and fodder rather than for bread. The wild ancestor of barley is now clearly established as being *H. spontaneum*, which is native to the eastern Mediterranean and adjacent Irano-Turanian regions. It appears that only one gene separates the two species, but it is critical for farmers since it controls the brittleness of the rachis, or stalk. Wild barley occurs in both primary and weedy habitats, and shows a remarkable amount of biochemical variation (Nevo *et al.* 1979). It is an aggressive colonizer of disturbed matorral and a common roadside and field weed that hybridizes freely with the cultivated species. The same situation is found in cultivated rye (*Secale cereale*), an important food grain throughout northern and eastern Europe, and its wild eastern Mediterranean ancestor, *S. montanum* (Fig. 8.4).

The genetics of oat (*Avena sativa*, including the cultivars commonly called *A. byzantina* and *A. nuda*) are no less complex than those of wheat, barley, and rye. The cultivated oat is closely related to an aggregate of wild hexaploid oats called *A. sterilis*, widely distributed in the Mediterranean Basin. Here again, sponta-

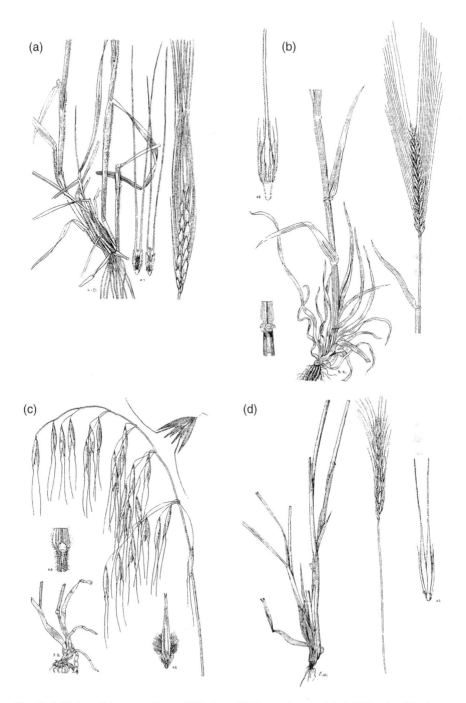

Fig. 8.4 Major wild progenitors of (a) wheat (*Triticum diococcoides*), (b) barley (*Hordeum spontaneum*), (c) oats (*Avena sterilis*), and (d) rye (*Secale montanum*) found in various habitats in the eastern Mediterranean and Middle East. (Reproduced with permission from Zohary and Feinbrun 1966–86.)

neous hybridization between the two species is common and much wild germplasm* remains to be collected and exploited by plant breeders. Two distinct modes of seed dispersal appear in the wild and weedy forms. One is the so-called *sterilis* form, which is synaptospermous, that is, the whole spike disarticulates upon maturity to form a drill-type dissemination device. In the so-called *fatua* form, by contrast, each floret shatters individually and is much harder to reap. As in the case of wheats and barley, domestication of oats required a modification of the wild mode of seed dispersal in order to obtain a non-shattering crop plant. Oats were brought into cultivation later than the above-mentioned cereals, some time between 3000 and 2000 yr BP, after which it also spread rapidly in Europe. This seems to agree well with the notion of Vavilov (1951) that oats (and many other early cultivars) should be considered as a 'secondary' or derived crop. That is, modern oat cultivars probably started out as weedy races infesting neolithic wheat and barley fields and only much later were they incorporated as a new grain crop sown separately (Zohary 1983).

Pulses

Numerous protein-rich pulses were domesticated in neolithic farming villages of the Near East, between 9000 and 8000 yr BP, and then accompanied cereals in their rapid spread throughout the Old World. These include peas (*Pisum sativum*), lentils (*Lens culinaris*), chick pea (*Cicer arietinum*), lupines (*Lupinus* spp.), and broad beans (*Vicia faba*). Wild progenitors of peas and lentils have been found in the eastern Mediterranean, namely *Pisum elatius*, *P. humile*, and *Lens orientalis*. It seems clear that domestication of pulses began simultaneously with that of cereals. This is supported by archaeological evidence as well as the shared geographical distribution and ecological patterns of the two groups. Priority was placed on the production of easy-to-store and highly nutritious seeds with a dual utilization of grass kernels rich in starch and leguminous seeds rich in protein. However, chick pea, broad bean, and other pulses were also grown as fodder, coffee substitutes, and green manure from very early times. Just as in the cereal grasses, most Mediterranean pulses form part of a bewildering array of feral types of cultigens, weedy races, natural hybrids, and cultivated derivatives of all of the above. Two examples that have been studied in detail are the *Vicia sativa* group (Zohary and Plitmann 1979) and the 12 species of inter-fertile lupines occurring in the eastern Mediterranean and North Africa (Plitmann 1981).

Many more arable crops

A great many vegetables were domesticated and cultivated in the eastern Mediterranean from very early times. These included beets, leeks, lettuce, cabbage, carrot, celery, radish, globe artichoke, and eggplant, as well as the subtropical melon and watermelon (Zohary 1983). Culinary herbs and spices of Near Eastern and Mediterranean origin constitute a list that is much longer still, and includes black cummin, borage, coriander, dill, fennel, laurel, mustard, oregano, rosemary, rue, tarragon, and thyme. Many of those elaborate plant chemical defence adaptations we discussed in Chapter 7 have been put to good use by humans in cooking, medicine, and rituals (see Table 8.2). The list of Mediterranean plants used in traditional healing runs to more than 200, and dozens of these are still included in pharmacopoeias.

Among wild and domesticated fodder legumes of the Mediterranean area, alfalfa (*Medicago sativa*) and annual medics (over 50 species), vetches (*Vicia* spp., with 50 annual and 25 perennial species), sang-foin (*Ornithopsis*), *Lathyrus* (35 species), *Lotus*, *Melilotus*, *Trigonella*, and some 110 species of trefoil (*Trifolium*) must all be mentioned. Palatable and nutritious grasses in the region that also underwent early domestication and improvement by farmers and pastoralists include orchard grass (*Dactylis glomerata*), fescue (*Festuca*), and ryegrass (*Lolium*).

Perfumes and oils

Sources of perfumes, balms, and religious–symbolic substances were also very numerous among early plant domesticates. Plant resins, gums, and volatile oils of all kinds were sought to mask body odours and to sweeten the ambience of tents, bathhouses, boudoirs, and palaces. Those which came into domestication include lavenders and myrtles (several species or chemotypes of each), lentisk, jasmine, henna, and olive, as well as various 'balsams', including frankincense and myrrh. Wild populations of lentisk, rock-rose, storax and other aromatics were managed in wild stands as well, or else exploited out of existence. Some of the earliest vegetable oils in history were obtained from safflower (*Carthamus tinctorius*) and flax (*Linum usitatissimum*). Safflower was formerly a very important source of natural dye, and flax gave the fibre of choice for the manufacture of fine textiles. Some 16 species of *Carthamus* occur in the Middle East, but the origins of their cultivation are still unknown.

Olive oil was so highly esteemed in ancient times that, in addition to its numerous uses as food and lamp oil, it was also used in medicine, for ritual offerings, and for the anointing of priests and kings. In Biblical times the olive tree was unquestionably one of the most important and valuable trees of the Hebrews. No other tree is mentioned so many times in the Bible, and olive orchards were clearly widespread throughout the Fertile Crescent and adjacent areas wherever climate and soils allowed it. In Greek and Roman mythology the olive was the symbol of Athena and Minerva, goddess of medicine and health.

Fruits

Trees producing fruit crops are another group of plants first domesticated in the eastern Mediterranean and the Middle East. The four classical fruit trees of the region are the olive tree, fig, grape, and date palm. The early domestication of date palm and grape are thought to go back as far as the Bronze (6500–7000 yr BP) or late neolithic ages (8000 yr BP). Zohary (1973) argued that the origin of the date palm (*Phoenix dactylifera*) was in the oases, wild canyons, and salt marshes of the Near and Middle East. The exact origin of wild grape (*Vitis vinifera* var. *silvestris*) is obscure, but early signs of cultivation have been found from the early Bronze Age in Israel, Palestine, Syria, Egypt, and the Aegean area. The world's most famous vine was cultivated in the Near East at least as early as the period covered in the Old Testament (*c.* 3800–5000 yr BP). In Genesis (8: 20) we read: 'Noah was the first tiller of the soil. He planted a vineyard.' The spies sent by Moses to explore the land of Canaan returned with 'a cluster of grapes and they carried it on a pole between them'.

A conspicuous feature of the domestication of grape was the change in sexual reproduction achieved through selection. All known wild forms are dioecious and fruit set thus depends on the transfer of pollen from male plants to female ones. In nearly all cultivated grape varieties, flowers are hermaphroditic and thus separate pollen donors are not needed provided the species is self-compatible (Zohary and Spiegel-Roy 1975).

Many other tree species have been domesticated for fruit crop from early times: carob, mulberry, hazelnut, pine nut, pistachio, pomegranate, and walnut, as well as dozens of stone fruits of Irano-Turanian origin. This latter group includes wild apple, pear, and hawthorn (*Malus*, *Pyrus*, and *Crataegus*) as well as medlar, mountain ash, quince, and service-berry (*Mespilus*, *Cydonia*, *Sorbus*, and *Amelanchier*), almond, prune, and apricot (*Amygdalus* and *Prunus*), three edible dogwoods (*Cornus* spp.), and one species of oleaster (*Elaeagnus angustifolia*) known as 'Russian olive' or 'Trebizond date'.

* * *

Having reviewed some of the 'ingredients' used by people to create a convivial *oecumene* in the Basin following the neolithic 'revolution', we shall now examine the various historical approaches to land use and resource management that provided a framework for the blossoming of Mediterranean civilizations. The first three of these—fire, grazing, and improved water management—all represent double-edged swords, so to speak, since they can all easily produce as much harm as good for biota and ecosystems when ill-managed. The crucial point is in finding the 'middle road' of what is now called 'intermediate disturbance', which is the 'golden rule' for Mediterranean agro-pastoral or agro-silvo-pastoral ecosystems.

How did humans maintain their living space?

Fire

Fire was the very first tool used by people, as testified by the remains of ash and charcoal found in palaeolithic archaeological sites in Spain, Greece, and Israel dating back some 20–25 millennia. In fact, the use of fire by palaeolithic hunter-gatherers appears to go back at least 500 000 years in parts of the Mediterranean (Naveh 1974). However, people only learned to kindle and control fire around 300 000 years ago (Trabaud *et al.* 1993) when it began to be used for a variety of purposes. By the Iron Age, 2600 yr BP, shepherds and farmers alike were using fire to attain more and better pasture and cropland. Many passages in the Bible show the importance of fire in the life of the Hebrews. In the first century BC Virgil also wrote of 'fires lit by shepherds in woodlands, when the wind is favourable' (*Aeneid*, X), testifying to the very ancient use of fire for improving pastures. Kuhnholtz-Lordat (1938, 1958) emphasized that until quite recently most human-set fires in modern times in the Mediterranean area were ignited for agricultural or pastoral purposes. In many developing countries of the Basin this practice is still in use.

Grazing and range management

The combination of forest cutting and fire lighting was a prerequisite for the development of grazing and browsing areas by domestic ruminants. In the Mediterranean area, livestock husbandry and breeding have been of enormous importance to humans and ecosystems at least since the neolithic revolution, and hence are amongst the most important forces shaping Mediterranean landscapes. Some pastures in the eastern Mediterranean have been continuously grazed by domestic ruminants for more than 5000 years.

Domestic grazing regimes and pastoral systems under mediterranean-climate conditions can take three different forms, depending on resource availability, social structure, and local and historical factors. The first is sedentary livestock raising, involving both stall feeding and free grazing, which is by far the most common form today. Second, there is true nomadic pastoralism where the whole household moves with the herd or flock, and which has now more or less disappeared from the Basin. Third, between these two extremes is 'transhumance' where only the herder moves with the stock. This is a remarkably well-adapted form that involves biannual movements of flocks between high summer pastures and winter grazing grounds at lower altitudes. It also allows cultivation of cereals or other crops as a supplementary source of revenue. The strategy of keeping mixed herds of small and larger ruminants was an important 'drought-insurance strategy' for these latter two forms (Le Houérou 1985).

The high plateaux and mountains of the Mediterranean have traditionally served as an 'escape zone' for animals, a space where herds and flocks could find food, water, and refuge from the dry, scorching Mediterranean summer. In such areas, continental and oceanic influences confer more reliable and higher rainfall than in the littoral, and throughout the year. This is well reflected in pasturelands, which are more productive, diverse, and actively growing throughout summer. Thus there evolved the semi-nomadic system of 'transhumance', which seems to date back at least to the Bronze Age (see Box 8.2). Some authors have ventured to suggest that the routes followed by shepherds in the different regions were often

Box 8.2 Transhumance

The word transhumance derives from the Latin *trans* (beyond) and *humus* (land), and thus means 'beyond the land of origin'. Employed as a verb in Spanish (*transhumar*), the term entered French in the early nineteenth century and came to replace the former term 'aestiver', meaning 'to spend the summer'. Transhumance covers both phases of a semi-nomadic system, which consists of biannual movements of herds and flocks between a lowland area and a mountain area of contrasted climate. In the most widespread form, 'ascending transhumance', flocks spend the autumn, winter, and spring in the lowlands, and move to the cooler mountain areas in summer, where pastures stay green and where drinking water is available (Fig. 8.5). 'Descending transhumance' describes the system whereby flocks leave their usual high mountain homes in winter to pass the coldest season of the year lower down, where the climate is milder. With time, the term came to incorporate burning of the lowland areas in the late spring to 'renew' pastures against the return of the flocks later in the year. This practice makes sense from the shepherd's point of view, and can be repeated more or less indefinitely if all resources are properly managed.

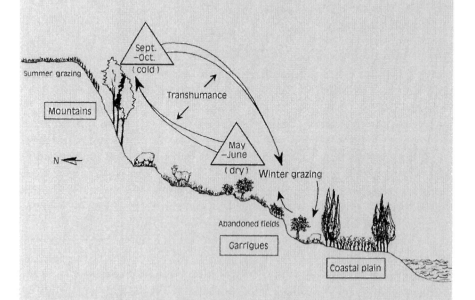

Fig. 8.5 Outline of a typical 'ascending' transhumance trail in southern Europe, wherein lowland flocks move up to the mountains in early summer, and return with dropping temperatures in autumn. (After Gintzburger *et al.* 1990.)

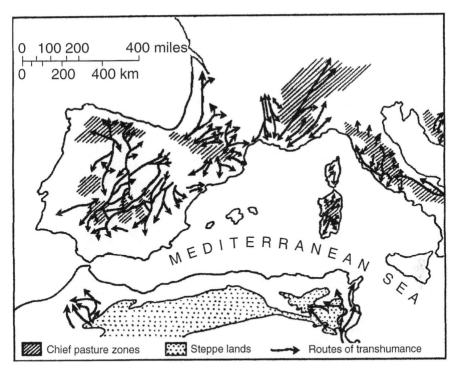

Fig. 8.6 Major routes of transhumance in the western Mediterranean. (After Houston 1964 and Brisebarre 1978.)

those established by migration routes of wild animals, especially the large herds of deer. Numerous petroglyphs and sculptures of large mammals in the mountains of North Africa and the central Sahara also suggest a very early origin for transhumance in those regions.

The distances covered in the biannual movement of herds in search of summer pastures are typically no more than 100–300 km in the moister areas of southern France and northern Italy. By contrast, in the semi-arid regions of southern Spain, southern Italy, and North Africa, each seasonal 'drive' could be as much as 700–1000 km and traverse a huge number of villages and property lines. From very early times, highly elaborate and ritualized networks of trails and footpaths (called *tratturi* in Italian, *cañada* in Spanish, and *drailles* in French) were employed for moving large herds between the Apulian plains and the Apennines, and the plains of southern France and all the nearby mountain ranges (Pyrenees, Massif Central, Alps). Similarly, in interior Spain, with its huge ancestral emphasis on livestock, the transhumance routes were far reaching and well maintained over a great many centuries. Seasonal movements of livestock in central Spain involved 4–5 million sheep, in the heyday of the early sixteenth century (Fig. 8.6).

In southern Europe generally, all forms of transhumance are on the decline today. For example, in the Department of Hérault, France, an area of 622 700 ha, there were approximately 400 000 head of sheep in 1894, and only 88 000 in

Fig. 8.7 A transhumant flock of sheep on the move in northern Spain.

1964. This figure has remained stable ever since (Gintzburger *et al.* 1990). In Morocco and parts of the Near East, however, transhumance and pastoralism continue to play an important role, but have undergone massive change as a result of the growing use of trucks to transport livestock, and subsidized feed (barley, bread, cottonseed, etc.).

Struggle for water

The struggle for water has always been a vital thread in the history of all Mediterranean peoples. When rainfall is scant or altogether lacking for several months of the year, sedentary people cannot survive for long without a reliable supply of drinking water for themselves and their livestock. The Persians and various other Near Eastern peoples built and maintained elaborate and nearly labour-free systems to collect and store rainwater in large quantities. Waterproof mortar, first invented around 3300 BP, allowed the construction of permanent water cisterns. In the steppes and desert regions of Syria, Iraq, and Iran, an ingenious system of well-chains called *foggara* or *qanats* was developed: long interconnected series of wells allowed underground (evaporation-free) transportation of water over many kilometres (Reifenberg 1955).

Ancient peoples built dams across wadi beds in the desert to catch floodwaters that might only occur once a year or less. Using rocks, gravel, and mud they also designed and built extensive but simple systems to divert floodwaters into city and farm reservoirs and to directly irrigate cultivated fields down-slope of the catchment areas. These works had to be capable of resisting the exceptionally heavy floods that might occur once every ten years or so, yet also be sensitive enough to utilize the smallest possible surface run-off.

In the Near East, sophisticated water-management and engineering practices date back at least to the chalcolithic period, 4500 to 5000 yr BP. The Nabateans, starting around 2500 BP, were particularly organized and determined in the practice of 'rainfall harvesting', as testified to by the extraordinarily large cities they built in some of the most arid and desolate regions of the Near East. These included the ancient Biblical city of Sela, in southern Jordan, renamed Petra by the Nabatean Arabs when it became their capital for seven or eight centuries. A series of Nabatean waterworks of varying dimensions and purposes were also laboriously excavated in Israel. They have been restored to working order by a small team of scientists over a period of 30 years (Evenari *et al.* 1982). The results of this work have helped to bring into attention the many known, but unexplored, archaeological sites in the 15 steep wadi-beds that plunge into the eastern shore of the Dead Sea on the Jordanian side. Innumerable remains of run-off agriculture and waterworks also exist in the desert fringes of North Africa. Both the Phoenicians and Carthaginians were well versed in sophisticated techniques long before the arrival of the Romans. The Phoenician colonists who brought the eastern art of writing may also have brought to southern Europe and northern Africa the art of irrigation (Reifenberg 1955). A few centuries later, the conquering followers of Mohammed, also brought many innovations and new crops to Iberia during their four-century long occupation of the Spanish Levant.

It was the Romans who refined run-off agriculture and water diversion systems into a high art. With their vast, centralized government, and large pools of forced labour, the Roman rulers were especially well placed to undertake large-scale construction, water transport, and irrigation systems under a variety of conditions all around the Basin. The Pax Romana provided favourable conditions

Box 8.3 Coping with silt

This is an essential aspect of water engineering in the drier parts of the Mediterranean and Near East. During the winter storms so typical of the eastern Mediterranean, desert wadis that are bone dry in summer suddenly fill with muddy torrents, which deposit boulders and stones but carry silt far away onto the plains, where it can harm crops and agricultural lands. At Kurnub (Mamshit), in the central Negev, one of the most astounding examples of ancient engineering can be seen in sharp detail, carefully mapped and interpreted by archaeologists. Although the region has only 75 mm mean annual rainfall, a large caravan station flourished here in Nabatean and Byzantine times, thanks to dams and canals that diverted floodwaters into underground reservoirs. Based on experimental research carried out at Kurnub and two other Nabatean sites in the Negev, scientists were able to confirm the hypothesis that Nabatean engineers not only knew how to capture sufficient amounts of rain and floodwater to support large populations, but also how to trap silt to prevent it from clogging and choking up fields (Evenari et al. 1982).

for intensive cultural exchanges that favoured the flow of know-how throughout the entire Mediterranean world. In coastal areas where monocultural production of cereals and other rainfed crops dominated, the mastery of hydrology and building attained by the Roman engineers is visible in the preserved remains of their monumental and remarkably astute hydraulic works. Some of our favourite specimens in southern France include the thousand-year-old aqueducts at the Pont du Gard, the industrial mills at Barbegal, and the public baths of the Emperor Constantine in Arles. In half a dozen other countries around the Basin, and several of the large islands as well, similar marvels exist, particularly near to cities. In Italy, the spectacular waterwork remains also reveal the intricacies of Roman farming and land-use planning in all its fascinating detail.

Terraces

From very early times, the history of Mediterranean peoples, both in the mountains and at the border areas between hills and plains, has been heavily marked by cultivated terraces. Hand-built stone terraces permitted cultivation on slopes ranging from 20 to 75%, and this sometimes required carrying soil up from the valleys on the backs of people or animals (Lepart and Debussche 1992). The construction of tailor-made water distribution systems added to the quality of terrace cultivation. Mainly fruit trees, vines, cereals, flax, and vegetables were grown on terraces. This management of hilly terrain is both a means of fighting against erosion and a water-saving device preventing water run-off. As such,

Fig. 8.8 Phoenician-built terraces are still in use in Lebanon and Israel, supporting mixed crops of grapes, olive trees, almond, figs, and, occasionally, cereals. Spontaneously sprouting oaks and other forest remnants are left growing when they spring up in or very near the terrace walls.

terraces are a popular and domestic counterpoint to the monumental works of the Romans and their vast imperial cities and farms. Until the early twentieth century, terrace cultivation remained a hallmark speciality of Mediterranean landscapes, from the mountains right down to the coast (Fig. 8.8).

On a small scale, wherever steep hillside terraces are abandoned the processes of erosion and gully formation soon set in and strip the slopes of their topsoil. All around the Mediterranean, vast hilly regions occur where formerly terraced and intensively farmed hillsides near human settlements are overgrown with neighbouring remnant seed-bearers such as oaks, rock-roses, maples, and junipers. In other sites, formerly terraced hills are now entirely denuded of woody vegetation despite having been totally abandoned for many decades. On a hillside near the former settlement of La Lauze (Hérault, southern France) for example, a Visigoth community built an impressive series of terraces in the fifth and sixth centuries AD. This site was definitively abandoned sometime in the twelfth or thirteenth centuries and farming never resumed. As early as the fifteenth century these hills were known in land tenure documents as *Monts-Chauves* or the 'Bald-hills', and even now woody vegetation has still not recovered. This is partly a testimony to the ill-managed use of fire and grazing since the 1600s, but ultimately the desertification was caused by the construction and then abandonment of the terraces.

Multiple uses of forest

Wild fruits

From at least the second millennium BP, Greek landowners cultivated a wide range of fruit trees, including pear, pomegranate, apple, and fig, all found in the gardens of Alcinoos in Homer's *Odyssey*. They also cultivated quince, walnut, almond, and stone pine as well as a number of Mediterranean trees that have more or less disappeared from modern usage in farms and gardens, such as the Christ's thorn (*Paliurus spina-christi*), which was widely used then as a 'living fence' (Amigues 1980). Theophrastus, Theocritus, and the epigrams of the Palatine Anthology, all report that poor people and shepherds ate a wide range of fruits, nuts, and mushrooms.

Cropping in 'wild fruit forests' implies a prehistoric practice of managing natural forests for the optimal production of semi-domestic fruits and nuts. While no solid data or cave paintings support this hypothesis, it makes sense in light of many landscape formations common in the north-east quadrant of the Basin and adjacent regions. For example, extensive groves of pistachio (*Pistacia vera*) grafted on wild *P. atlantica* trees occur throughout Iran and, to a lesser extent, on *P. palestinus* in various parts of the eastern Mediterranean. Very early in history, western and eastern Mediterranean peoples grafted scions of productive almond trees onto non-productive ones or those producing bitter-tasting fruit. Wild olive and carob trees are also grafted in Turkey, Crete, and Cyprus, and in inner Anatolia the most common wild trees over large areas are a hawthorn (*Crataegus lacinata*) and a pear (*Pyrus elaeagrifolia*). Along with apples and almonds, these trees are generally left when forests are cleared for farming or grazing purposes, and frequently serve as stocks on which to graft other cultivated varieties (Zohary 1983). How ancient those semi-cultural forests are is not known, but there is no reason to doubt that the fundamental horticultural skill of grafting was discovered as early as the domestication of tree crops itself. Cultivated olives, after all, require grafting; ungrafted, they quickly revert and produce a small worthless fruit just like wild types. Similarly, pistachio nut trees (*Pistacia vera*), are far more productive and long-lived when grafted onto a stock of the more vigorous *Pistacia atlantica*.

Semi-wild 'forests' or woodlands were perhaps maintained for a variety of products and purposes other than fruits, oil, and nuts (see Box 8.4). The valonia oak (*Quercus macrolepis*) was for centuries widely planted in Anatolia, where it is native. In some areas of south-west Turkey, wild stands were, until the 1950s, still taken care of and treated as if they had been planted. The main revenue of this tree comes not from the wood but rather from the galls, and acorn burs from which a highly sought-after tanning agent is derived, but the acorns were eaten by livestock and some wood products were provided from regular pruning. Similarly, carob trees were once planted very extensively in warmer parts of both the southern and eastern shores of the Mediterranean, primarily as a forage and fodder tree in association with sylvo-pastoral systems.

Box 8.4 The dye-producing kermes oak

The kermes oak, whose scientific name *Quercus coccifera* means 'beetle-bearing oak', forms a dense shrubland in many parts of the Mediterranean (see Chapter 4). This knee-high formation was, in many cases, intentionally maintained by people to serve as an outdoor 'factory' for a strong, colour fast red dye that was very highly prized in ancient times. This substance was produced by females of the cottony cochineal insect, *Kermococcus vermilio*, whose sac-like bodies contain carminic acid in high concentrations. As the female attaches herself permanently to a kermes oak, and remains immobile throughout her life, feeding upon the oak through sucking mouthparts, this carmine-red and bitter substance presumably serves as a defensive warning to potential predators. When mature, she lays a number of eggs and dies, after which the young emerge and immediately begin feeding on the host plant.

The use of these scale insects and the harvesting of the dead females, dropping them in vinegar and then letting them dry to obtain the means for dying wool and other fabrics bright red, was widespread by Old Testament days. Kermes was so highly prized by the Romans that it formed part of the tribute exacted from a conquered nation. Feudal lords and monasteries in the thirteenth and fourteenth centuries also accepted it as partial payment of taxes and rent.

Unfortunately, we know nothing about how the kermes shrublands were managed to optimize scale insect proliferation, but fire was certainly applied regularly. What is most provocative and stimulating to imagine, however, is the countless small and large hands that must have spent hours collecting the pea-sized bodies of dead females of *Kermococcus* by the million under the Mediterranean sun, prior to the invention of aniline dyes in the late nineteenth century.

Sacred trees

One important factor determining how early humans reshaped and 'redesigned' Mediterranean forests and landscapes in ancient times was the role of certain 'sacred' trees and groves. Both oaks and terebinth were sacred in the belief systems and ritual practices of the ancient Near East, since both were associated with Abraham. Both sacrifices and the burial of loved ones took place under these trees that, accordingly, were not to be cut or burned. Both olive and styrax trees were also considered as sacred in the eastern Mediterranean, but some individual trees were considered more 'sacred' than others.

Box 8.5 Sacred pomegranate

The pomegranate (*Punica granatum*) was long thought to constitute not only a genus but a separate family. Quite recently, a second species, *P. protopunica*, was discovered on the island of Socotra, off the coast of north-east Africa (Cronquist 1981). According to Zohary (1973), *P. granatum* 'grows fairly abundantly in the jungles [sic!] of the Caspian coastline' as well as along rivers of this same area. Where the sacred pomegranate of the ancients was first domesticated is anybody's guess.

The thick, fleshy fruit of pomegranate, with its unusually deep tones of reddish purple, was highly prized for the edible pulp around its seeds, and the delightful fermented beverage made from it. The skin was also used for tanning the finest leathers in ancient Egypt. This plant was so highly thought of that a detail of its fruit shape was reproduced by the designers of King Solomon's crown, and later imitated in the crowns of many European monarchs. That hard, fleshy pulp made it possible to transport the fruits long distances without damage, a trait of economic value. It was carefully tended, cultivated, and selectively improved.

Throughout southern Europe and the Near East, both pomegranate (*Punica granatum*) (Box 8.5) and hackberry (*Celtis australis*) were considered as 'safe' trees to sleep under, since no evil spirits could abide there. It is not unlikely that the 14 cedar groves still surviving in Lebanon were protected as being sacred or at least as marking important burial spots.

Religious uses of plant parts, particularly those that give off good smells when burnt, were legion in ancient times. While 'sacred' trees were spared the woodsman's axe, many other species were systematically exploited for the 'services' they could provide. The word 'perfume' derives from the Latin *per fumum* ('in the smoke'), and rituals involving burnt fragrances and volatile oils of aromatic plants are still found in mountain villages of Anatolia, North Africa, Albania, and elsewhere in the Basin.

In ancient and medieval times, several species of resin-bearing rock-roses were so highly valued that they were managed as wild, or semi-wild sources of the raw material of incense and perfumes. As the leaves of these species exude their resin in daytime, especially in summer, shepherds in Crete, Corsica, and elsewhere are reported to have combed the beards of goats each evening in order to harvest the resin of rock-roses as a supplement to their meagre daily income. When collecting firewood, these people harvested other shrubs and left the rock-roses to grow.

Other services and products

The services and products that forests and woodlands offer to people have always been much more varied in the Mediterranean area than further north in Europe.

Rural peoples in all regions of the Basin have managed them for a wide range of purposes: livestock grazing, bee-keeping, wood cutting, hunting, and for gathering useful plants, fruits, and fungi. A fine balance was often achieved among woodlots, pastoral grasslands, shrublands, and open spaces reserved for cultivation. The resulting mosaic greatly contributed to the biological diversity of Mediterranean landscapes. The goals sought, but not always achieved, were diversity and sustainability.

Forested areas in all Mediterranean cultures were used for wood and charcoal. With an average wood production usually not exceeding 0.5–1.5 m^3 ha^{-1} yr^{-1}, annual productivity in most Mediterranean forests is low compared to that of lowland forests elsewhere in the temperate zones. Only in some particularly favourable locations does timber production reach 8 m^3 ha^{-1} yr^{-1}. To give a comparison, production values worldwide vary from 1.5–4.6 m^3 ha^{-1} yr^{-1} in forests where an oak species is dominant to over 10 m^3 ha^{-1} yr^{-1} in coniferous and beech wood temperate forests (Llédo *et al.* 1992).

More than anywhere else in Europe, Mediterranean woodlands have traditionally been used for many other services and products. These include food (game, fruits, mushrooms, honey), shelter, medicinal plants, cork, tanning agents, and resins. Some fruits are of commercial importance, such as the large edible seeds (pine-nuts) of the Italian stone pine (*Pinus pinea*) which, to this day, provides a larger return than the timber itself in some countries such as Tunisia. Similarly, the maritime or cluster pine (*Pinus pinaster*) has long been cultivated in the western Mediterranean for its copious production of resin, which is tapped after the fashion of rubber trees. For centuries, this was the main source of turpentine for western Europe, as well as providing the tar used for caulking the hulls of ships during Roman and subsequent eras. *Pinus peuce* takes its name from its presumed identity with the *peukê*, as the ancient Greeks called the tree from which they made resinous torches and signal 'lamps' to provide light at night.

In southern Europe, until the end of the eighteenth century, the downy and holm-oaks were preferentially cut for charcoal (Box 8.6), industry (glasswork, metallurgy), and domestic heating and cooking. Trees were traditionally exploited according to a 20-year forest clear-cut rotation scheme. Because the downy oak resprouts less well from stumps than evergreen oaks, huge areas formerly dominated by this deciduous tree were gradually transformed into more or less depauperate forests dominated by evergreen oaks and other stump-sprouting sclerophylls. Since no other woody matorral species can compete with oaks for calorific value, these 'lesser' species were generally neglected or eliminated by woodcutters except when local needs or market conditions provided an incentive. Examples include cade, boxwood, and mountain ash.

Collecting firewood was a time-consuming priority for villagers until the use of fossil fuel became generalized after the Second World War. Wood collecting from orchard prunings and woodlot coppices was accompanied by a number of other activities. For example, in the north-west quadrant the bark of young branches of holm-oak was widely exploited for the fabrication of tannin. A 20-year cycle of clear-cutting was found to be optimum for this purpose as well as for fuel-wood

Box 8.6 Charcoal production

Since the Iron Age, charcoal has been the main source of energy for humans in the Mediterranean area, as illustrated by the innumerable ancient charcoal production sites, up to 40 sites per hectare in some cases, still visible in many holm-oak woodlands. These sites are small circular plots, c. 5.5 m in diameter, sometimes sustained by a wall or embankment in hilly terrain. The soil is dark black to this day, and covered with tiny pieces of residual charcoal. Old iron ovens and the ruins of charcoal makers' huts are still visible in many areas. Charcoal production greatly increased in the Middle Ages when proto-industrial activities such as glassworks and iron metallurgy required charcoal, which has much higher calorific properties than wood. Charcoal production sites are thus a vivid testimony of the past and are often used as archaeological sites to reconstruct the history of forest and its inhabitants (Bonhôte and Vernet 1988). Vegetation throughout the Basin has been profoundly marked by charcoal manufacture practices, both in its floristic composition and structure as shown by anthracological* studies that compare extant vegetation with that which prevailed in earlier periods on the basis of dated charcoal remains combined with other techniques. Generally speaking, charcoal production over millennia resulted in the homogenization of woodlands over large areas by favouring evergreen oaks at the expense of other less resilient species. Cyclical clear-cutting of these holm- and *calliprinos* oaks over small areas of a few hectares every 20 years resulted in a perpetual rejuvenation of coppices. In fact, the young stems, which do not exceed 5–6 m in height at the time of cutting, resprouted vigorously from centuries-old underground root crowns. Although human action was severe and considered as leading to a progressive degradation of the former mixed evergreen/deciduous forests, charcoal production was nevertheless a sustainable forest-use practice that resulted in a stable ecosystem (Vernet 1973). Charcoal production in southern Europe sharply decreased in the nineteenth century and completely disappeared in some countries in the twentieth, except in Portugal. In parts of North Africa and the eastern Mediterranean, however, the practice still exists.

production. Thus, in many areas matorrals were managed for a double production of tannin and wood. Additionally, small branches were collected for kindling, and leaves may also have been used for a variety of purposes. While woodcutting was underway, mushrooms, berries, nuts, medicinal plants, resins, gums, oils, stimulants, dyes, and other forest 'by-products' were also assiduously sought and collected by villagers, hunters, and woodcutters. Most of these activities are now disappearing, as evoked in Fig. 8.9.

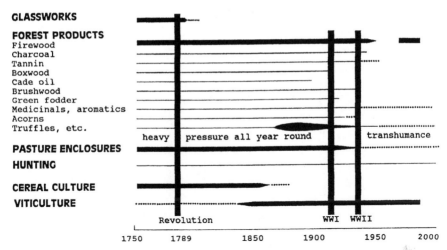

GLASSWORKS

FOREST PRODUCTS
Firewood
Charcoal
Tannin
Boxwood
Cade oil
Brushwood
Green fodder
Medicinals, aromatics
Acorns
Truffles, etc.

heavy | pressure all year round   transhumance

PASTURE ENCLOSURES

HUNTING

CEREAL CULTURE

VITICULTURE

Revolution   WWI WWII

1750   1789   1850   1900   1950   2000

Fig. 8.9 Evolution of the utilization of Mediterranean forests and matorrals in southern France over the past 250 years. Dotted lines indicate a gradual increase or decrease of utilization. Thickness of lines indicate relative importance of each activity—WWI and WWII: the two world wars. (Based partly on de Bonneval 1990 and Soulier 1993.)

Traditional landscape designs

Traditional Mediterranean life styles were based for centuries on self sufficiency at both family and village levels. This was largely imposed by the isolation of the many small river basins and other habitable sectors offered by a mountainous terrain in epochs when transportation systems were limited and slow. Over time, a large number of local land use systems, more or less clearly delineated in space, arose in the various Mediterranean hinterlands. They vary greatly according to regions and ethnic groups. Among them, the two best known are the *ager-saltus-sylva* and the *dehesa–montado* systems. They differ in the spatial organization of the three main activities, cultivating, grazing, and forest-product harvesting. These activities occur in separate areas in the first system and are combined in a single area in the second. These two systems have different consequences on biodiversity patterns as will be shown in Chapter 9 (see Fig. 9.3).

Ager-saltus-sylva

Perhaps the most influential of all ancient land use systems in the Mediterranean area was the triad called *ager-saltus-sylva* (field–pasture–woodland). Following the precepts of Theophrastus, published in 313 BC, this system sought to optimize microclimatic and edaphic variations in each farmlot or landscape. As the possibilities for overland transportation were limited most villagers were self sufficient, that is, their needs were all primarily fulfilled by their own growing, hunting, and breeding practices. The plot (or field) was the smallest section of rural space used by farmers. It was also the basis for cadastral registration and taxation (Lepart and Debussche 1992). For each of the different crops in use, specific landscape units were identified as being the best suited. Therefore, plot limits often coincided with geomorphologic limits. Rocky areas were often covered with forest to be used for wood, charcoal, etc., while stony plateaux were used primarily for extensive grazing of livestock. On thinner soils, at the foot of the hills, olive groves were recommended, with vineyards to be planted either below, on stony ground where olives fare less well, or else at higher elevations where grapes are more resistant than other crops to the frequent winter frosts. Cereal crops, by contrast, were cultivated on the deepest, most fertile soils, in the plains or in intermontane valleys, as well as a variety of fruit trees, textile plants (hemp, flax), and vegetables. The most energy-demanding crops, the ones requiring irrigation, fertilizers, and a large amount of manpower, were grown closest to the villages. Over the years, land management led to the installation of intricate networks of linear elements such as fence walls, sustaining walls, hedges, ditches, embankments, irrigation and drainage canals, and paths. Many Mediterranean landscapes, especially in Italy, are so indelibly marked and shaped by these artificial constructs, that parts of the Roman cadastre are still visible today from aerial photographs (Chouquer *et al.* 1987). Clearly, one goal of this management approach was to avoid the 'ups and downs' of *ager* or *saltus* expanding out of control at the expense of *sylva*. It also helped avoid the potential conflict between nomadic or semi-nomadic shepherds and farmers or foresters. That is, traditional common grazing lands permitted extensive livestock raising and the biannual transhumance of large herds of sheep and goats, in conjunction with the ancient rotation of cereal cropping and fallows that dates back at least to the days of Homer. In some regions, however, huge areas of forest or grazing lands belonged to a king, a lord, or an abbey, who allowed villagers to use them for grazing or collecting timber or firewood under carefully controlled conditions (Lepart and Debussche 1992).

Dehesas and montados

In some parts of the Basin, the *ager-saltus-sylva* triad is replaced by one or more 'agroforestry' or agro-sylvo-pastoral systems, wherein the three main rural activities of wood-gathering, livestock husbandry, and agriculture are pursued conjointly in a single space. Under this scheme, livestock grazes acorns or chestnuts and grass under open forest or woodland cover, while annual or perennial crops are sown between planted or protected fruit or forage trees where they take advantage of the shade provided in summer. Many other possibilities for mutual benefits among soil, crops, and animals are also generated by agro-sylvo-pastoral systems that mimic natural ecosystems more than *ager-saltus-sylva* which segregates each economic pursuit into separate landscape units. They appear to contribute in maintaining ecological equilibria as well as herbaceous plant diversity at ecosystem and landscape levels of integration.

Some examples of ancient agroforestry in the Mediterranean Basin are found in Sicily and the Calabrian Mountains of southern Italy, central Corsica, and the

Cévennes Mountains of southern France. In all three regions, both chestnut and mulberry tree-based systems were maintained over several centuries with a range of small livestock, subsistence crops, and that remarkable cash crop, the oriental silkworm (*Bombyx mori*), all providing income and sustenance for local populations. Either coppiced for wood production, or left to grow tall and produce annual crops of carbohydrate-rich food and fodder, the chestnut tree formed the 'backbone' of these mountain systems until a rust disease devastated most populations in southern Europe.

More spectacular still are the 'dehesa' and 'montado' systems that cover more than six million hectares in the plains of the southern Iberian peninsula (Joffre *et al.* 1988, see Fig. 8.10). These, and related systems in Sardinia, Greece, and elsewhere, combine extensive grazing of natural pastures with intermittent cereal cultivation (oats, barley, and wheat) in park-like woodlands of cork oak, holm-oak, and smaller numbers of deciduous oaks (*Q. faginea* and *Q. pyrenaica*). These savanna-like formations result both from selection and protection of superior, well-shaped trees occurring among natural stands, and, in some regions, from the intentional planting of acorns chosen from selected trees. Tree density is maintained at 20–40 ha^{-1}, and mature trees are pruned regularly to remove infested branches, broaden their canopy cover, and increase acorn production. Various wood products are also collected in the process, including timber for various uses, charcoal, tannin, and natural cork. A well-managed cork oak dehesa with 40–50 mature trees ha^{-1} produces, on average, 600–1000 kg of cork every 9 years (Joffre *et al.* 1988). In addition, cereal crops are produced, and a variety of grazing animals raised, including pigs, sheep, goats, cattle, and bulls. A special breed of semi-feral 'Iberic' pigs, which was in use by the end of the Middle Ages, lives primarily on acorns between October and February and achieves the remarkably high weight increase of 1 kg for every 9 kg of acorns eaten.

Joffre *et al.* (1988) report that herbaceous plant assemblages in Andalusian dehesas differ depending on whether they are under tree canopy or not. Under tree cover, there is greater moisture and more organic elements in the soil from leaf shedding and animal excretion. As a result, there is twice as much potassium, phosphorus, nitrogen, and carbon as in soils in the same field not under tree canopy. Consequently, a mosaic-like structure of herbaceous plant assemblages is promoted, typified by increased botanical diversity, longer growing seasons, and greater pasture productivity. This enhances animal productivity, which in turn probably favours long-term growth and vitality for the trees. Hence a mosaic-like structure in a two-tiered system is maintained.

Although the Spanish dehesas and Portuguese montados are of anthropogenic origin, they show remarkable stability, biodiversity, and sustained productivity as a result of their balanced two-tiered vegetation structure, heavy incorporation of animal husbandry, and botanically rich mosaic-like herbaceous plant layers (Joffre and Rambal 1993). These are examples of systems, developed over centuries, that have proved to be adapted to local constraints. These systems are being increasingly altered or abandoned today, however, because of many things. A decrease in agro-sylvo-pastoral activities due to rural depopulation is a

Fig. 8.10 A typical dehesa near Sevilla, southern Spain.

contributing factor, but increasing deforestation and semi-industrial clearing in order to extend mechanized cropping lands are also creating conflicts, largely driven by the impetus of global and European economic pressures. In many regions, dehesas and montados are being replaced by large stands of eucalyptus. Yet dehesa-like systems are increasingly being recognized as potentially well-adapted and economically viable multiple-use agro-ecosystems suitable for promoting sustainable development in many farming areas of the Mediterranean Basin.

A final example of an ancient agroforestry system well adapted to the hottest Mediterranean life zone is the argan 'forests' of the Sous region in south-western Morocco (Benchekroun and Buttoud 1989). Until the mid-twentieth century over one million hectares in this dry corner of the Basin were occupied by a centuries- or millennia-old agroforestry system based on open parklands featuring the endemic argan tree (*Argania spinosa*), as well as local varieties of olive and almond trees, and an endemic acacia tree (*A. gummifera*) whose large pods were used for animal fodder. What allowed the system to survive is the argan tree, whose hard seeds yield a culinary oil more highly prized locally than even olive oil. In this semi-arid region where no permanent water is available for livestock or irrigation, this dehesa-like system made life possible for local populations and their herds. In favourable years, when rainfall was adequate, the ground between trees was lightly ploughed and short-season cereal crops were sown. In wet and dry years alike, the argan and olive trees produced edible oils of high commercial value. Foliage of the argan trees and the acacias provided supplementary fodder to goats and other small livestock, which were present nearly year-round. Once again, a system 'mimicking' nature was created, incorporating a multi-tiered vegetation structure and firm integration of plants and animals, albeit in lower densities than in the Iberian systems. From over one million hectares originally, only about 600 000 ha are left today. Like the dehesas and montados of Andalusia and Portugal, the argan agro-forests are now in danger of collapse. This unique and ecologically well-adapted poly-cropping system built on endemic plants is gradually being replaced by intensive, irrigated farming methods. The trees themselves are falling to the axe and bulldozer one by one.

Summary

The long-lasting management of Mediterranean space, and the ongoing but fluctuating exploitation of all available resources, has not always resulted in a calamitous decrease in biodiversity. Exceptions are those few animal and plant species that directly competed with humans or else were too heavily harvested or killed at one period or another. Overall, it appears that human activities have in fact been beneficial for many components of biological diversity in the Basin. Gomez-Campo (1985), Pons and Quézel (1985), and Seligman and Perevolotsky (1994) have all argued that the highest species diversities in the Mediterranean area are found in frequently but moderately disturbed sites. The endless redesign of Mediterranean landscapes through dozens of varied land-use practices that did not involve the massive use of chemical fertilizers and pesticides were probably of great benefit to biological diversity in the Mediterranean area over several millennia. Thus, one major risk to Mediterranean biodiversity today derives from extensification of disturbance regimes in the north-west and north-east quadrants, and excessive intensification of land use systems in the southern quadrants related to growing human populations and prevailing rural poverty. In the next chapter we look at past and current trends in biological diversity for several groups of organisms.

9 Biodiversity ups and downs

In view of the alarming figures concerning the ongoing mass extinction of plants and animals everywhere in the world, one might expect a large decline of biological diversity in a region that has been for so long managed, modified, and degraded by humans as the Mediterranean Basin. It is unfortunately extremely difficult to assess the actual rate of diversity decline, if any, because there are almost no references in time against which we can compare the present situation. However, a large body of historical records, both palaeontological and archaeological, shows that human-induced decline of biological diversity started many thousands of years ago in most parts of the Mediterranean. We will start by telling the sad story of mammals on the larger islands.

Mass extinction: the Holocene holocaust of insular mammals

During the Quaternary, most of the larger Mediterranean islands were populated with odd assemblages of animals including tortoises, giant rodents, flightless owls, dwarf deers, hippos, and elephants. Many of these endemic animals arose from archaic Tertiary lineages and evolved striking adaptations such as gigantism or dwarfism. The absence of land connections with the nearby mainland since the Messinian Salinity Crisis indicates that most mammals probably reached the islands by swimming. Various palaeontological and archaeological data starting from the neolithic (7000 yr BP) show that these highly endemic and disharmonic* upper Pleistocene faunas of Mediterranean islands were ravaged by humans and more or less completely replaced by present species assemblages. Extinction of these mammals was long thought to have resulted from climatic changes of the upper Pleistocene. However, evidence is accumulating that if climate changes may have precipitated extinction in some cases, most extinction events were directly or indirectly caused by humans. A large series of archaeological sites discovered in several islands, for example the famous 'Eagle Cliff' in Cyprus, revealed that human settlements occurred there before 10 500 yr BP (Simmons 1988). This dating proves that invasion of Cyprus and presumably other islands by humans dates to soon after the end of the last glaciation, when human societies were based on hunting and gathering. Actually, the recent and unexpected discovery of human fossils in Sardinia, dating back to 20 000 yr BP, indicates that this island at least was inhabited by humans during the upper Pleistocene

Fig. 9.1 Dwarf elephants and hippos, which were widespread on some of the larger Mediterranean islands before the arrival of humans. Note the tiny size of the mammals relative to that of the land turtle and swans. (Reproduced with permission from Blondel 1995.)

(Sondaar *et al.* 1995). At the Eagle Cliff site, the remains of over 200 different animals have been discovered, including dwarf hippos and elephants (Fig. 9.1). All the bones were broken and bore marks of cooking, which indicates these animals were hunted for food. This deposit also yielded a large number of manufactured tools and necklaces, as well as piles of discarded shells of edible sea molluscs (more than 20 000) that could only have been transported there by humans. The site also included many bones of large birds such as great bustards, geese, doves, and ducks. Dating of this material by radiocarbon methods gives an approximate age of 10 465 yr BP.

The story of Corsican mammals

The story of mammals on Mediterranean islands can be briefly summarized using Corsica and Sardinia as a case study (Vigne 1992, Blondel and Vigne 1993). These two islands share the same history because they were connected during much of the Pleistocene, and the shallow channel separating them today is no more than 60 m deep. The Corsico-Sardinian complex is a 'continental microplate' that definitively parted from the European plate in the Oligocene–Miocene (30 Myr BP) (see Chapter 1). It is the largest island complex of the Mediterranean area, extending over more than 32 000 km², that is, some 7000 km² larger than Sicily, and nearly four times as large as Crete or Cyprus. It is also the nearest to the mainland, which explains some faunal specificities, such as the absence of the very unusual 'elephant–deer' assemblages (Sondaar 1977) that characterize other large islands such as Cyprus. Some endemic taxa survived until the end of the Pleistocene, including a mole (*Talpa*), a giant shrew (*Episoriculus*), a rabbit-like lagomorph (*Prolagus sardous*), two vole-like rodents (*Tyrrhenoglis*, *Rhagapodemus*), several ungulates (*Nesogoral*, *Sus nanus*), and a medium-sized monkey (*Macaca majori*). Later, some species invaded the island during a marine regression and produced local endemic species: a carnivorous dog (*Cynotherium sardous*), a vole (*Tyrrhenicola henseli*), a deer (*Megaloceros cazioti*), and even a mammoth (*Mammuthus lamarmorae*).

Some of these taxa became extinct during the climate changes of the upper Pleistocene, with the result that at the beginning of the Holocene, just before humans invaded the island, the insular mammal fauna included a very small number of endemic taxa. This highly endemic Pleistocene fauna is now entirely extinct since not a single non-flying mammal of the present fauna has been found in fossil deposits older than the lower Holocene. Hence, a complete turnover has been achieved in the mammalian island fauna since the Pleistocene.

As soon as humans invaded these islands, some 9000 years ago, hunting pressure rapidly led the *Megaloceros* deer, and probably the carnivorous dog, to extinction. Some mammals succeeded in resisting hunting and competition with new invaders for over eight millennia, but they finally went extinct as well under the combined effects of human predation, competition with new immigrants, and vegetation changes arising from pastoralism and agriculture. Some species, especially small rodents, persisted until the Roman invasion. Others, such as the rabbit-like *Prolagus*, survived much longer, up to the eighteenth century in the predator-free islet of Tavolara off the coast of Sardinia, as reported by the great Italian naturalist Cetti in 1777. Starting around 7000 yr BP, people progressively introduced all the extant mammal species on these islands. Some were intentionally introduced as game species (e.g. deer and fox), while others were brought as domestic animals (sheep, goat, pig, horse, donkey, cattle, and dog). Many small mammals (e.g. *Crocidura*, *Suncus*, *Apodemus*, *Mus*, *Rattus*, *Eliomys*, and *Glis*) were presumably introduced inadvertently as stowaways on ships.

Although the diversity of carnivores on Mediterranean islands is fairly high (Cheylan 1984), such large species as the wolf, the wild cat, and the lynx were never introduced, probably because they would have been perceived as potential competitors to humans. Introduced predators were small species, such as the fox (sixth millennium BC), the dog (fifth millennium BC), the weasel (*Mustela nivalis*), and the cat (*Felis catus*). Most of the 24 species of mammals introduced in Corsica are ecological generalists strongly associated with humans from both ecological and cultural points of view (Vigne 1990). Overall extinction–immigration processes resulted in a three- to five-fold increase of mammal species diversity but a dramatic loss of indigenous species and genetic diversity.

Some large species became an important part of the present fauna through feralization*, as exemplified by the mouflon (*Ovis ammon musimon*), which was absent from western Europe during the Pleistocene. The present Corsico-Sardinian mouflon is nothing other than a relictual neolithic domestic sheep that escaped human control before the late neolithic (Vigne 1988). A similar process of feralization probably occurred in the Corsican boar and the wild cat (*Felis silvestris libyca* var. *reyi*). In addition, competition involved habitat losses through forest clearing and direct competition from introduced ungulates such as sheep, goats, and cattle (Vigne 1983).

The story of mammals on Corsica may be extrapolated to other Mediterranean islands. In Cyprus and, to a lesser extent in Crete, mesolithic people progressively hunted large mammals to extinction. The situation is slightly different in the more remote Balearic Islands where human colonization was delayed until the neolithic, some 8000 yr BP. Domestication of an endemic antelope (*Myotragus balearicus*) apparently saved it from extinction until the Bronze or Iron ages (Alcover *et al.* 1981). Fossilized bones of humans have been found together with remains of this antelope, which suggests the species was protected, managed, or else definitely under domestication. The replacement of an impoverished, endemic, and disharmonic mammal fauna by a super-saturated* and monotonous fauna during the Holocene occurred on all the larger Mediterranean islands and even smaller islands. The only relictual Pleistocene endemic species on islands in the whole Mediterranean are three shrews, and a rodent (see Chapter 3). Even the Cretan goat (*Capra aegagrus cretica*), long thought to be endemic, is probably a feral population of the domestic goat (*Capra hircus*).

Effects of the upper Pleistocene–Holocene upheavals have been less severe in bird faunas. However, in the mid-Pleistocene (350 000–150 000 yr BP) avifaunas of Corsica included several species that are now extinct, such as a giant barn owl (*Tyto balearica*) and a dwarf great owl (*Bubo insularis*), both of which persisted into the Holocene. Almost every large Mediterranean island had its own endemic species of little owl: *Athene angelis* in Corsica, *A. cretensis* in Crete, *A. noctua* in the Balearic Islands, and one still undetermined species in Sicily (Mourer-Chauviré *et al.* 1997). They persisted until the upper Pleistocene but ultimately went extinct as well.

Box 9.1 The Corsican mouflon

The Corsican mouflon (*Ovis ammon musimon*) has long been considered native in Corsica and Sardinia. In fact, archaeo-zoological studies have revealed that it is descended from a sheep that was domesticated in the Middle East about 8500 yr BP. The domestic sheep in question (*O. aries*) was rapidly spread throughout the Mediterranean Basin and, in some places, especially on large islands such as Corsica, a few individuals escaped human control. These feral animals sometimes gave rise to genetically altered populations through a process whereby ancestral characters progressively revert in the absence of human selection. Much later (twentieth century), the Corsican mouflon was introduced for game purposes into many parts of Europe, North America, and even to remote places such as the Hawaii and Kerguelen Islands. This mouflon is a rather small subspecies weighing from 30 kg (females) to 40–45 kg (males). Adult males have spiral horns that vary greatly in shape and size. Surprisingly, some females have horns as well. The coat is brownish with large white patches near the feet, on the head and back, and on a large part of the belly. The animal lives in small herds of varying size. During the autumn rutting period, a rigid social hierarchy exists among males, resulting in fights that may be fierce; the dominant male mono-polizes most of the females. Around 400–600 individuals presently live in Corsica, in two disjunct areas of the main mountain range. They move up and down according to seasons grazing in a variety of habitats, from the top of the mountains (1600 m) in summer to lower altitudes in winter. They feed upon a large variety of annual plants, including the 'mouflon herb' (*Armeria multiceps*) and the leaves of various shrubs and small trees.

Fig. 9.2 The Corsican mouflon.

The decline of diversity in recent times

Except for large mammals, which paid a heavy tribute to human persecution in preceding centuries, few extinction events have occurred recently for most groups of Mediterranean plants and animals. For example, in reptiles and amphibians only two extinction events have been reported in the Mediterranean this century. The first is that of the crocodile (*Crocodilus niloticus*), which disappeared from Israel and southern Morocco at the beginning of this century and in the 1950s, respectively. The second is that of the endemic black-belly toad (*Discoglossus nigriventer*), which formerly occurred in a marshy area at the Israel–Syrian border. Among birds, very few species have disappeared from the Basin in historical times, perhaps no more than three large species in North Africa: the lappet-faced vulture (*Torgos tracheliotus*), the helmet guinea fowl (*Numida meleagris*), and the Arabian bustard (*Choriotis arabs*). The demoiselle crane (*Anthropoides virgo*), a bird of steppe regions and high valleys of central Asia, used to breed in some parts of Turkey and the Middle Atlas of North Africa up to the 1930s. Only a few pairs still breed occasionally in these regions. Another threatened species is the bald ibis (*Geronticus eremita*), which used to breed in many parts of the Basin but is restricted today to small scattered colonies.

The fact that few species became extinct does not mean, however, that decline in biological diversity did not occur. Actual loss of a species is one thing, impending declines in distribution and population sizes, and threat of extinction is another, for which we have far less reliable information. What is much more certain is that anthropogenic changes have had major consequences on the dynamics and distribution of biodiversity over the past several millennia everywhere in the Basin.

There is also evidence that marine biodiversity is presently facing changes that can be related to increasing sea water temperature (Francour *et al.* 1994). Among the most conspicuous decline of diversity in sea species is the regression of *Posidonia* beds. *Posidonia oceanica*, a sea grass endemic to the Mediterranean, covers large meadows of prime importance for fishes which find food, shelter, and spawning grounds there. Most of these meadows are declining in importance and vitality, especially in the northern part of the Basin, presumably because of pollution (Bianchi 1996).

Dummy extinctions among plants and invertebrates

Despite a large number of endemics that are narrowly distributed in a single or few localities, no more than 33 Mediterranean plants are considered to have become extinct this century, of which 31 are species and two are subspecies. This is a mere 0.11% of the recognized taxa for the region (Greuter 1994). Such a level of extinction is perhaps not significantly higher than during geological times. However, Leon *et al.* (1985) estimated that nearly 25% of the Mediterranean flora may be threatened in the decades to come. Mathez *et al.* (1985) reported that a great many plant species known to occur in Algeria have not been seen for at least

Box 9.2 The decline of the black truffle and other fungi

The black truffle, a highly perfumed hypogeal fungus known as the 'black diamond', used to occur abundantly in woodland clearings and other habitats on calcareous soils all around the Mediterranean Basin. In the second half of the nineteenth century there were huge areas of central and southern France where this fungus grew wild or was semi-cultivated. In the twentieth century, however, truffle production has declined dramatically: only 12 tons of black truffles were sold in all of France in 1991, as compared to the 1370 tons sold in 1889 (Callot and Jaillard 1996).

Given the remarkably high prices of fresh black truffles (550 Euros kg^{-1} in 1992–93), how can this decrease be explained? During the 'golden age' of black truffle production (1880–1915), thousands of farmers, villagers, and land owners all over France planted fields, field edges, and diseased vineyards with acorns from oak trees known to have borne truffles. By 1900, resistant grape rootstocks were imported from California in response to the phylloxera root rot epidemic that attacked vineyards in the early 1870s, and most people went back to growing vineyards, which are much more reliable, if less lucrative, than truffles. More importantly, huge areas where natural stands of truffles had been harvested annually became largely unsuitable for truffles in this century because of rural depopulation. One direct result of this has been the rapid dwindling in number and size of the vast flocks of sheep that used to keep the understorey of Mediterranean forests clear. In the absence of those herds and flocks—which also fertilized the forest, and browsed the lowest branches of trees—thick, coarse, shrubby vegetation quickly took over, thereby creating an environment unsuitable to truffles.

Jaenkie (1991) showed that populations of many ectomycorrhizal fungi (i.e., they have a symbiotic relationship with the roots of a plant), as well as the average size of individual edible mushrooms harvested, have decreased significantly throughout Europe over the past four decades. The decline is apparently due to indirect effects of air, soil, and water pollution, for example massive increases in nitrates, all of which have led to a decrease in forest tree vitality. Fungi may be an early warning signal of generalized decrease in biological diversity.

20 years. An alarming species decline is also under way in the Canary Islands and Madeira where 24% and 20%, respectively, of native plant species are considered to be threatened with extinction. Very little is known, however, about the decline in population sizes of plant populations. One example in mushrooms is that of the black truffle (*Tuber melanosporum*) (Box 9.2).

In fact, for many plants and invertebrates most presumed extinctions are surrounded with considerable uncertainty. Several supposedly extinct plant species may in fact survive in unnoticed localities, or else as part of dormant seed banks. This is especially true in the Mediterranean area where few species are monitored adequately in the field. In Greece, for example, no less than three species reported as being extinct have been rediscovered in the last few years (Greuter 1994). One example is the Peloponnesian pheasant's-eye (*Adonis cyllenea*), recently rediscovered 130 years after its first discovery on Mount Killini in 1854. Each year botanists find 'new' species for a given region or else species that had not been noticed for several decades and were thought to be extinct. The same is true for many groups of invertebrates and micro-organisms. From a 5-year intensive sampling of ants in a small area of 400 m^2, Espadaler and Lopez-Soria (1991) found as many as 40 species, an exceptionally high figure for a Mediterranean habitat. Many of them are considered by myrmecologists as 'interesting' and rare species that are unusual in published faunistic accounts. Actually they are probably present in many habitats but remain unnoticed because a much higher sampling effort than is achieved in most faunistic surveys would be required to detect them. These authors argue that 'rarity' in Mediterranean ant species may be explained by insufficient sampling of all existing microhabitats. In other cases, however, the situation is unambiguous. About 57 million orchid tubers, in 38 different species, are harvested each year in Turkey alone (Sezik 1989) to prepare 'salep', a milk-based drink of great popularity throughout the eastern Mediterranean. In this case, there is no doubt whatsoever that many wild species are threatened with local extinction.

The high cost to mammals

Much more is known about the conservation status of higher groups of vertebrates than of plants and invertebrates. The combination of habitat changes and direct persecution in recent times has doomed many large species of mammals to extinction, so that the present-day mammal fauna of the Basin is but a weak reflection of what it was in upper Pleistocene times (see Chapter 2, p. 42). Many Greek and Roman artworks bear witness to the incredible richness of mammal faunas from which animals were captured in Spain, the Balkans, and North Africa for Roman games and spectacles. Some species disappeared long ago, for example the elephant, which was still abundant in North Africa in Roman times. Hannibal used it to help transport his armies through the Pyrenees in 218 AD and then the Alps, where most of the 37 individuals pressed into service died from starvation and cold. More recently, several large species such as the ass (*Equus asinus*), the gazelles (*Gazella rufina*, *Oryx dammah*, *Alsephalus busephalus*), and the lion (*Panthera leo*), have been eradicated from the Mediterranean fauna. In many areas of North Africa, local and visiting European hunters have also doomed to local extinction large mammals such as porcupine (*Hystrix cristata*), Cuvier's gazelle (*Gazella gazella*), and the mouflon (*Ammotragus lervia*).

Others, such as the brown bear (*Ursus arctos*), are scattered in small disjunct populations in Spain, the Pyrenees, Italy, Greece, and Bulgaria. Several of them, including the panther (*Panthera pardus*) and the monk seal (*Monachus monachus*), are now on the verge of extinction.

Although this book deals mostly with terrestrial diversity, we will take the monk seal as a case study of a declining mammal because its status is well known and much effort is currently devoted to saving it from extinction. This seal is a large species, of *c.* 3 m length and weighing 240 kg (males) to 300 kg (females). Its original range extended throughout the whole Mediterranean area, and the Black Sea, the Atlantic coast of Mauritania, and Macaronesia. The monk seal was intensively hunted in classical Greece after huge herds were described by Homer. It was even still fairly common in the Mediterranean in the mid-nineteenth century, until a burst of seal-hunting caused all populations to dwindle to mere smatters. Ongoing fragmentation of populations can mostly be attributed to persecution by fishermen who consider the monk seal as a competitor for fish and damaging for their fishing gear.

The monk seal is now listed as one of the world's six most threatened mammals by the World Conservation Union (IUCN), since no more than 600 individuals still survive, or survived in the recent past, in very small disjunct populations in Greece (80–400 individuals), along the Turkish Mediterranean coast (50–100), along the north coast of Africa (50–100), in a few sites in the Black Sea and Madeira, and in the north Atlantic coastal regions of Mauritania (Panou *et al.* 1993). In the summer of 1997, a severe disease, presumably caused by the combination of saxitoxins produced by a dinoflagellate bloom and/or a morbillivirus, resulted in mass mortality of seals in the colony of coastal Mauritania. It has been estimated that only around 90 individuals from the former 300 seals have survived. This means that nearly 25% of the extant world population of this species at that time disappeared within about two months (Reijnders 1997).

The Mediterranean monk seal, which probably inspired the Greek legends of mermaids, depends on coastal caves for resting and pupating. Many variables seem to determine cave suitability, such as size and shape, the presence or absence of an inner sand beach, relative protection from wave action and a safe distance from people when they are trying to breed and raise their young. Panou *et al.* (1993) described one frequently used cave that was 100 m^2 in size, with a broad entrance 7 m in diameter, a low ceiling, a suitable beach, and good protection against storms and people. Availability of suitable caves of this type, sufficiently remote from humans for the seals to feel at ease, is certainly a limiting factor to keeping viable populations alive today.

In spite of much effort to protect the last surviving populations in the Northern Sporades, Greece, and in the Ionian Sea, the chances of saving the Mediterranean monk seal seem very slim indeed. Simulation models show that, as is often the case in large long-lived animals, the most critical factor for population dynamics is assuring adult survival. More than half of reported adult and juvenile mortality is caused by fishermen and hunters deliberately killing animals. An additional one-quarter die accidentally in fishing nets. Normal birth rates are insufficient to counterbalance such heavy losses. The only hope for conservation of this species lies in a drastic reduction in human-related mortality.

Population decline in birds

But the main concern is a general trend of sharp decline in population sizes among nearly 25% of European bird species (Tucker and Heath 1994), especially large raptors and long-distance migrants (Berthold *et al*. 1998). Even if only a few species of birds have become extinct in recent times, many formerly widespread species are now threatened and confined to small, scattered populations. For birds, as well as for many insects and reptiles, the main consequences of human-induced changes in Mediterranean habitats have not been so much a decrease in overall species richness at a regional scale as a tremendous advantage for species adapted to drylands and shrublands, as opposed to forest dwelling species. As birds are known to be good indicators of environmental conditions, the picture they provide of the state of the environment for wildlife presumably reflects what is happening for many other life forms as well. Although there is a general agreement that habitat loss is prominent among threats to birds and other animals, few studies have demonstrated the precise mechanisms responsible for population decline. Pollution is certainly a major cause of decline and the general effects of pesticides on birds, especially raptors and aquatic predators, are well known. Fortunately, the most harmful pesticides are now banned in Europe so that many species, for example the peregrine falcon, are steadily recovering in the Basin as elsewhere in Europe. But pesticides are still a threat for many migrant species that overwinter in tropical Africa because of their widespread use in many countries south of the Sahara desert.

This may be the direct cause of a sharp population decline of the lesser kestrel throughout the Mediterranean. This species declined in Spain from *c*. 100 000 breeding pairs in the 1970s, to below 5000 in 1988, coincident with a period of heavy pesticide use both in Spain and Africa (Gonzalez *et al*. 1990). For most migrant species, pesticide use in Africa to fight against locusts might affect birds through reduction in prey availability rather than through acute or sub-lethal poisoning. Other causes of decline of bird populations in the Mediterranean might be climatic changes, especially drought that affected Sahelian wintering grounds for several years, and through reduction of bodies of surface water as a result of drainage and irrigation works. Such causes have been suggested to explain population decline in several species, for example the white stork (Kanyamibwa *et al*. 1990) and the purple heron (Bibby 1992). As reported by the European Union for Bird Ringing (EURING), changes in the survival rates of several species have been shown to be strongly linked with the severity of drought in the Sahel region of western Africa. In years following severe African droughts fewer birds are counted in European breeding areas, and survival rates measured by ringing are much lower than in years of normal rainfall.

The Mediterranean area is visited twice a year by billions of migrant birds that shift seasonally between the Palaearctic region and their Mediterranean and

Box 9.3 Shooting and trapping of migratory birds

The reputation of Mediterranean peoples as indefatigable hunters and poachers is not unjustified. Recent work published by the British Trust for Ornithology, based on the analysis of recoveries of ringed birds, strongly suggests that a decline in survival rates of many migratory birds throughout Europe may be directly related to excessive hunting and trapping pressures occurring in Mediterranean countries (Crick and Jones 1992). Such pressure is particularly high on islands such as Malta, Cyprus, and most of the Aegean islands. Magnin (1987) estimated that approximately two million birds are killed annually by illegal mist-netting and liming in Cyprus alone. In Malta, the figure is estimated at three million, including well over 5000 birds of prey. Due to intensive bird shooting in recent decades, Malta lost all its breeding birds of prey, that is, peregrine falcon, kestrel, and barn owl. The Balkans and coastal countries of the Middle East (except Israel) remain hot regions for the hunting of migratory birds, particularly Cyprus and Lebanon. Areas of intensive hunting closely correspond with important migration routes, especially 'bottlenecks' where geographical features such as narrow straits and mountain ranges cause birds to be channelled through small areas, for example on both sides of Gibraltar, both ends of the Pyrenees, and eastern flyways such as the Bosphorus straits and coastal areas of Lebanon. From a detailed analysis using recovery data from 19 European bird ringing schemes for 20 migratory species (5 raptors and 15 passerines), McCullogh et al. (1992) showed that the majority of the populations of these species are subject to considerable hunting pressure during migration and wintertime in the Mediterranean. However, from hunting logs it appears that from 1980 onwards there has been a decline in hunting of most species except for legitimate quarry species. It may be, however, that the decline is only apparent as a result of a deliberate under-reporting of birds taken by illegal hunters faced with the growing public opposition to hunting and new regulatory policies of the European Union.

Other threats to raptors include illegal traffic for falconry and for exhibition centres, leading to a sharp increase in prices. The cost of a young peregrine falcon may reach 4500 Euros and that of an imperial eagle or lammergeyer 10 500 Euros (Kurtz and Luquet 1996).

tropical winter quarters (see Chapter 6, p. 162). This traditionally represents a manna for Mediterranean peoples who have developed a panoply of systems for catching them. All species, irrespective of their size and protection status, are caught, cooked, and eaten including the tiny kinglets! Woldhek (1979) pointed out that indiscriminate shooting and trapping of protected species occurs on a wide scale in all Mediterranean countries (see Box 9.3). Recent estimates suggest that up to 1000 million birds are killed annually in the Basin (Magnin 1991).

It has often been argued, and even demonstrated for some game species, that mortality due to hunting pressures has no significant effects on population densities. This is because increased mortality due to hunting is compensated for by a decrease in natural mortality rates as a result of density-dependent processes of population regulation. However, critical parameters of population dynamics are too poorly known for most species to allow an accurate estimation of the rate of annual harvest of populations that would be compatible with the long term health of migratory birds.

One well-documented example of the decline of bird diversity has been provided by Brosset (1990), who worked for several years in eastern Morocco in the 1950s, and then returned to the region in 1989. The survey of the most important habitats revealed a tremendous decline of large raptors and the bald ibis (*Geronticus eremita*) 35 years after the first survey. The two most badly degraded habitats in terms of bird species diversity were the Alfa steppe and the large cliffs that characterize the huge mountain ranges of *c.* 15 000 km^2 that stretch between Tlemcen and Debdou, near the Algerian–Moroccan border.

The Alfa steppe in this region was characterized by large isolated betoum trees (*Pistacia atlantica*) that formerly served as nesting sites for a large series of birds including brown crow (*Corvus ruficollis*), lanner (*Falco biarmicus*), kestrel (*Falco tinnunculus*), hobby (*Falco subbuteo*), black kite (*Milvus migrans*), long-legged buzzard (*Buteo rufinus*), short-toed eagle (*Circaetus gallicus*) and, more rarely, African eagle owl (*Bubo ascalaphus*). However, most of the betoum trees have been cut down in recent decades and all the above-mentioned birds have disappeared from the region.

The cliffs in this region used to be one of the most famous habitats in the western Palaearctic for large raptors and the bald ibis. By 1989, most large bird species had disappeared or became rare in eastern Moroccan cliffs (Table 9.1). Surveys of the status of large raptors at broader scales of space suggest that their decline is generalized throughout eastern Morocco and indeed in most parts of North Africa as well. In contrast, none of the small passerines that are linked to cliffs and rocky habitats, for example Moussier's redstart (*Phoenicurus moussieri*), black wheatear (*Oenanthe leucura*), trumpeter finch (*Rhodopechys githaginea*), crag martin (*Hirundo rupestris*), and rock bunting (*Emberiza cia*), shows any sign of decline. This is a clear indication that the main cause of decline among large raptors has not been climate or habitat change, but rather direct persecution by humans.

Freshwater fish

Although only four species of freshwater fish have gone extinct in the Mediterranean area in recent times, more than 75% of the indigenous fish species are considered threatened by the World Conservation Union (Crivelli 1996). The main threats to them are habitat destruction through draining,

Table 9.1 Decline of raptors, corvids, and the bald ibis in the Zekkaras cliffs and Djebel Mhasseur of eastern Morocco, between 1953 and 1989

Species	Number of pairs in			
	1953–59	1966	1975	1989
Golden eagle	1	0	0	0
Booted eagle	1	1	0	0
Bonelli's eagle	1	1	0	0
Black kite	3	2	1	0
Long-legged buzzard	1	1	0	0
Peregrine falcon	1	1	1	0
Lanner	1	1	1	1
Kestrel	4	2	2	0
Lesser kestrel	5	5	13	11
Egyptian vulture	2	0	0	0
African eagle owl	1	0	0	0
Raven	2	1	1	0
Jackdaw	80	0	0	0
Alpine chough	85	5	10	6
Bald ibis	20	10	0	0
Total	208	30	29	18

Source: after Brosset 1990

pollution, and the introduction of exotic species. For example, in the Mikri Prespa and the Megali Prespa lakes, at the border between Greece and Albania, 8 of the 20 fish species present have been introduced. Two of these were introduced from North America (*Lepomis gibbosus* and *Onchorhynchus mykiss*), five from Asia and one from central Europe, the sheatfish (*Silurus glanis*). All these species are potentially a threat for native species, which include two endemics, a barbel, and a trout. Based on experience elsewhere, these introduced species will compete with native species for food, and spawning grounds, or else will prey directly upon them.

Invertebrates

Almost nothing is known about the distribution and abundance status of most invertebrate species in the Mediterranean Basin due to a dearth of long-term monitoring schemes that would allow an investigation into the magnitude of changes. However, there is a growing body of evidence for an ongoing decline in several groups (Collins and Thomas 1991; Pullin 1995) although we lack data on its extent. A dramatic decline during the past few decades is obvious for large conspicuous insects such as butterflies, large bees, dragonflies, and many groups of Scarabeids, Cerambycids, and large moths, including the flamboyant peacockof-the-night, *Saturnia pyri*. Reduction of species diversity and population abundance in most groups of large insects reflects a parallel decline of large insectivorous birds such as roller, little owl, and all species of shrikes. Several species of maybugs, including the beautiful *Polyphylla fullo*, have also become extremely scarce in recent years. On the other hand, there does not seem to be any decline in the cicadas, of which seven species occur in southern France alone (e.g. *Tibicen plebejus*, *Cicada orni*, and *Tettigetta pygmaea*). Long-term monitoring programmes are badly needed to evaluate the extent of the decline in invertebrates.

Earthworms, whose importance to ecosystem functioning was discussed in Chapter 7, is one example of declining invertebrates. Over the past decades a marked impoverishment of earthworm communities has been documented, especially in forest or formerly forested ecosystems throughout Europe. This has been attributed to leaching and nutrient export by logging, fire, acid rains, heavy metal pollution, and conifer plantations leading to litter and top soil removal (Granval and Muys 1992). The combination of soil acidification and earthworm decline has resulted in a decrease of biological activity and compaction of soils, reduced formation of humus and natural regeneration of trees, fragilization of trees due to superficial rooting, and accelerated erosion processes (Hildebrand 1987). Current levels of heavy metal soil contamination may also have important consequences for major physical and biogeochemical ecosystem processes since they adversely affect earthworms. Given their ecological function in ecosystems, especially in respect to physical properties of soils, the decline of earthworm communities contributes to an increase in soil erosion and in the occurrence of severe flooding that periodically plagues various Mediterranean regions. This is especially true in agricultural areas, such as vineyards, where copper is widely used as a fungicide, or in mining areas where various heavy metals can be distributed in water courses and ground water reservoirs. In this context, the findings of Abdul Riga and Bouché (1995) are alarming: these authors have discovered a regional eradication of the genus *Sclerotheca* from many areas of southern France. These impressively large species (> 25 cm long) are active ground feeders and play an important role in water filtration and decomposition of organic matter. They are much more sensitive to heavy metal pollution (Cu, Zn, Pb, Mn, Cd, and Fe) than any other co-occurring group of earthworms. The entire genus has been exterminated from many polluted areas of southern France.

On a more positive note, earthworm reintroduction experiments are increasingly undertaken, and some of these have demonstrated an increased primary productivity after earthworm introduction in depauperate soils, and an increase of more than 50% in forest productivity after enrichment of soils with fertilizers in the presence of anecic earthworms, that is litter-feeding burrower earthworms that are sensitive to acidity (see p. 194) (Toutain *et al.* 1988). Moreover, liming and/or fertilizing degraded forest soils proved to be more efficient in the presence of *endogeic* (deep dwelling soil species) and *anecic* earthworms (Granval and Muys 1992). In most degraded forest soils, earthworm reintroduction appears to be the best way to promote bioturbation and the regeneration of soil activity.

Subtle evolutionary and ecological changes

Subtle changes in biodiversity may result from hybridization between closely related species that secondarily come in contact as a result of human-induced changes in the environment. For example, Mediterranean orchids have a wide range of isolation barriers that lower the risks of interspecific recombination under normal circumstances. But in many regions of the Basin the range of isolating barriers may be weak because of millennia of human history that have altered landscape structure and created innumerable disturbed sites conducive to hybridization (Kullenberg and Bergström 1976). Infra- and inter-generic hybrids are common within the orchid genera *Ophrys*, *Orchis*, *Dactylorhiza*, and *Serapias*. For example, at least 100 hybrids involving the 50 species of *Ophrys* have been documented and no less than 15 hybrids between *Serapias* and *Orchis* have been recorded so far (Baumann and Kuenkele 1982). However, the behaviour of the dominant insect pollinators helps prevent hybridization. For example isolation of the *Ophrys fusca*/*O. lutea* complex from other species of *Ophrys* may result from the fact that pollinia are deposited on the abdomen of the pollinating insect in the former whereas they are carried on the insect's head in the latter (Dafni and Bernhardt 1990).

Other threats to biodiversity may arise from new species interactions resulting from introductions of alien species. We will discuss this point in the section on biological invasions but one example worth mentioning here is that of the accidental introduction of a lizard onto an island. When it was recently introduced on the island of Minorca, the Sicilian lizard *Podarcis sicula* came in competition with the local endemic species, *Podarcis lilfordi*, which soon disappeared from the island.

Loss of cultivars and human-selected species

An additional aspect of biodiversity decline to consider is that of the 'regression' of species of economic value that are now neglected after centuries of selection and attention. In Greco-Roman times, Barbary thuja (*Tetraclinis articulata*) wood was so highly prized that, by the first century AD, Pliny reported that Cicero had paid a sum equivalent to 275 000 Euros for a table made from this wood, and that many wealthy Roman nobles competed to have their own thuja coffee table 'trophy', for its beautiful and unusual 'burl' wood produced by the fire-resistant ligno-tubers and root crowns of this tree (Amigues 1991). As a result of this fad, large stands of *Tetraclinis* in coastal Libyan foothills and various oases of central Egypt (sic!) had already been wiped out by Pliny's day. The surviving populations in Tunisia and Morocco, and the extreme south of Spain, are but tiny, scattered fragments of formerly extensive mixed forests in those countries. Similarly, the cork oak (*Quercus suber*) is now entirely restricted to the western Mediterranean, but it formerly grew in many parts of Greece until exploited out of existence within historic times. For both these trees genetic diversity has no doubt been reduced to a mere fraction of its former extent.

Not so long ago a huge range of Mediterranean fruit, nut, herb, and also fodder plant varieties were propagated, cultivated, and continually 'improved' according to the needs and tastes of growers and consumers. Over the centuries, hundreds of varieties of olive, almond, wheat, and grape, etc. were passed down and preserved in gardens and orchards. In all too many cases, the horticultural and agricultural selections and discoveries of past generations have been lost in very recent decades. At the turn of this century, 382 named cultivars of almond were in use on the island of Mallorca alone (Socias y Company 1990). How many can be found there today? Five? Ten? Throughout the Mediterranean area, the great majority of 'minor' fruits, nuts, vegetables, and other plant varieties selected in the past are extinct and lost forever. The loss of these ancient varieties is a problem of great concern and much effort is devoted today to restoring and preserving them from extinction (see Chapter 10).

Pros and cons of human impact

Not all groups of animals and plants of the Basin are in decline, however, and we must be careful when formulating a global assessment of the status of Mediterranean biodiversity. There are many examples of spectacular population recoveries and sudden range expansions of formerly rare or highly localized species.

Long-term effects of land use practices

Over the past dozen millennia biological diversity in the Mediterranean Basin has depended on the many interactions between humans and living systems. Human activities have had positive as well as negative effects on biodiversity. For example, it is likely that the two types of traditional land use systems described in the previous chapter, *ager-saltus-sylva* and 'dehesa–montado' both had positive, albeit different effects on various components of diversity. Clearing forest on a large proportion of an area to plant pastures and crops allowed many species of shrubby and grassland habitats to colonize the area, and thus increase biological diversity at the scale of landscapes, the so-called gamma diversity, in a most dramatic way. One may imagine that the other two components of diversity (alpha and beta) differed greatly between the two systems as a response to a different distribution of habitat patches—'coarse-grained' in the *ager-saltus-sylva* triad, and 'fine-grained' in the 'dehesa–montado' system. The former was presumably characterized by moderate alpha diversity and a high beta diversity whereas dehesa–montado was presumably characterized by still higher alpha diversity but much lower beta diversity (Fig. 9.3). The collapse or near collapse of these systems undoubtedly had or will have consequences on biodiversity at the scale of landscapes and entire regions. Thus, the most dramatic consequences, albeit poorly known, of long-term human design of landscapes have been changes in the distribution and dynamics of biological diversity.

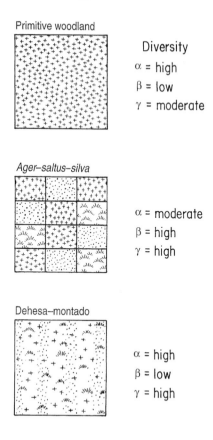

Primitive woodland

Diversity

α = high

β = low

γ = moderate

Ager–saltus–silva

α = moderate

β = high

γ = high

Dehesa–montado

α = high

β = low

γ = high

Fig. 9.3 Hypothetical changes in values of the three components of diversity (see Fig. 4.12 p. 107) as a result of land management of former holm-oak woodlands in the *ager-saltus-sylva* and 'dehesa–montado' systems.

Even if forest recovery as a result of recent abandonment of traditional land use practices in most Euro-Mediterranean countries is beneficial for forest-dwelling species, it results in habitat reduction for species of open habitats. In less than thirty years we have witnessed no less than six species of birds recolonizing newly forested areas of the Mt. Ventoux region, including the buzzard, honey buzzard, black woodpecker, marsh tit, and song thrush (see Chapter 10, p. 279). However, population recovery by these species has been counterbalanced by a retreat of species of shrubby and open habitats such as warblers, pipits, and buntings.

Thus, habitat changes have not the same effects on all groups. Although intensive use of herbicides and insecticides has caused a reduction in bee species diversity in many parts of the Basin (Dafni and O'Toole 1994), forest clearing, agriculture, and even the establishment of sand pits, roads, and paths have had beneficial effects for many species confined to open habitats. Bees find a new range of nest sites in these new habitats as well as in dwarf shrub communities such as low matorrals, bath'a, and phrygana (Table 9.2).

As a general rule, the periodic changes in vegetation cover that result, over the long term, from changes in land use patterns by people, have produced alterations in the distribution patterns and populations sizes of species, depending on whether they are 'at home' in closed woodlands or in open habitats.

Spectacular increases in biodiversity

Increases in biodiversity have sometimes been observed in habitats that are considered as highly threatened everywhere in the Mediterranean, such as wetlands. For example, Table 9.3 looks at population sizes of the most common bird species in the Camargue since 1956. The overall numbers of breeding pairs for all species combined increased between the 1950s and the 1990s, mostly as a result of sharp increases in some species. Examples of increasing species are the

Table 9.2 Effects of forest clearing and management of Mediterranean ecosystems on the diversity and abundance of solitary bees.

Habitat type	Structure of the vegetation	Habitat characteristics	Diversity of bees
Oak woodland	Dense vegetation, many small low-reward flowers, few annuals and geophytes	Scarcity of nest sites, low diversity of plants as well as pollinators	Few solitary bees
Matorral	Sparse vegetation, high insulation and degree of patchiness	High availability of nest sites, high diversity of annuals and geophytes	High diversity and dominance of solitary bees
Traditional agricultural habitats	Farmland	Frequently disturbed habitats, high availability of nest sites, spread of ruderals and segetals	Proliferation of solitary bees
Industrial agriculture	Crop fields, orchards	Use of biocides, apiculture, reduced floral diversity	Few solitary bees, competitive exclusion of solitary bees as pollinators

Source: after Dafni and O'Toole 1994

slender-billed gull (*Larus genei*), the Mediterranean gull (*Larus melanocephalus*), which did not occur at all in the Camargue as breeding birds in the 1950s, and the sandwich tern (*Sterna sandvicensis*). There is virtually no significant decline trend for any species, except perhaps the avocet (*Recurvirostra avosetta*), where a huge interannual variation of population sizes renders the observed variation of numbers ambiguous or insignificant.

The slender-billed gull, the Mediterranean gull, and the sandwich tern have also increased elsewhere in the Mediterranean, for example in the Pô delta, Italy, the Ebro delta, Spain, the Evros delta, Greece, and in the salt pans of Sfax, Tunisia. Other gull species that were considered threatened some decades ago have undergone a spectacular demographic increase and have recently colonized many new sites. This is the case for Audouin's gull (*Larus audouinii*), which forms large colonies in Corsica, the Ebro delta, the small Chaffarina Islands off the Mediterranean coast of Morocco (more than 3000 breeding pairs), and some other areas. Absent from the Ebro delta in the 1970s, it began colonizing the area in the 1980s with *c.* 200 breeding pairs in 1987 and 11 000 in 1997! The precise demographic mechanisms responsible for the spectacular increase of these species are still obscure.

Table 9.3 Numbers of pairs of the main bird species breeding in the lagoons of the Camargue in 1956, 1976, and 1991.

Species	Year		
	1956	**1976**	**1991**
Black-headed gull	2900	8000	5200
Yellow-legged gull	300	2700	4000
Slender-billed gull	0	12	200
Mediterranean gull	0	7	120
Common tern	3000	1500	1100
Sandwich tern	15	1000	1430
Little tern	400	450	370
Gull-billed tern	250	200	340
Avocet	750	850	500
Total	7615	14719	13260

Source: after Sadoul 1996

The rates of increase in the Camargue of the slender-billed gull (22% per year between 1967 and 1994) and the Mediterranean gull (20% per year between 1969 and 1994) are too high to result only from local demographic processes because local recruitment of offspring is not enough to account for this large increase. Modelling population dynamics of these species, Sadoul (1996) showed that such spectacular population increases would imply an average yearly production of viable offspring of 1.42 and 1.60 young per breeding pair for the slender-billed gull and Mediterranean gull, respectively. Yet, the observed offspring production of these species in the Camargue is usually less than 0.5 offspring per nest! One explanation for the discrepancy between reproduction levels and population increase of these two species is immigration from eastern populations, possibly those of the badly-polluted Black Sea where populations have sharply declined between 1985 and 1994 from *c.* 30 000 to 6000 and from 300 000 to 50 000 breeding pairs for the slender-billed gull and Mediterranean gull, respectively (Sadoul 1996). It is presumed that some of these birds migrated to the Mediterranean area in search of better habitats. The flamingo also showed population increases and colonized new sites in Spain and Sardinia during this same period, probably from the core colony of the Camargue, which has been protected and highly productive for the last thirty years.

Another example of spectacular increase in local diversity is that of the Moulouya River mouth in northern Morocco. Some decades ago this was a small marshland of 50 ha at most with no more than four breeding species of birds. The size of this marshland has increased at least tenfold over the past few decades thanks to the building of large reservoirs in the upper course of the river and the expansion of irrigated farmland that raised the water-table. Today this area is one of the best breeding sites in North Africa for many rare bird species such as purple gallinule (*Porphyrio porphyrio*), marbled teal (*Anas angustirostris*), pratincole (*Glareola pratincola*), several species of herons, and many marsh warblers such as the rare Savi's warbler (*Locustella luscinioides*). The growing concentrations of birds has of course attracted raptors, for instance the black-winged kite (*Elanus caeruleus*). This is also an excellent stopover area for thousands of migratory waders and ducks (Brosset 1990). The large improvement of this marshland for wildlife is an indirect, unintentional effect of human action, but it demonstrates that biodiversity can sometimes recover quickly once favourable conditions reappear. For example, as a result of more stringent regulation policies and active action plans launched by BirdLife International for the protection of raptors, many species are now in a phase of population recovery, especially in Euro-Mediterranean countries where there is a growing concern for nature conservation. Species that are increasing in numbers include the lammergeier (*Gypaetus barbatus*), griffon vulture (*Gyps fulvus*), golden eagle (*Aquila chrysaëtos*), Spanish imperial eagle (*A. adalberti*), Eleonora's falcon (*Falco eleonorae*), and several others (Muntaner and Mayol 1996).

Several large mammals, notably ungulates (deers, wild boar, chamois, ibex), are also increasing in population sizes and range in many parts of the Basin, notably as a result of increasingly stringent hunting regulations and conservation policies. For example, the population of the mountain gazelle (*Gazella gazella*) in Israel, which had sharply decreased by the 1950s as a result of severe poaching, built up again from a few hundred to about 10 000 individuals in the 1980s. In some districts, the density may now reach 35 animals km^{-2}. Incidentally, the grazing habits of this gazelle has resulted in a recovery of plant species diversity, including species that are heavily grazed (Kaplan and Gutman 1989) (see Chapter 10, p. 280). Other mammal species have spontaneously returned thanks to generalized forest recovery and agricultural land abandonment. One spectacular example in France is that of the wolf, which colonized the National Park of Mercantour, in the southern Alps, from adjacent Italian mountains where its populations are large and in good health. Eradicated from the French Mediterranean region in the nineteenth century, it was first observed in this park in 1992. Five years later, there were at least 19 individuals in four groups. Traces of wandering individuals have been observed at least 300 km further north. Hopefully other Mediterranean regions further west will be colonized in the near future. This is definitely a happy event even if predation pressures on sheep raise problems for livestock managers and legislators.

Biological invasions

We consider a biological invader as any species of plant, animal, or micro-organism that colonizes new territories as a result of intentional or unintentional introduction by humans. Basically the term does not include the natural colonizations of areas that occurred on long scales of time as a result of large geographical upheavals of the Miocene–Pliocene and climatic fluctuations during the Quaternary. Nor does it include processes of range shifts of native species even if these shifts are more or less influenced by human activities. The introduction of exotic species into natural ecosystems may cause considerable disruption to native ecosystems (Drake *et al.* 1989) so that it is biologically unsound to consider introduced species as adding to the biodiversity of a region.

How important are invading species in Mediterranean habitats, and to what extent do they become harmful for native species assemblages and ecosystem functioning? The geographical location of the Basin and its long tradition of marine and terrestrial trade with all parts of the world make it particularly vulnerable to biological invasions. Humans are by far the most effective dispersal agents that ever existed on Earth. Since the earliest times they have intentionally introduced plants and animals over great distances and created new synan-thropic* environments. Thousands of plant species have been introduced as medicinal herbs, crop plants, or horticultural subjects while dozens of animal species have been transported as livestock, game, or pet species. Many invading species, both plants and animals, have probably been introduced unintentionally in ship ballast, and as hitch-hikers in bags of grain and fertilizer, or bales of wool, cotton, straw, and hay. Humans and their animals have carried even more invaders in the form of bacteria, viruses, and parasites. By introducing alien species humans have strongly influenced natural distribution patterns of many species, with unfortunate consequences for native communities and ecosystems in many cases, and profound impacts on human health and economy.

What makes a successful invader?

To be successful, a biological invader must have ecological, physiological, genetic, and morphological characteristics that promote long-distance dispersal of offspring and propagules, rapid colonization rates, and high competitive ability. In plants, attributes of obvious importance are morphological character-istics of seeds that facilitate their transportation by wind or animals (di Castri *et al.* 1990). Long ago Baker and Stebbins (1965) identified a series of biological attributes likely to facilitate the invasiveness of a species. However, reviewing the physiological, demographic, and genetic attributes of invaders, Roy (1990) showed that the invasive flora of a country is composed of a large array of plant types with a wide range of traits. He was unable to find any particular set of traits that may favour the invasiveness of species.

In fact, despite a growing concern about the negative effects of invasions, surprisingly little is known of the determinants of distribution, abundance, and population dynamics of invading plants. It has been recently proposed that the fate of introduced species mostly depends on context-specific processes within the target community, such as interspecific competition, herbivory, and resource availability in space and time. If such is the case, invading species–community interactions determine invasion success. Therefore, it is likely that small biological differences among species interact with habitat characteristics to produce distinct patterns of distribution and abundance. One example is provided by two closely related species of annuals that have invaded a huge variety of biogeographic regions throughout the world: *Conyza canadensis* and *C. sumatrensis* (Box 9.4).

The few plants

Surprisingly, invading plant species are not a real threat in Mediterranean ecosystems. Of the 25 000 or so plant species existing today in the region, no more than 250, that is, around 1%, are considered non-native and very few may be considered as harmful for natural communities. Even the two very aggressive invasive species of *Conyza*, which invaded the Old World some 150 years ago, do not escape from heavily disturbed habitats. One hundred years ago for example, a time when there was much international trade in wool and other sheep products, the local flora in our region contained hundreds of recently adventitious species as testified by the famous 'Flora Juvenalis' of Montpellier (Thellung 1908–10). Not one of them succeeded in massively invading the region. In succeeding decades none has penetrated far from roadways, ditches, and cultivated fields. Indeed, most plant invaders in the Mediterranean Basin survive only in unstable and human-created habitats such as cultivated fields, old fields, orchards, urban and peri-urban wastelands, and roadsides. One exception is the riparian forests, which are often invaded by woody alien species such as *Robinia pseudoacacia*, *Fraxinus ornus*, *Acer negundo*, and *Lonicera japonica*.

The low invasiveness of Mediterranean ecosystems contrasts with the situation in many other regions in the world where invading species have displaced whole cohorts of native species and disrupted ecosystem dynamics. Between 57 and 82% of plant species introduced to the four other mediterranean climate regions of the world originated in southern Europe or the Near East with the percentage of strictly Mediterranean species being highest in the mediterrranean-climate region of Chile. In California, as much as 20% of the extant flora consist of invading species, mostly of Eurasian origin (Raven and Axelrod 1978). In the South African Cape Province, an equally massive but dual invasion has taken place, with trees and shrubs coming from Australia, and many annuals and biennials arriving from Eurasia. As well as sharing a mediterranean-type climate, California and Chile were first settled by people from the Mediterranean itself. Thus, most early introductions came from the Iberian peninsula. The success of introduced plants from the Mediterranean to other mediterranean-type regions of the world is due as much to their 'preadaptation' to a set of factors such as fire, low soil nutrient levels, and grazing pressures operating in their new environments, as to their response to a particular climatic regime.

The reasons why the Mediterranean is relatively resistant to invasion are still obscure and largely conjectural. Several hypotheses have been proposed to account for the susceptibility of ecosystems to invasion by alien species. One possible explanation relevant to Mediterranean ecosystems is that they have been

Box 9.4 The invading *Conyza*

In the Mediterranean region, *Conyza canadensis* is restricted to recently perturbed areas while *C. sumatrensis* typically invades old fields in early stages of secondary succession. Thébaud *et al.* (1996) experimentally demonstrated from cross-habitat transplant experiments that a variety of factors, including interspecific competition, availability of soil nutrients, and water resources, all influenced the demographic performance of the two species. Their rates of survival and reproduction differed sharply depending on these factors. The relative performance of the species during the experiment nicely matched their regional distribution, with *C. sumatrensis* doing better in most habitats. The higher reproductive effort but lower competitive ability of *C. canadensis* explains why this species invades only the most recently perturbed areas, where their seedlings are not confronted with resource limitations. A greater competitive ability allows *C. sumatrensis* to invade and persist in more 'advanced' communities as well. This illustrates how subtle biological differences between closely related species can determine the patterns of distribution and abundance of invaders in a recently invaded area. This elegant experiment shows that invasion failure or success may depend on many traits interacting with many features of the target community. This supports the view that species–community interactions can determine success or failure for invading organisms.

subjected to continuous disturbance of fluctuating regimes and intensity over many millennia, accompanied by thousands of spontaneous colonization events. According to this view, these ecosystems have become progressively more resistant since 'old invaders' prevent access to potential 'new invaders' (Drake *et al.* 1989). Fox and Fox (1986) also argued that mature communities characterized by complex interspecific interactions among a full set of native species should be less susceptible to invasion because all available resources will be optimally used.

Most invasive plant species of the Mediterranean area are from temperate climatic zones rather than from the other mediterranean-type regions. Apart from ornamental plants such as *Pelargonium*, *Arum*, and *Impatiens*, which just barely escape cultivation in some regions, only two invading species in the Mediterranean come from the Cape Province of South Africa: the succulent creeper called ice plant (*Carpobrotus edulis*), which may extend over large areas on dry coastal rocky substrates, and the Bermuda buttercup (*Oxalis pes-capra*), which may become a weedy pest in gardens and farmland. One accidental introduction from Chile is the spiny-fruited *Xanthium spinosum*, which has become somewhat naturalized in the Basin. From Australia, many species of *Atriplex* and nitrogen-fixing *Acacia* have been voluntarily introduced, but, here again, none has succeeded in invading forest or matorral. The only exception is *Acacia dealbata*, the 'mimosa', which has long been cultivated along the Côte d'Azur for its fragrant and beautiful flowers and sold throughout Europe in early spring. This tree is now escaping in some areas of southern Europe.

Several plant invaders come from temperate North America but few of them have had any major impact. Examples are the pigweeds (*Amaranthus albus* and *A. retroflexus*), false indigo (*Amorpha fruticosa*) in wetlands, black locust (*Robinia pseudoacacia*), evening primrose (*Oenothera biennis*), and the *Conyza* spp. mentioned above. Some succulent species introduced from the Americas for ornamental purposes occasionally escape cultivation along the coasts, including a few agaves (e.g. *Agave americana*) and several prickly pear cacti (*Opuntia* spp.).

The only ecological 'disaster' related to non-native plants introduced into the Mediterranean is the large-scale plantations of eucalyptus (*Eucalyptus camadulensis*, *E. gomphocephala*) that cover well over a million hectares, especially in Spain (400 000 ha), Portugal (700 000 ha), North Africa (280 000 ha), and Israel. Although they may be economically very rewarding in the short term, with timber production reaching 5–27 m^3 ha^{-1} yr^{-1}, such plantations are biological 'deserts' and may have dire consequences for native communities and landscapes over the long term (Quézel *et al.* 1990). For example the utilization by birds and insects of eucalyptus leaves, flowers, and fruits is minimal outside Australia, probably because native Mediterranean faunas are incapable of using these resources. In Portugal no more than 13 species of birds breed regularly in eucalyptus plantations, as compared to 30–35 species recorded in oak forests nearby. Additionally, eucalyptus plantations in the Mediterranean and elsewhere are infamous for extracting all available water and nutrient resources in a given site.

The even fewer birds

In most groups of animals, too, the number of invading species is quite low. There is hardly any species of bird from other continents that regularly occurs as an established breeder in the Mediterranean. Among birds, even the pheasant, which was introduced by the Romans, has nowhere succeeded in establishing long-term self-sustaining populations in the wild. It occurs only in localized lowland areas where humans continuously breed it for hunting. The many attempts to introduce species as game birds regularly fail. Hunters repeatedly try to introduce exotic species in hopes of increasing their game scores. Examples in the western Mediterranean are the Californian quail (*Colinus virginianus*), the pheasant (*Syrmaticus reevesii*), the black francolin (*Francolinus francolinus*) from the Middle East, and some others.

Exceptions to the rule are three cases of intentionally introduced birds that escaped human control. The first is that of very small free-living populations of the ring-necked parakeet (*Psittacula krameri*) which breed in some Mediterranean cities such as Nice. The second example is that of the tiny, red avadavar (*Amandava amandava*) from Asia, which has developed a rather large free-breeding population of more than 1000 individuals in the Spanish province of

Extramadura (de Lope *et al.* 1984). There exists a risk that this escaped granivorous cage bird might become a pest in field crops, but thus far it has not. The third case is that of the blue magpie (*Cyanopica cyanopica*). This beautiful bird lives in small flocks in the open forests of southern Spain and Portugal, in populations totally disjunct from the bulk of the world populations of this species, all of which occur in China and Manchuria. Explaining such a pattern has long been a popular brain-teaser for biogeographers. The most likely explanation is that the species was brought from Asia to Spain sometime during the seventeenth century by European navigators or missionaries. Even for conservation purists, who disapprove in principle of all displaced, non-native species, the blue magpie can perhaps be forgiven its presence and considered as contributing to regional biodiversity. In addition to its beauty, it apparently never competes with any native species of bird.

Fishes: a more severe threat

The only group of vertebrates with successful and sometimes harmful invaders in the Mediterranean area is the fishes. Many freshwater fish species have been intentionally introduced in Mediterranean streams and marshes more or less recently. Several of them, such as carp (*Cyprinus carpio*), pike (*Esox lucius*), and pike-perch (*Stizostedion lucioperca*) became naturalized and are now well-integrated components of extant communities. These introduced fish have been shown to result in the decline or extinction of several native species, including some endemics. For example, in Spain, no less than 19 introduced fish species are considered as threatening native endemics. The most harmful are large predators such as pike, sheatfish (*Silurus glanis*), pike-perch, bass (*Micropterus salmoides*), and perch (*Perca fluviatilis*).

The mosquito-fish (*Gambusia affinis*) from south-eastern North America has been introduced in almost every Mediterranean wetland area, such as lakes, lagoons, and lowland rivers, where it was hoped it would control mosquitoes. Several species of carp have also been introduced in many parts of the Basin for commercial purposes. But even when introduced in closed fish ponds, alien species may have detrimental effects when individuals escape human control. For example, when Israelis introduced the silver carp (*Hypophthalmichthys molitrix*) from the Far East for commercial production in aquaculture systems, they were running a great risk. This is an extremely aggressive species with a high potential for escape and invasion. The introduction in 1983 of herbivorous carps in Lake Oubeira, Algeria, resulted in the rapid destruction of half the reed-beds present (Pearce et Crivelli 1994), with dire consequences for the native biota. Fortunately, Mediterranean wetlands have not to date experienced the catastrophic consequences of invasion by exotic fish species such as those that have so badly ravaged cichlid fish communities in the great lakes of East Africa.

A famous historical case of invasion by fish is the so-called 'Lessepsian migrations' of species that invaded the Mediterranean from the Red Sea when the Suez canal was completed in 1869 (Ferdinand de Lesseps was the engineer responsible for the canal). This involved no less than 41 species (Ben-Tuvia 1985), not to mention the many species of parasites they brought with them. These invading fishes gradually invaded coastal waters of Israel, Lebanon, and Egypt, and then expanded as far west as Libya and Tunisia. In 1953 'Lessepsian' red mullets (*Upeneus asymmetricus* and *U. moluccensis*) constituted 10–15% of mullet catches (the remainder being the indigenous mullets *Mullus barbatus* and *M. Surmuletus*) off the coast of Israel and up to 30% in 1973 (Maillard and Raibaut 1990).

Bombus terrestris: a threat to native bees

There are extremely few cases of documented invasions by insects. One worth mentioning is that of the bumble-bee *Bombus terrestris*. This species first invaded Israel in the early 1930s and then spread over most of the country. Increasing numbers of this bumble-bee have been associated with a reduction in honey bees and several species of solitary bees, hence representing a threat to the indigenous bee fauna (Dafni and Shmida 1996). *Bombus terrestris* is a generalist species able to harvest floral resources from a large spectrum of unrelated plant species in various habitats. It is active early in the morning and reduces nectar availability, which results in food shortage for other bee species. This is especially true in late spring when the pollen market switches from a surplus of flowers to a surplus of bees (Shmida and Dafni 1989) (see Chapter 7, p. 185). Bee censuses conducted on several core species of phrygana in Mt. Carmel, Israel, have shown a reduction in all other species of bees as *Bombus terrestris* numbers have increased. Although honey bees have shown a constant retreat due to another invader species, the parasitic mite *Varroa*, observations on bee activity revealed a regular exclusion of solitary bees from several plant species in the presence of *Bombus*, suggesting competitive exclusion (Dafni and O'Toole 1994). Harnessing bumble-bees for commercial pollination in agribusiness may be a real threat for native bees.

Close companions of humans: alfalfa and house mouse

We will close this chapter by considering two species of Asiatic origin that became commensal with humans for completely different reasons. The first, alfalfa, is used as hay and fodder for animals by almost all peoples around the Basin whereas the second is an undesirable but inescapable 'fellow-traveller' with people.

Alfalfa

One of the most widely cultivated forage plants in the world, alfalfa (also known as lucerne), is a complex of related species and hybrids. The perennial alfalfa was apparently domesticated in the Middle East sometime between 6000 and 8000 yr BP, and from there was carried west to all parts of the Mediterranean Basin (Fig. 9.4). Along the way, hybridizations, autopolyploidy, and introgressions* took place with related perennial species, including *Medicago falcata* in northern

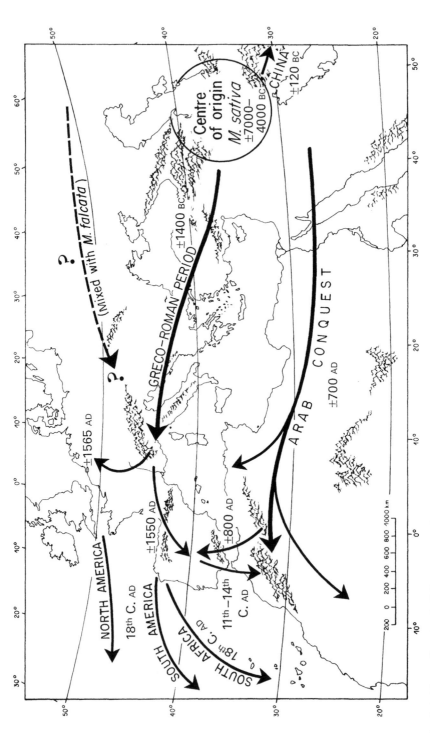

Fig. 9.4 Nine millennia of alfalfa migration (*Medicago sativa*), from its Near Eastern centre of origin, across the Mediterranean area and western Europe, and then to the New World and South Africa in the last two centuries. (After A. Zoghlami, and P. Guy, in Prosperi *et al.* 1995.)

Europe, *M. glutinosa* in the Caucasus, and *M. coerulea* and *M. glomerata* in the Mediterranean Basin (Lesins and Lesins 1979). In addition, at least two natural hybrids (*M. media* and *M. varia*) are accepted as 'good' species, arising from cultivated alfalfa back-crossing with *M. falcata*, no doubt in the vicinity of cultivated alfalfa fields.

The house mouse

A large body of palaeontological and archaeozoological data make it possible to reconstruct the spatio-temporal 'conquest' of western Eurasia and North Africa by the house mouse (*Mus musculus*) complex (Boursot *et al.* 1993). According to the current systematic status of this complex of small mice, four species occur in western Eurasia and North Africa (see Fig. 9.5). Three of them occur as wild-ranging populations: *Mus spretus* in the western part of the Mediterranean (Iberian peninsula, south of France, and North Africa), *M. spicilegus* in the southern parts of central Europe around the northern part of the Caspian Sea, and *M. macedonicus* in the Balkans and the Middle East. The fourth species, *M. musculus*, is mostly commensal, living in close contact with human settlements. Two subspecies or semi-species* of the house mouse have been identified, the long-tailed house mouse, *M. m. domesticus*, around the Mediterranean and in western Europe, and the short-tailed house mouse, *M. m. musculus*, in northern and continental Eurasia (Bonhomme 1986). They are mostly allopatric but freely hybridize along a narrow zone that extends across Europe from Denmark to Bulgaria.

The house mouse belongs to a lineage that most probably originated in south-west Asia, with the first *Mus* fossil, found in Pakistan, dating back to 7 Myr BP. By the end of the Pliocene, the genus was established in North Africa where it subsequently differentiated, some 1–3 Myr BP, to yield the two extant species, *Mus musculus* and *M. spretus*, which co-occur today in North Africa and south-western Europe. Arriving from the east at the very end of the Pleistocene, the house mouse is present in the Near East with two species in Israel, the house mouse (*M. musculus*) and a short-tailed wild mouse, *M. macedonicus*. The first records of the house mouse in Israel date from the Natufian culture, *c.* 10 000 yr BP (Auffray *et al.* 1988), when the first permanent settlements were built. It may be that sedentary villages and later cities in the Fertile Crescent provided a new ecological niche for this animal and allowed it to avoid competition with the already established *Mus macedonicus*.

Invasion of the Mediterranean and western Europe by the house mouse from the Middle East proceeded very rapidly. From 10 000–4000 years BC it occurred only in the Middle East, and then spread into the western Palaearctic through two main routes of colonization—a 'land route' to the north (L on Fig. 9.5) and a 'sea route' to the south (S on Fig. 9.5). The first corresponds to a Danubian neolithic human culture whereby mice followed humans along a 'northern route' to the north of the Caspian and Black seas (Auffray *et al.* 1990). Spread of colonization through this route was rapid since Belgium was invaded as early as the fifth millennium BC. The second route, further south, appears to have brought the

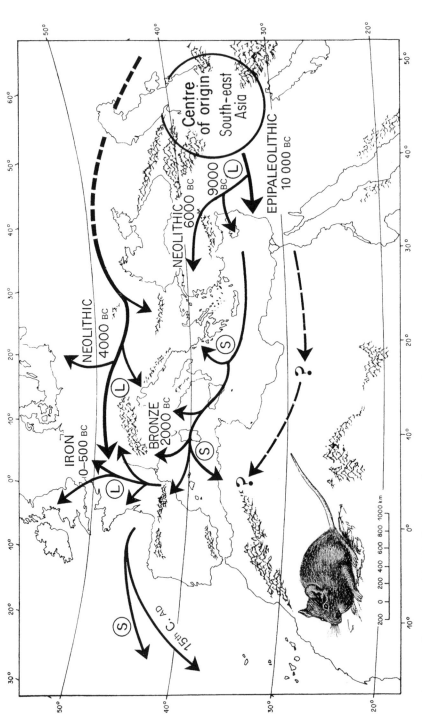

Fig. 9.5 Patterns of invasion of the Mediterranean Basin and western Europe by the house mouse. S = sea transportation, L = land invasion. (After Auffray *et al.* 1990, and personal communication).

mice across the Mediterranean Sea during the Bronze Age, 4000 BP, a time of extensive navigation all over the Mediterranean. This epoch corresponds to the beginning of the brilliant Aegean civilisation and the Middle Empire of Egypt, which were characterized by a boom in sea trading and the development of large ships that provided many opportunities for mice to travel as stowaways. Thus, within a very short time in the Bronze Age the whole Mediterranean Basin was invaded by house mice. Genetic evidence supports the view that the *M. musculus musculus* subspecies of north-eastern Europe corresponds to invasion along the northern terrestrial route while the *M. musculus domesticus* subspecies of western and southern Europe, whose populations exhibit very low genetic differentiation rates (Britton-Davidian 1990), corresponds to a rapid invasion via the Mediterranean sea route.

10 Challenges for the future

It would be a hopeless dream to think that the pre-historic 'First Eden' of the Mediterranean Basin, so vividly described by Sir David Attenborough (1987), will ever be regained; 'paradise' is definitively lost. Current trends in the Mediterranean do not lead to optimism about the future of biodiversity in general in spite of much recent effort to preserve it. The deep-rooted Mediterranean mindset that 'Culture' is mainly a humanistic affair and that Nature is second in importance has resulted in the past in too little attention being paid to environmental quality and 'services', and also to short-sighted resource management. After 10 000 years or more of 'co-habitation', most Mediterranean ecosystems are so inextricably linked to human interventions that the future of biological diversity cannot be disconnected from that of human affairs. Biological diversity is intimately linked to all the other components—economic, political, and sociological—that constitute the greatest challenges for the decades to come: promoting sustainable development and controlling population growth. For achieving the first of these goals, many argue that biodiversity has a vital role to play. But biological diversity is but one of the many factors that humans are faced with. The first and most dramatic objective should be to alleviate the poignant discrepancies that make increasingly different the northern and the southern banks of the Basin, with its many consequences for biological diversity.

A microcosm of world problems

Today, some twenty to twenty-five states totalling 380–400 million people constitute two sharply contrasting worlds within the Basin, each with its separate histories. To the north-west, five countries totalling 165 million inhabitants belong to the European Community, and enjoy an average yearly income of about 17 000 Euros per capita. In contrast, the average income of some 235 million people of the eastern and southern shores of the Mediterranean (with the exception of Israel) is four times less, and barely reaches 3700 Euros per capita per year. Emerging slowly from the recent wars in the north-east quadrant, the five redefined Balkan states, and neighbouring Albania, all resemble North Africa and the poorer Near Eastern countries, economically speaking.

Population density also varies widely in the Basin, from 1080 inhabitants km^{-2} in Malta, the highest national density anywhere except for Monaco and

Table 10.1 Population size, rate of increase, and yearly income in some countries of the northern, southern, and eastern parts of the Mediterranean (Population et Sociétés 1997).

Country	Population size (inhabitants $\times 10^6$)		Current rate of increase (%)	Yearly per capita income (Euros in 1997)
	1997	2025[a]		
Spain	39.3	39.0	0.0	13965
France	58.6	62.7	0.4	19563
Italy	57.4	54.8	0.0	18624
Greece	10.5	10.2	−0.1	11136
Algeria	29.8	47.7	2.4	3185
Morocco	28.2	39.9	2.0	3628
Lebanon	3.9	6.1	2.2	4146
Israel	5.8	8.0	1.5	14500

[a] Estimates based on conservative 1997 projections (Population et Sociétés 1997).

Singapore, to only 28 inhabitants km^{-2} in Corsica. Demographic trends in the rich north-west quadrant are, however, diametrically opposed to those in the other three. In 1950, the 140 million inhabitants on the northern bank represented 66.7% of the Basin's population. By 1990, with some 190 million, those same countries accounted for only 50% (di Castri 1998). Since their independence in the 1960s, North African countries have averaged between 2% and 3.5% annual population growth, which is comparable with the highest national growth rates in the world.

These discrepancies are aggravated by the fact that human populations in the poorer Mediterranean countries will be growing even faster in the near future, since they are much younger than those of the richer, northern shores. According to di Castri (1998), the ratio of 15–24-year-old to 55–64-year-old people is about 2.5 times higher in the southern Mediterranean than in the Euro-Mediterranean countries. Conservative analysts have suggested that by the year 2025 the population of southern Europe will stabilize at around 170 million, while that of the Near Eastern and North African countries will have increased by about 70%, to surpass 400 million (Table 10.1).

The northern shores

Along the north bank of the Mediterranean human pressure on many ecosystems has steadily decreased because of agricultural abandonment and rural depopulation, dating back to the end of the nineteenth century but accelerating greatly since the Second World War. Across the entire range of life zones and habitat types we described in Chapters 4 and 5, a progressive recovery of forest and matorrals is proceeding at a rate of 1–2% per year. Abandonment of fields and pastures favours the recolonization and spread of plant species that were formerly scattered in the landscape. For example the surface areas covered by the anemochoric* Aleppo pine increased three-fold between 1878 and 1904 and by 2.6 times between 1904 and 1978 (Achérar *et al.* 1984). In a study area of *c.* 1000 km^2 in southern France, the amount of area occupied by vegetation over 2 m or more high increased from 7% in 1946 to 49% in 1979 (Lepart and Debussche 1992). In the same period cultivated and grazed areas declined from 22% to 11%. Other expansive plants that colonize many old fields are bird-dispersed species such as *Pistacia, Rhamnus, Phillyrea,* and junipers. These processes of plant recovery lead to a decrease in habitat patchiness and the typical 'moving mosaic' landscapes that are so characteristic of the Mediterranean area and beneficial for biological diversity (see Chapter 8, p. 229). Certain species are expanding their ranges whereas numerous other are disappearing locally from many regions. Since the potential productivity in most Mediterranean lands is insufficient to justify a reallocation of abandoned land to forestry production, more and more inland areas are increasingly becoming abandoned.

On the other hand, there is a growing local demand for tertiary activities, especially near the coast, from promoters, speculators, and entrepreneurs of all sorts. From one end of the Basin to the other whole regions and landscapes are losing their age-old configuration and contours. So a gloomy dichotomy emerges: far from the coast there are deserted fields, orchards, and pastures, progressively encroached upon by shrublands, and increasingly dense, unproductive, and ill-managed woodlands. Along the coasts, in the densely urbanized and homo-genized industrial zones that continue their inexorable sprawl, all ecological and cultural contact with the Mediterranean past is abandoned and lost.

Well over 8800 km of Mediterranean seacoasts (19%) are now occupied by tourist installations, concrete structures, and networks of roads (Henry 1977). In Italy, France, and Spain, including the Balearic and Canary Islands, large parts of Mediterranean coastal areas have disappeared under concrete and macadam. On the island of Mallorca (Balearic Islands), for example, 48% of the coastline has been irreversibly 'artificialized' in this fashion. In the north-east quadrant of the Basin, in Greece, Turkey, Cyprus, Lebanon, and Israel, the same process is under way. Huge and growing human agglomerations combined with concentrated industrial zones result in high levels of sea, lagoon, and estuary pollution. The Cousteau Foundation and many others warn of increasing marine pollution as the Mediterranean Sea becomes the dustbin for ever-expanding large cities along the coasts and on the islands, only few of which are equipped with adequate waste and sewage treatment systems. This pollution is bacteriological, organic, metallic, radioactive, chemical, and more. When flying over the coasts in summer one sees a brown ribbon several hundreds metres wide, all along the coast, accompanied by dense crowds of people basking in the sun or occasionally 'enjoying' the dirty waters nearest the beaches. In spite of all this, the Mediterranean Basin is still the number one tourist area in the world, currently drawing 30% of annual tourist trade, mainly from countries of northern Europe. Tourist influx is also expected to increase in the decades to come with an estimated 250–300 million visitors each year for the entire Basin (Ramade 1997). In Greece, for example, 90% of tourists remain exclusively along the coasts, where 75% of the local population also live.

According to demographic predictions of the Blue Plan (an Action Plan launched by the United Nations for Environment Programme), permanent human populations will rise from 82 million inhabitants in 1985 to 144–170 million in 2025. The unbelievably rapid growth of tourism and urban development in Mediterranean coastal areas is a real threat to many habitats of high biological value.

The southern shores

In the over-populated southern and eastern banks of the Mediterranean, the situation is just the reverse. Disturbance regimes in farm, pasture, and forest areas are moving towards still greater intensity of land and resource exploitation for the short-term survival of local people. In the absence of sustainable land use systems updated to meet the needs of young and growing local populations, ongoing forest clearing and outright desertification are proceeding at a rate of about 2% per year (Marchand 1990), which is just about the same rate at which agricultural lands are being abandoned in southern Europe!

A visit to any mountainous areas of North Africa today reveals that demographic pressure, combined with a highly conservative rural economy still largely disconnected from outside markets, results in very low crop yields and overall productivity. This leads to the all-too-familiar cycle of increasing ploughing and grazing areas followed by soil erosion, and then new clearing elsewhere. Contrary to the situation in the industrialized Euro-Mediterranean countries (and Israel), where economic input comes mostly from service industries, here in the south people and national economies still depend almost entirely on local resources. In the Ouarsenis region in Algeria, for instance, and in the Rif Mountains of Morocco, slopes of up to 50° are increasingly ploughed and cultivated without any special measures being taken, resulting in catastrophic gully erosion and soil loss. In Algeria and Morocco forested areas now cover less than 30% of their potential territory (Marchand 1990). The rest has been permanently and, in all probability, irreversibly removed. Of the remaining 30%, the quantity of wood and grazing material harvested annually far exceeds primary production. Table 10.2 provides some revealing and frightening statistics from a fairly representative region of Morocco.

In 1971, wood consumption for cooking and heating was estimated at 55 million m^3 per year in North Africa, that is, 0.5 m^3 of wood per capita per year. This represented 41% of the total energetic consumption of the region. As in much of Africa, this situation has changed little in the intervening years. Wood requirements for domestic use (cooking and heating) in these countries are still roughly 0.5–1 m^3 per capita per year. Marchand (1990) has estimated that more than 140 million people in the Basin will be affected by fuel-wood shortages in the

Table 10.2 Statistics relevant to forest destruction and overgrazing in the Azizal Province of Morocco (after Marchand 1990).

Size of the area:	1 000 000 ha
Annual human population growth:	1.5%
Remaining forested area:	340 000 ha
Primary annual wood production:	230 000 m^3 (0.7 m^3 ha^{-1} yr^{-1})
Annual rate of wood harvesting:	c. 490 000 m^3
Yearly loss of woody 'capital':	c. 260 000 m^3, i.e. 1.7% of that remaining
Territory grazed:	50%
Sustainable carrying capacity:	0.8 head ha^{-1}
Current grazing load:	1.6 head ha^{-1}

year 2000. The consequences for standing woody biomass, and on ecosystems in general, are easy to imagine.

The situation is similar in the steppe areas of the south where bush-dotted grasslands represent the primary resource. Elimination of alfa grass by over-grazing from the Alfa steppe, which covers more than three million hectares in Algeria and similar areas in both Tunisia and Morocco, can have profound implications for ecosystem stability and potential productivity. The sheep population in Algeria increased from 3.8 million animals in 1963 to 16.1 million in 1987 (Aidoud and Nedjraoui 1992). As a result, most of the steppe regions have been so badly degraded that livestock breeders have increasingly reverted to stockyard breeding and the motorized transport of animal feed and water. Table 10.3 summarizes the dramatic changes that have resulted from overgrazing and systematic over-cutting of woody plants in an Alfa steppe zone of central Algeria.

Summarizing the overall situation, one Tunisian government minister recently said 'the northern [Mediterranean] countries can build upon rich and fertile plains whereas we have only the desert to build upon. Where is sustainable regional development to come from when you have huge natural resources and over-industrialization on one side, and sparse resources and over-population on the other?' This situation makes of the Mediterranean Basin a striking microcosm of current world problems.

Conservation

In looking to achieve goals related to biological conservation and sustainable development, a useful approach is to analyse the feedback mechanisms that keep ecosystems 'running'. Studies of positive or negative feedback cycles at local or regional levels may be a crucial long-term strategy for planning new management policies for ecosystems and biological resources. The ultimate goal would be to find and restore an optimally 'running' system, involving as many as possible of the native constituents of living systems, from populations to whole

Table 10.3 Recent evolution of the Alfa steppe, based on long-term studies in a series of permanent study plots in central Algeria (after Aidoud and Nedjraoui 1992).

	1976	1989
Bare ground (%)	16	83
Vegetation cover (%)		
Ephemerals	11	9
Perennials	39	6
Alfa	34	2
Total	84	17
Phytomass (kg dry matter ha^{-1})	2100	750
Production (FU ha^{-1} yr^{-1})[a]	130	60

[a] FU = Food units (1 FU = 1650 Kcal for ruminants).

communities. However, compromises must be made between fully preserving biological diversity and native ecosystem functioning on the one hand, and satisfying the immediate and long-term needs of local human populations on the other. This idea is illustrated in Fig. 10.1. Geographic Information Systems (GIS) are a good tool for achieving this goal since they help to manage the wealth of climatic and environmental data and integrate it with databases on species distribution and ecology (Scott *et al.* 1991). By storing, and then allowing the manipulation of, many types of mapped data on soils, vegetation types, species distribution, etc., GIS can highlight correlations among elements of the landscape, and help in planning the management of landscapes, ecosystems, and whole regions.

As we argued in Chapter 8, the highest biological diversity in Mediterranean ecosystems probably does not occur in woodlands, even the pristine ones depicted by the first circle in Fig. 10.1, but rather in the various agro-sylvo-pastoral systems such as the traditional *ager-saltus-sylva* and 'dehesa–montado' systems combined with transhumance (Fig. 10.1, second circle upper left). Therefore, relying on the 'potential natural vegetation' (PNV) for devising conservation strategies would probably not be the best approach in the Mediterranean where ecosystems and landscapes have been transformed and redesigned by humans over several millennia. The PNV system was defined by Küchler (1964) as 'the vegetation that would exist today if humans were removed from the scene and if plant succession after its removal were telescoped into a single moment.' A strategy based on PNV is a useful approach in many countries that have only recently been transformed by humans, such as most parts of North America, but certainly not in the Mediterranean area where the series of vegetation that would lead to potential natural vegetation is difficult or impossible to reconstruct. It is within this very general framework that action plans should ideally be devised to promote the conservation and restoration of

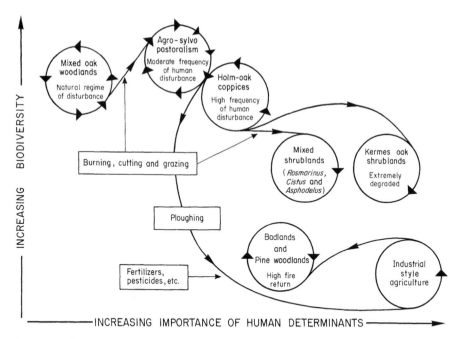

Fig. 10.1 Schematic representation of some human-induced changes in mixed oak woodlands in the Mediterranean Basin. Under pressure from prolonged disturbances of different kinds, systems appear to 'shift' across ecological thresholds from one trajectory to another. This causes changes, in the dynamics of energetic, hydraulic, and elemental fluxes, and to species diversity. Numbers of arrows on the circles indicate the relative richness of ecosystem dynamics (inspired from Woodward 1993), as indicated by 'eco-diversity', that is, number of functional groups, interspecific interactions, etc. (Modified from Blondel and Aronson 1995.)

biological diversity at the levels of species, habitats, and landscapes.

Broadly speaking, biological conservation relies on two strategies. The first consists in assisting threatened individual species to survive in at least some localities. The second aims at the conservation of habitats, ecosystems, or larger areas. The best strategy for the long-term protection of biological diversity is the preservation of natural communities in the wild, that is, *in situ* or *on-site* preservation. A relatively important amount of scientific work and money has already been devoted to biological conservation of selected species in the Mediterranean and many new projects are getting underway.

Conservation of species

Except for some groups of plants and large animals such as birds and mammals, we lack information on the distribution and abundance of most species in the Mediterranean. Species to be protected are often chosen more on the basis of aesthetic or cultural criteria, such as size, colour, or notoriety rather than on scientific grounds. Many protected areas has been created to protect 'charismatic

Box 10.1 Preserving species: the bad side of the coin

As the saying goes, the road to Hell is paved with good intentions. The inclusion of rare or endangered species in 'red books' can have detrimental effects by attracting clandestine collectors and promoting illegal trade in specimens whose price on the black market is directly related to the species' growing rarity. For several species of rare carabid beetles and butterflies, for example, recent price hikes have prompted local collectors, under pressure from unscrupulous dealers, to search ever more avidly for this unexpected source of income. Some local populations of the rare butterfly called Spanish festoon (*Zerinthia rumina*), which is closely linked to its patchily distributed host plant (an *Aristolochia*), have been hunted to the last individual by collectors who are not always aware of the gravity of their actions. A similar threat exists for several other showy species such as Corsican swallowtail (*Papilio hospiton*), southern swallowtail (*P. alexanor*), and the superb moth *Graellsia isabelae*.

megavertebrates' that capture public attention, have symbolic value, and are crucial to ecotourism' (Primack 1993). This is the case for many bird reserves. Even for certain insect species much effort has been made. For example, in Italy and France attempts are in motion to conserve the beautiful carab beetle *Carabus olympiae* but there is complete inaction in regard to their much rarer and more threatened relatives *C. clathratus* and *C. alysidotus*, both of which are rather drab and inconspicuous (Baletto and Casale 1991). Nothing at all has been done so far to preserve the legions of minuscule and microscopic organisms that make ecosystems 'turn'.

One problem is that many rare, endemic species occur in localized areas that are not likely to be included in large reserves; conservation of such species requires specific action plans. For example, in a detailed study on insular plants both in the western and eastern Mediterranean islets, Lanza and Poggesi (1986) showed that several species occur only on small satellite islets off larger islands. They have either gone extinct on the large island, as a result of changes in their environment, or else they never occurred there because the species in question live only in the relatively inhospitable islets where few competitors occur. In this latter group are a wild onion (*Allium commutatum*), a shrubby mallow (*Lavatera arborea*), and a perennial grass, *Parapholis marginata*, of a genus endemic to the Mediterranean. Strikingly, there were strong similarities in the ecological distribution of these species on islets off Corsica and those found on islets near several of the larger islands of the Aegean Sea. The showy hairy pink, *Silene velutina*, disappeared from Corsica in recent times but still persists in a few small nearby islets. This gives an idea of the complexity of species conservation in the Mediterranean, in spite of great efforts to adopt stringent conservation measures in the framework of international conventions (e.g. CITES and the Bern convention).

Species reinforcement and reintroduction

The survival of very rare and threatened species often requires active intervention, including reinforcement of populations, translocation, reintroduction, and captive breeding. Indeed, the smaller a population becomes, the more vulnerable it is to the so-called 'vortex effect' (Gilpin and Soulé 1986), that is, demographic variation, environmental stochasticity*, and genetic factors that tend to drive the population to extinction. Saving very small populations is a difficult task because the success of reintroduction depends on population viability, and this involves complex genetic, behavioural, and demographic processes. The success of reintroductions relies on careful studies of such issues as the genetic consequences of inbreeding, behavioural consequences of captivity for those species that are released after having been bred in captivity, and the population consequences of infectious diseases (Sarrazin and Barbault 1996). In some cases the consequences of outbreeding depression* must also be considered, because hybrid offspring born from divergent genotypes or populations may no longer have the precise mixture of genes that allows individuals to survive in a particular local environment. This often raises difficult problems. For example, if we seek to reinforce the fast-dwindling population of the brown bear in the Pyrenees, where this species is on the brink of extinction, should new animals be reintroduced from Italy, Greece, or Scandinavia? The Italian and Greek bears live in ecologically similar conditions, but the Scandinavian population is genetically closer to the indigenous Pyrenean stock (see Fig. 2.6). Outbreeding of populations of different geographic origin may cause genotypes and co-adapted genes that evolved under certain local conditions to disappear altogether.

Hundreds of species could be reinforced or locally reintroduced in the Mediterranean Basin but reintroduction policies are faced with two problems. First, if properly carried out on scientific grounds, they are extremely expensive and cannot be applied to all species. They require a careful preparation of public awareness. Second, economic or political criteria very often prevail over scientific (and moral) arguments for biological conservation. Whatever their motivations, conservationists prefer to choose 'flagship' species such as large birds or mammals rather than the cohorts of disappearing invertebrates. In this context, several projects of reintroduction and reinforcement of raptor populations are in progress for eleven species in four Mediterranean countries (France, Italy, Spain, Israel). Fifteen of the projects involve large vultures (griffon vulture, black (cineraceous) vulture, lammergeyer) and may be very successful with indirectly beneficial effects for other species. For example, a project of reintroduction of the griffon vulture started in 1968 in the Cévennes, France, where it had been hunted out by the beginning of this century. After a long process of captive breeding, the first individuals were released into the wild in 1981. Nowadays a nearly self-sustaining population that includes more than 160 free-living individuals produces more than 30 young yearly (Terrasse 1996). Such successful projects should prompt ecologists to develop action plans for many other species.

For species that become extremely rare and endangered, off-site preservation facilities include zoos, game farms, aquaria, and captive breeding programs. However, certain animals, particularly marine mammals, are so large that the facilities necessary for maintaining and handling them are extremely difficult and prohibitively expensive. This is why projects for reinforcing the Mediterranean population of monk seal have not yet given significant results.

A guild* of species that currently receives a certain amount of restoration and revised management effort is that of large ungulates. This prominent group in the Mediterranean includes several species of deer (roe, red, and fallow deer), and several subspecies of ibex, goats, and mouflon. In addition to their aesthetic and cultural values, they are of great hunting interest and clearly play important roles in many functional processes in forest, shrubland, and steppe ecosystems. Thus, a sustainable management scheme for ungulate populations should strive to conserve genetic diversity while also making rational use of this resource. Most regions, especially in the northern part of the Basin, clearly have a great potential for increasing population density and spreading the range of various ungulates.

Preserving ancient varieties of domesticated plants and animals

How are we to preserve the extraordinary variety of ecotypes and gene pools of cultivated plants and domesticated animals selected over millennia by traditional agriculturists and pastoralists? The 'option value' of a species or an ecotype is its potential to provide an economic benefit to human society at some point in the future. The growing biotechnology industry is finding new uses for many species and varieties, yielding a wide range of economic benefits within the context of traditional cropping and livestock raising practices, but also in medicine and pharmacology, as well as biological control. Resistance to a particular disease or pest is often found in just one or a few varieties of a crop that is grown in only small areas.

Preserving this genetic variability is especially critical in the Mediterranean Basin where traditional farmers are widely abandoning their local varieties and land races. Thus, an important challenge is to identify and save wild, ancestral gene pools in the Mediterranean before they are irretrievably lost along with their habitats. In the past 30 years there has been a burgeoning of interest in the need for, and value of biological diversity of ancient races and cultivars by UN agencies (FAO, UNEP, UNDP, UNESCO) and other governmental and private institutions. The Mediterranean Basin includes 45% of the bovid varieties and 55% of the goat varieties of Europe and the Middle East. Of 145 varieties of domesticated bovids and 49 varieties of sheep that occur in the Basin, 115 and 33 species, respectively, are considered as in danger of extinction by the FAO World Watch List of Domestic Animal Diversity (Georgoudis 1995) mostly as a result of the diffusion of European dairy breeds that dramatically reduced the number of local breeds. Currently there is a great deal of effort by international, governmental, and private organizations to conserve the genetic resources of 'old native' breeds of several species such as the Mediterranean buffalo, cattle, sheep, and goat. There are many examples of well-managed 'herd books' and programmes based specifically on the storage of semen. This is especially important in the framework of sustainable development because the disappearance of local races may seriously limit the exploitation of marginal lands where only locally selected breeds can thrive in the harsh environments and prolonged drought. In several Mediterranean countries, active conservation programmes employ pedigree, progeny testing, and data recording for preserving the local diversity of goats.

Many plants of economic interest occur in small fluctuating and poorly dispersed populations, such that they are at risk of extinction or severe genetic loss. Hundreds of programmes in progress aim at conserving genetic resources of fruit trees, grape, field crops, forage species, vegetables, and ornament plants (Charrier 1995). For cultivars that are no longer used by rural farmers but which constitute highly valuable genetic resources, as well as for wild relatives of cultivated species that are really threatened in the wild, one strategy is to preserve them. Facilities for preserving plants *ex situ* or off-site include botanical gardens, arboreta, botanical conservatories, and seed banks, all of which have great potential for preserving genetic variability for the future (see Box 10.2 and Avishai 1985). Integrating research on the current status of a species with conservation *in situ* and cultivation *ex situ* provides good tools for preserving biodiversity. Several botanical gardens have developed gene banks specifically designed for the storage of germplasm of wild plants. Some of the world's most ancient botanical gardens were created in the Mediterranean during the sixteenth century, mostly for the conservation of herbs and medicinal plants. Today there are approximately 100 botanical gardens and arboreta around the Basin (42 in Italy alone) (Du Puy and Jackson 1995). Most of them are established in Euro-Mediterranean countries with the notable exception of the Jardin d'Essais du Hamma in Algeria, which maintains an important collection of 8000 taxa of native Algerian plants. A great many botanical gardens today consider conservation as a major theme in their activity and maintain living collection of economic plants such as fruit tree cultivars and medicinal and aromatic plants. For example the beautiful botanical garden of Gibraltar established in 1991 already contains an impressive collection of succulents, and aims to become an important Mediterranean centre for native plant conservation and environmental education.

One promising initiative is the International Centre for Advanced Mediterranean Agronomic Studies (CIHEAM), which includes the Mediterranean Agronomic Institutes of Bari (Italy), Chania (Greece), Montpellier (France), Zaragoza (Spain), and Meknès (Morocco). This Institute has established a working group on 'under-utilized fruit crops' such as figs (*Ficus carica*), loquats (*Eriobotrya japonica*), Japanese persimmons (*Diospyros kaki*), pomegranates (*Punica granatum*), and Barbary figs (*Opuntia ficus-indica*). These species have often been considered as marginal crops but are of great value in localized situations, especially as a result of an increasing demand for their fruits in industrialized countries and the local market. The fig tree and the pomegranate are two of the seven Biblical plants (along with wheat, barley, olive, date, and

Box 10.2 The botanical conservatory of Porquerolles

This conservatory, established in 1979 on the island of Porquerolles, France, intends not only to inventory and monitor wild plant species that are rare or threatened in any part of the Basin, but also to intervene directly to assist them in their survival. Achieving this goal is pursued through cytotaxonomic and ecological research on target species. Conservation programmes include both *ex situ* conservation of genetic resources of wild species and cultivars in the form of a seed bank that currently holds more than 2000 species, and *in situ* conservation of rare and endangered populations of plant species. In the case of populations on the verge of extinction, this can take the form of reinforcement, but where local eradication has already taken place, attempts at reintroduction are undertaken, with attention being paid to the dangers of 'genetic pollution' in a regional context. In both pre- and post-eradication situations, it is the *ex situ* seed bank and living collections that make the *in situ* efforts possible. This conservatory also maintains a project to document, record, and conserve traditionally grown fruit trees.

grape) representing the ancient Israelite agriculture. A network, called the 'Identification, Conservation and Use of Wild Plants in the Mediterranean Region' (MEDUSA), was formally established at the Mediterranean Agronomic Institute of Chania (Greece) (Skoula *et al.* 1997). MEDUSA aims to identify and conserve native Mediterranean plants of economic interest, and to propose methods for the economic and social development of rural areas using management systems based on alternative crops, wild relatives of existing crops, and the rational use of wild, uncultivated plant resources. Extensive field surveys are now underway, within the framework of MEDUSA, on the ecological and conservation status of, and ethno-pharmacological knowledge concerning, medicinal and aromatic plants in Greece, Tunisia, and Turkey. At a number of other CIHEAM centres, especially those in Zaragoza, Spain, and Meknès, Morocco, a great deal of genetic and applied horticultural research is underway on the olive tree, pistachio, and almond tree, among others. This research involves both *in situ* and *ex situ* study of wild relatives and local races of these traditional Mediterranean fruit and nut crops whose economic future in the region depends entirely on maintaining a high level of research and conservation commitment at national and international levels.

Hotspots

In his worldwide analysis of biodiversity, Myers (1988, 1990) rated the Mediterranean Basin as one of the world's 18 hotspots. However, the Basin is far too extensive and heterogeneous to be treated as a single hotspot area.

Identifying key sectors or 'regional hotspots' that warrant special treatment (see, for example, Fig. 3.1 for vascular plants) would be a first step to developing conservation strategies at a regional scale. However, to be meaningful as candidates for special attention, these regional hotspots should present two characteristics: they should be rich in species for many different groups of organisms (Gaston 1996), and they should also be areas with a large number of rare, endemic species.

The first condition is rarely met, as shown by Williams and Gaston (1994) and Prendergast *et al.* (1993), who demonstrated that areas of maximal richness for different taxonomic groups rarely coincide. For example, in the Mediterranean area the regions of greatest species richness are the eastern Mediterranean and Iberia for reptiles, and the western Mediterranean for amphibians (Meliadou and Troumbis 1997). Nevertheless, examples of obvious candidates for regional hotspot areas in the Mediterranean are the Baetic Cordillera of south-eastern Spain and the mountains of southern Greece. In these regions there are exceptionally high concentrations of endemic species among plants, fishes, reptiles, and insects.

Protected areas

Preserving habitats and their biological communities is the most effective way to preserve biological diversity. Except for some well-known threatened species such as the monk seal or pelicans, communities and ecosystems should be the target of conservation efforts because conservation of communities can preserve large numbers of species in self-maintaining units. Much effort has been made recently to raise the number of preserves in the Basin, with the result that the total area under protection increased by 26.7% between 1985 and 1996. The total coverage of protected areas was 4.3 million hectares in 1996, including 3.4 million of terrestrial ecosystems, 0.9 million of coastal areas, and 0.5 million hectares of marine ecosystems (Ramade 1997). Protected areas range from minimal to intensive use of habitats and resources. Following the classification of the World Conservation Union (IUCN 1985) they range from strict nature reserves to multiple-use management areas that are not managed primarily for conservation but still may contain most of their original species. Such multi-use management areas may be particularly significant, especially in the Mediterranean, since they are often much larger than strict nature reserves. In an area like the Basin, characterized as it is by highly dissected landscapes with thousands of small private properties, it is difficult to establish reserves of appropriate size. Ideally, the size of any reserve should be big enough to make the targeted species' survival viable. It is also important that reserves not stand out too emphatically with respect to their surrounding environment, like oases in a desert.

In this respect a good strategy is to embed strict nature reserves in a matrix of managed areas, as is done in biosphere reserves and regional natural parks. The UNESCO's Man and Biosphere reserves, of which there are currently 36 in ten countries of the Mediterranean Basin (mostly in the western part, with 11 in

Table 10.4 Protected areas in some Mediterranean countries. (After Ramade 1997.)

Country	Size of Mediterranean area ($\times$ 10^3 ha)	Protected inland and coastal areas ($\times$ 10^3 ha)	%
Spain (79, 3, 11)[1]	40000	1604	4.0
France (39, 2, 2)	5800	498	8.6
Italy (75, 2, 2)	20000	415	2.1
Malta (1, 0, 0)	32	0.3	0.9
Croatia (16, 4,1)	2500	199	7.9
Greece (28, 4, 2)	10000	107	1.1
Turkey (35, 9, 0)	48000	517	1.1
Cyprus (8, 1, 0)	925	102	11.0
Lebanon (3, 2, 0)	1040	4.8	0.4
Israel (8, 2, 0)	2077	20	1.0
Tunisia (9, 6, 4)	10000	35	0.3
Algeria (13, 6, 1)	30000	202	0.7
Morocco (6, 0, 1)	30000	52	0.2

[1] The three figures in the first column correspond to: total number of reserves, national parks, and Man and Biosphere reserves, respectively.

Spain alone; see Table 10.4), have been conceived precisely for this purpose. The biosphere reserve concept aims to coordinate the functions of the conservation of biodiversity, the long-term monitoring and study of changes in ecosystems, and the contribution to sustainable development of local human populations. Thus, they are protected areas flexible enough to cope with the varied ecological and socio-economic conditions that prevail in each particular region. Their aim is not so much to preserve large areas of the most common habitat type as it is to include representatives of all habitats on a regional scale. Unfortunately, except for strict nature reserves, many 'protected areas' exist only on the map and receive little protection from habitat degradation, over-harvesting, and pollution.

Establishing large multi-use managed areas should be especially rewarding in the Mediterranean because of the huge potential for ecotourism based on activities such as hiking and bird- , dolphin-, or fish-watching. Ecotourism may provide one of the most immediate justifications for protecting biological diversity in the framework of sustainable development. Although not widespread as yet in the Basin, ecotourism can represent a considerable amenity value in most Mediterranean tourist areas with exceptional scenic beauty and archaeological sites. Ecotourism is a dramatically growing industry in many countries, earning approximately 11 billion Euros per year worldwide (Primack 1993). There is a large potential for ecotourism in islands and coastal areas that often include extremely valuable habitats in terms of biological diversity, but which are the most vulnerable ecosystems in the Mediterranean. The coastal fringe is the area of most human pressure with urbanization, industry, agriculture, transportation installations, and tourism. In some countries large efforts are made at a governmental level to preserve as much as possible of coastal ecosystems. For example, the French government created a 'Conservatory for Littoral and Lacustrine Shores' in 1975, which has already purchased 53 000 ha of coastal properties (10% of the total French coastline) for the express purpose of nature conservation.

Reintegration of fragmented landscapes

Besides action plans specifically devised to protect species and communities, the most important goal for putting into practice the concept of sustainability is to reconsider land use planning in the broadest sense. For Mediterranean ecosystems that have been subjected to long periods of human perturbations, there exist three possible alternatives to continued degradation or abandonment. These are restoration, rehabilitation, and reallocation (Aronson et al. 1993a). All three should be planned in the context of what Hobbs and Saunders (1992) have called the 'reintegration of fragmented landscapes'. Restoration and rehabilitation of ecosystems both aim at recreating self-sustaining ecosystems characterized by autogenic succession in plant and animal communities and sufficient resilience to repair themselves following natural or moderate human perturbations. Restoration of an ecosystem aims at the complete (though admittedly virtually impossible) return to a pre-existing state in terms of species composition, that is, the re-establishment of all indigenous species and the elimination of all exotics. Rehabilitation, by contrast, concentrates on repairing damaged or blocked ecosystem functions, with the primary goal of raising ecosystem stability and productivity. It is primarily practised in economically disadvantaged regions where rural populations still depend on local resources for their livelihood and survival. However, it is also highly pertinent to conservation aims since increasing productivity and profitability for local people in certain regions should ultimately reduce pressure on protected areas set aside for nature conservation or restoration.

In a semi-arid region of Tunisia, experimental attempts to regain former levels of pastoral productivity through rehabilitation efforts have been in action for some time. A primary goal was the reintroduction of nitrogen-fixing legumes and other 'keystone' species that have become rare or were eliminated altogether as a result of overcutting and overgrazing (Floret and Pontanier 1982). Table 10.5 shows four ecosystem indicators that reflect rapid and dramatic changes, firstly in the course of ecosystem degradation covering several decades, and also in the initial period following rehabilitation-oriented interventions. Note the dramatic change in all four indicators during the very brief period (3 years) of the rehabilitation experiment. These data strongly suggest that simple techniques can be found to reverse degradation-related trends, and to 'pilot' Mediterranean and arid zone ecosystems into a new trajectory exemplified by greater biodiversity and productivity in a relatively short period.

Table 10.5 Evolution of four ecosystem indicators, in late spring, of the typical steppe, 50 km north-west of Gabes, Tunisia, in various stages of degradation, and 3 years after the beginning of an experimental rehabilitation (Aronson *et al.* 1993c; Le Floc'h *et al.* 1995)

Ecosystem indicators	Stages of ecosystem development				
	RS1	RS2	RS3	SP1	Rehab.
No. of annual species	41	13	6	2	21
No. of perennial species	23	18	3	1	12
Total plant cover (%)	60	20	2	0.5	37
Above-ground biomass in spring (kg dry matter ha^{-1} yr^{-1})	1800	600	200	100	*c.*1200

RS1, *Rhanterium suaveolens*-dominated steppe on deep sandy soils in relatively undisturbed conditions.
RS2, same steppe moderately degraded through overgrazing and woodcutting.
RS3, same steppe badly degraded and with truncated soils.
SP1, *Stipagrostis pungens*-dominated steppe, with all shrubs removed.
Rehab., rehabilitated steppe after reseeding of selected native species, including legumes, grasses, and the dominant shrub, *Rhanterium suavcolens*.

We will now look at some more examples of restoration or rehabilitation that have been done, or should be done, in the three most important types of Mediterranean landscapes, namely forests, matorrals, and wetlands.

Forests

As explained in Chapter 8, Mediterranean forests traditionally provided a wide range of products and 'services'. From the time of the Roman emperors to the present, many attempts have been made to restore forests following periods of heavy exploitation, especially in mountainous areas. The French programme 'RTM' (Restoration of Mountain Lands) was an early and exemplary initiative to modernize forestry, and to develop new techniques for restoring heavily degraded mountain slopes in the Mediterranean area. Sufficient time has now elapsed to allow a thorough analysis of the methods used and results achieved.

Nearly 140 years old, the effort at Mt. Ventoux, southern France, is worth summarizing as an example of the numerous RTM programmes undertaken around the same time. In the middle of the last century nearly all the southern slopes of the Mt. Ventoux range, just like those of most other mountains in southern Europe, were completely deforested as a result of several centuries of woodcutting and overgrazing. In the mid 1850s, the French botanist Martins wrote that 'most of the massif is the sole realm of thyme and lavender, almost devoid of trees'. A stony desert, with only scattered sub-shrubs and large expanses of bare ground exposed to wind and surface erosion, was all that was left of the dense forest mantle that had formerly included stands of pines, oaks, beech, maples, and many others according to the life zones occurring from about 350 m to 1800 m (see Chapter 4). Devastation was the only word to describe the scene. The second Empire was a period of great prosperity in France and, among other monumental public works, the decision was taken in 1861 to reforest Mt. Ventoux. The main reason given was to stop soil erosion and, if possible, to 'restore' some of the lost soil. An enormous amount of money was devoted to this project. Between 1861 and 1873, 2500 ha were planted with corns and tree saplings on eroded and rocky slopes. Workers had to climb up each day from their base camps at the foot of the mountain, carrying with them an armload of plants and jerrycans of water. The sequence of tree species planted ranged from Aleppo pines and holm-oaks at the foot of the mountain, to mountain pines near the timberline, at 1500 m altitude, with downy oaks and beech in between. Moreover, many non-native species were experimentally introduced, such as atlas cedar, spruce, larch, cluster pine, and black pine. Remarkably this huge and still ongoing project has succeeded in reconstituting dense forest cover of mixed ages, even on the most exposed southern slopes of the mountain, and autogenic succession has clearly been reinstated. The original plantations of native trees, complemented by Atlantic cedars of Algerian origin, have contributed to the gradual reconstitution of soils and the reestablishment of a mixed canopy under which native trees, shrubs, and herbs have been able to recolonize.

Since the main aim of the programme was to protect soils against erosion, management did not involve any kind of major interference with the development of vegetation. Careful studies conducted in the 1970s (du Merle 1978) have shown that plant and animal communities of the reconstituted forests do not differ greatly from those that occurred in the few places where the forest cover had not been destroyed. Bird communities of the planted areas in particular are now very similar to those of relatively undisturbed forest patches used as a reference, as shown by in-depth quantitative studies. The RTM reforestation programme has led in this case to the return of a European woodland avifauna and the replacement of Mediterranean species that had secondarily occupied these areas after destruction of the primeval forests.

A certain number of special cases might also be usefully considered in the context of conceptual approaches to restoration, such as isolated forest patches long preserved in specific sites for religious or other cultural reasons (see Box 10.3).

Matorrals

The longstanding traditional uses of matorrals allowed the conservation of considerable biodiversity but, in recent decades, especially on the southern rim of the Basin, many have become heavily degraded. Yet grazing, even heavy grazing, is not necessarily a threat to biodiversity and may even increase it. The high degree of resilience of Mediterranean matorrals, especially in the presence of a

Box 10.3 The Sainte-Baume relictual forest

The massif of Sainte Baume lies about 50 km south-east of Mt. Ventoux. This is a relictual forest fragment that is certainly not pristine, but can be called 'primary-type' forest. This 136-ha site was spared from woodcutting and forest exploitation in the past because of the legend that Mary Magdalene spent the last 33 years of her life here.

The cave located about halfway up the steep northern face—*baoumo* means cave in Provençal—has been a celebrated stopping-place for pilgrims on their way to Santiago de Compostella (Europeans' first popular pilgrimage site) from the early Middle Ages to this day. Accordingly, just as with the eastern Mediterranean 'sacred trees' mentioned in Chapter 8, the woods around the Sainte Baume site have been spared for centuries. Even though the site is only 15 km inland, huge beech, maple, and lime trees abound, with thick ivy vines, yews, and hollies in the understorey, such as can rarely be seen elsewhere in the north-west Mediterranean quadrant. This reflects the northern exposure of this slope and protected inland position of the site. It also testifies to the conservation-minded management it underwent as a result of being a pilgrimage site. Accordingly, it could serve as an 'ecosystem of reference' for forest restorationists working in similar conditions. Further south, and in drier conditions, a similar kind of model can be seen in the 60 ha of forest at the Bussako Palace, 50 km north of Lisbon. This property was for centuries attached to a monastery, before being taken over and preserved as a royal hunting reserve.

balanced load of grazers and browsers, either domesticated or wild, can result in a dynamic co-existence of living systems characterized by great stability, diversity, and productivity (Etienne *et al.* 1989). Comparing two islands off Crete, one heavily grazed by the Cretan wild goat (*Capra aegragus cretica*) and the other ungrazed, Papageorgiou (1979) demonstrated that plant species diversity is much higher in the former than in the latter, despite the evil reputation of goats as 'four-legged locusts'. Moreover, maintaining herds of sheep, goats, or other ungulates in matorrals and forestry production zones is perhaps the best way to reduce the risk of catastrophic wild fires.

Among unwanted side effects of conservation and restoration policies are the tendencies of some populations of some wild grazers and browsers to get out of control. For example, several decades of stringent conservation policies in Israel allowed the formerly vulnerable mountain gazelle to build up such large populations by the 1980s that they became a nuisance to farmers. By limiting population sizes to 15 individuals km^{-2} a new management policy not only allowed the harvesting of 1000–1800 animals each year for meat, but also resulted in a significant increase in plant species diversity of the rangelands used by the gazelles (Kaplan 1992). The dramatic conclusion here is that a moderate grazing intensity should be recommended to conserve the full potential of species diversity and also optimize ecosystem productivity. Seligman and Perevolotsky (1994) obtained similar results in their review of both sheep and cattle grazing systems in the eastern Mediterranean. The dehesa system we highlighted in Chapter 8 is an example of how long-term productivity can be obtained by mixing tree crops, annual cereals, and livestock. Needless to say, the socio-economic problems far outnumber the technical or purely ecological problems involved in combining sustainable economic development with the growing need for conservation in the Mediterranean area. One challenge will be to learn how to manage matorrals in the near total absence of short-term economic interest for the state or private individuals.

Wetlands

Mediterranean wetlands are among the most threatened ecosystems on Earth. They formerly extended over very large areas but have been widely drained during the last few millennia. Major wetland drainage begun in Italy during the Etruscan period (fifth century BC) and accelerated in the time of the Romans, expanding agricultural areas at the expense of riverine wetlands in all the major valleys of Italy and in their provinces in France, Spain, and North Africa. In Roman times, there were 3 million hectares of wetlands in Italy alone. At the beginning of the twentieth century, only 1 300 000 were left, a figure that had dropped to 300 000 ha by 1991. After the collapse of the Roman Empire wetlands recovered, but drainage was renewed again under the leadership of large and powerful monasteries in the Middle Ages. Wetland drainage increased still further after the Renaissance, often with the help of Dutch engineers. A further acceleration took place in the 1850s with the introduction of steam-powered machinery. A new impetus for drainage occurred at the end of the nineteenth and the beginning of the twentieth century to eradicate malaria, which was endemic in all Mediterranean countries.

No more than 28 500 km^2 of wetlands are left in the Basin today (Pearce and Crivelli 1994), including 6500 km^2 of coastal lagoons, 12 000 km^2 of lakes and natural mashes, and 10 000 km^2 of artificial wetlands, mostly dammed lakes. Salinas are present in all Mediterranean countries with a total coverage of 621 km^2 but the largest ones occur in the more industrialized countries of southern Europe (570 km^2 versus 51 km^2 in North Africa). In spite of much warning against the destruction of wetlands over the last 50 years, and despite several projects aiming at conserving and restoring surviving Mediterranean wetlands, dredging and draining continued unabated, until a decade or so ago, to make new space for agriculture, aquaculture, intensive grazing lands, salt pans, and tourist installations. In the Camargue, one of the best protected wetlands of the whole Basin, no less than 40% of natural habitats have been lost in the last 50 years (Tamisier and Grillas 1994). There is now a decelerating trend in wetland destruction in most parts of the Basin but the main pending problem is mismanagement and overexploitation of many of them. Additional threats are sedimentation, siltation, episodes of hypertrophic anoxia, and sea level elevation, which is occurring at a rate of 1–2 mm per year. The International Project on Climate Change forecasts that this will increase to between 3 and 8 mm per year by 2030. Many coastal wetlands will be at risk of over-flooding.

Yet, Mediterranean marshlands perform a range of functions that deserve conservation and better management. First, they play a role in regulating the often huge variations in annual rainfall. Coastal wetlands sponge up rapidly accumulating quantities of water during heavy rainfalls and then serve as a source of surface water during dry periods. Thus they can help buffer and alleviate the potentially devastating effects of storms, if properly managed. Second, wetlands are habitats for wildlife. Many of them are hotspots of diversity for many rare species of plants, insects, fishes, and birds. Nearly 50% of Europe's bird species and 30% of the plant species depend more or less exclusively on wetland habitats. Hundreds of rare and endemic species of insects in the Basin are characterized by at least one aquatic stage in their development. For example, among carab beetles not less than 500 species are exclusively wetland dwellers in Italy alone (Baletto and Casale 1991). Third, wetlands offer many services to humans, including hunting, fishing, reed cropping, animal grazing, and ecotourism. In most Euro-Mediterranean countries hunting is in fact a life-saver for wetlands. The estimated annual harvest of hunters in France alone amounts to 1–3.5 million ducks with an annual value of the carcasses equalling 10.5 million Euros. Many of the largest and most famous wetlands of southern Europe owe their existence to economic benefits derived from hunting and grazing by local land races of horses and cattle. For example, experiments in progress in the Camargue aim at rehabilitating abandoned rice fields through management of standing water to improve habitat quality for wildlife and grazing potential for horses and cattle. By monitoring water levels and salt concentration, Mesléard *et al.* (1995) found the best compromise for the development of an herbaceous plant cover that allows grazing by livestock and provides feeding and breeding habitats for birds. A long-term process of preserving and restoring Mediterranean wetlands began in the early 1960s with the IUCN's 'MAR' project. Another important vehicle for wetland conservation is the RAMSAR convention (see below). A symposium held at Grado, Italy, in 1991 defined an action plan to stop and to reverse the loss and degradation of Mediterranean wetlands. The Tour du Valat Biological Station is entirely involved in basic research and conservation biology in these habitats (Box 10.4).

Managing and living with fire

On average, 200 000 ha of forest and matorral are burnt each year in the Mediterranean Basin (Le Houérou 1981), but inter-annual and inter-regional variability is very high. The summer of 1989, for example, was exceptionally dry throughout the area, and nearly 2 million hectares of forest and shrubland burned down. In September 1994, over 150 000 ha burned in one week in northern

Box 10.4 The Tour du Valat Biological Station

For more than 50 years the biological station of La Tour du Valat, in the Camargue, has conducted ecological studies on the population biology of key animal and plant species and the ecosystem functioning of Mediterranean wetlands. Long-term studies have been carried out on birds, fishes, mammals, including semi-wild herds of horses, and plants in an attempt to integrate population dynamics in food webs at the level of ecosystems.

For stopping the loss of Mediterranean wetlands, rehabilitating degraded ones, and piloting rational use of the remaining wetlands across the whole Basin, a long-term action plan called 'MedWet' was recently launched under the auspices of the RAMSAR Convention supported by the European Community. This action plan, monitored by the Tour du Valat Biological Station, with partners such as the World Wildlife Fund (WWF), Wetlands International and the Greek Centre for Wetland Conservation, promotes international cooperation to preserve vital or key specimens of these pivotal habitats, while also updating resource management and environmental protection plans in the framework of sustainable development.

Spain. Natural or spontaneous fires account for only about 1–2% of the fires started in the Mediterranean today and in the historical past (Trabaud 1981). Since matorrals and woodlands combined include about 150 m^3 ha^{-1} of wood, a 10 000 ha fire destroys on average 1.5 million m^3, which is roughly the same amount of wood as is exploited commercially each year in the entire French Mediterranean region. Although only some 0.3% of the total forested area in southern Europe is affected annually by fire, in some areas the figure may reach 10% and more. In the so-called 'red belt' of southern France, for example, fires rage out of control every 25 years on average in any one place (Le Houérou 1990). When the 'return rate' of fires is too high, the degradation of vegetation leads to dwarf formations and asphodel 'deserts' that are of low diversity. When landscapes were intensively managed by dense rural populations, most fires were not excessively harmful since they were immediately attended to and quickly put out. In most Euro-Mediterranean regions today, rural depopulation combined with encroachment of woody vegetation result in the accumulation of inflammable material over large areas. Hence the risk of very large fires with catastrophic consequences exists over millions of hectares. Fire-control policies are difficult to implement, and the consequences of large fires are difficult to control.

Although most people consider fires as devastating scourges, it is unrealistic and biologically unsound to strive to totally prevent their occurrence. Fires are natural disturbance events that contribute to maintaining the 'moving mosaic' of communities and ecosystems at the scale of landscapes (see Chapters 6 and 7). As such, they participate to a large extent in maintaining biological diversity and ecosystem 'health'. Therefore they must be integrated in management policies and peoples must learn to live with occasional fires. But there is a need to devise techniques that control the frequency and intensity of fires. The best strategy would be to manage forests and woodlands so as to anticipate and reduce the risk of catastrophic fires *before* the next fire breaks out. Although the idea of intentionally setting fires is anathema for most people in the Basin, fire itself, if properly managed, has just as important a role to play as grazing in the rational maintenance of open spaces, and the management of mosaic landscapes that are biologically diverse.

Concurrently, given the large number of accidental or intentional fires that occur each year, and the too-rapid 'fire return' cycle this entails, it is vital to develop effective re-vegetation techniques for restoring burned lands. Otherwise, a desertification process is set up, especially in hilly or montane areas with shallow topsoils. In eastern Spain the shrublands and Aleppo pine forests around Valencia are among the most seriously affected by wildfires within the Mediterranean Basin, and this is one of the regions where the danger of fire-induced desertification is greatest. Over the past 30 years the yearly average has been about 2%, with the result that the portion of the province considered as 'badlands' has increased by 14% (Vallejo and Alloza 1998).

In response to this ongoing debacle a group of researchers has been studying post-fire regeneration, combined with field experiments consisting of actual interventions aimed at post-fire restoration on a large scale. Techniques based on direct seeding and planting of numerous native species are being studied, and species selection is underway for tree and shrubs adapted to colonizing the different soil types and climatic life zones where the last fire had occurred (Vallejo 1996). Furthermore, a 'toolbox' full of post-sowing or planting techniques is being developed, including mulching of newly planted areas to improve chances for plant survival and reduce erosion at the same time. This team is currently spearheading an international project to develop greater cooperation throughout the Mediterranean area by developing a restoration and rehabilitation network and database. This project is conducted in the context of the European Union's plans to promote large-scale reafforestation, and ongoing efforts to stem the tide of desertification on the southern and eastern banks of the Basin.

International cooperation

As a result of a growing concern about the continuing degradation of the Mediterranean Sea, which is a natural link and a common property for all Mediterranean peoples, all countries bordering the Mediterranean met in Barcelona in 1975 under the auspices of the United Nations for Environment Programme (UNEP) and launched an action plan, the so-called Barcelona Convention, for stopping and reversing the degradation of the environment. Initially designed to struggle against pollution of the Sea, this project was soon extended to terrestrial ecosystems because most pollution comes from the land. This initiative, known as 'Blue Plan', which is funded by the World Bank, is brought into play by all the Mediterranean countries and the European Community (Grenon and Batisse 1989). The Barcelona Convention appointed a Coordination Unit, which is located at Athens. Specialized institutions of the United Nations and international non-governmental organizations such as the WWF and World Conservation Union (IUCN) are involved in the Blue Plan. These organizations have an extremely active role as regards the biodiversity in the Mediterranean coastal zone: many hotspots of plant diversity, for example the Baetic cordillera of southern Spain, the mountains of central and southern Greece, north-east and south-west Anatolia, and several others, are among the 231 sites throughout the world identified as the most important centres of plant diversity requiring protection and proper management (Médail and Quézel 1997).

Another tool for the conservation of wetlands of international importance is the RAMSAR convention, established in 1971 to halt the destruction of Mediterranean wetlands and to promote instead their ecological, scientific, economic, and cultural value (Kusler and Kentula 1990). Up to now, 89 wetland sites totalling 46 000 km^2 in 12 Mediterranean countries, have been nominated for inclusion in this convention. Together with the UNEP Mediterranean Action plan and the UNESCO Man and Biosphere programme, this convention is actively concerned with promoting sustainable development of the wetlands that constitute the most important component of these sites of exceptional biological diversity. There are many other governmental and private initiatives to stop environmental degradation of both marine and terrestrial ecosystems in the Mediterranean. One of them is the Alghero Convention organized at Alghero (Sardinia) in 1995 by the association 'MEDVARAVIS'. It is supported by 35 Non-Governmental Organizations (NGOs) under the auspices of the Bern Convention, Council of Europe, IUCN, and the United Nations Environment Programme (Mediterranean Action Plan). More species-oriented projects include action plans designed for the protection and restoration of populations of large mammals and birds. For example, in 1992 BirdLife International launched a project to prepare action plans for all the globally threatened birds that occur in the Basin. As a result of these efforts the populations of numerous species of large raptors are recovering in many countries of the northern bank, sometimes at spectacular rates (Muntaner and Mayol 1996).

Alternative futures

The most promising development in the struggle to preserve biological diversity in the Mediterranean area is the realization that biodiversity is in fact a key component of sustainable regional development. Although sustainability has become a key concept, there is no generally accepted guidance on how to define and assess sustainability, still less on how to achieve it. Without such guidance, two problems arise. First, many activities will continue to contribute to the depletion of species and the degradation of ecosystems. Second, uses with social and conservation benefits have to struggle against hostile policies and private interests. Therefore the most urgent task is to develop and define guidelines on the sustainable use of wildlife and ecosystems. Such guidelines will have to be flexible enough to take into account the extraordinary diversity of cultures, traditions, and land use practices in the Mediterranean Basin. Any guideline on sustainability will be inefficient if local peoples, with their practical experience and cultural values, are not involved in its definition and do not agree with it. As regards development and environment, the Mediterranean stands among the worst regions in the world (di Castri 1996), which is due in large part to its incredible diversity in all biological and anthropological aspects. This results in a large number of frontiers that act as brakes to partnership, cooperation, and global management. As a result of this (and of the recent growth of several forms of fundamentalism, the opposite of what is needed for development and environment), there is much ingrained, irrational resistance and rejection to any action plan that would integrate the many regions of the Basin into a single unit.

One avenue that would be particularly promising for promoting the idea of sustainability would be to design agricultural systems in 'nature's image' (Dawson and Fry 1998). This ecologically based approach, which has recently gained momentum in other parts of the world, consists in mimicking the natural functions of the biota of any given region. The working premise is that natural ecosystems of any region are adapted to fluctuations in key resources as well as to the constraints imposed by the environment. They therefore provide regionally specific models for sustainability if well mimicked by new land use practices. Several experiments in progress in the United States and Australia should help in redesigning land use systems in structural and functional terms. In the Mediterranean, the 800-year-old 'dehesa–montado' system (see Chapter 8) is an example whose sustainability derives from sub-optimal production, an adaptation to the highly variable Mediterranean climate (Joffre and Rambal 1993). Successful mimic systems should look for complementary species according to the 'M5' golden rule: Making Mimics Means Managing Mixtures (Dawson and Fry 1998). Mediterranean habitat mosaics are an ideal place to put into practice these new models, which should be based on a scientific understanding of the functional characteristics of the main players. The problem is to identify plant and animal species that will provide a diversity of functional roles, accommodate environmental variation, and grow well in mixtures. They must also offer at least the hope of economic viability.

Conclusion

Promoting sustainability requires raising collective consciousness about the importance of preserving natural and cultural heritages that should be shared by all who live around this luminous Sea. Only in that way can we possibly create a common space of cooperation and trade, the kind of 'Mediterranean Cultural

Community' that Albert Jacquard (1991) has called for. Science can perhaps help us learn how to conserve Nature but the politics of nature conservation is a social and cultural affair to be taken up by Nations. Thus the ultimate challenge will be to take fully into account and preserve, insofar as possible, the common biological and cultural heritage shared by all the regions and nations within the Basin, including the regional and ethnic specificities that form the essence of the Mediterranean world. The Mediterranean has always been more than a simple geographical unit. For millennia, it was a humanist and spiritual forum, the home of Plato, Aristotle, and Hippocrates, Moses and Maimonides, Jesus and Mohammed, Augustine and Averroës. This spiritual and cultural heritage had, and still has, profound influence on living standards, economy, and relationships between humans and nature in all parts of the Basin. Long before the modern era the Basin had already undergone several periods of cultural and economic 'globalization'. The high levels of trade and cultural exchange pursued by the ancient Greeks, Romans, Phoenicians, Venetians, and others superimposed new living standards and cultural diversity without destroying pre-existing ones. To achieve this effect in our day a gigantic effort will be necessary because of the tremendous disparity in all that makes up the lives of the various societies around the Mediterranean. While it is true the Mediterranean no longer plays the pivotal role in international relations it once did in the age of the great seafaring ships, its long history as biological and cultural crossroads and melting pot, as well as its position as undisputed cradle of western civilization, surely makes of it as good a place as any to start this long process.

Despite 10 000 years of intense human presence, the Mediterranean is still one of the most spectacularly rich biological hotspots in the world. Coming back home, at the end of this long journey, we trust that the keen naturalist we imagined in the first page of this book has enjoyed this guided tour of Mediterranean biodiversity, even if the long excursions we have taken along dusty and rugged paths were sometimes exhausting.

Afterword

Richard Hobbs, CSIRO Wildlife and Ecology

Mediterranean-type climates, characterized by a predominantly winter rainfall regime, exist in five regions of the world: parts of California, South Africa, Chile, southern Australia, and the Mediterranean basin. While all these regions share broad climatic conditions and contain ecosystems which have similar structures and dynamics (Hobbs *et al.* 1995), each region has a unique history of human habitation and use. The Mediterranean basin is probably the most difficult region to develop an overview of because of its geological, biological and socio-political diversity. Its rich and prolonged history, involving the rise and fall of civilizations, conquest and colonization, and reorganization of state boundaries make the current political map of the region a complex tapestry of nations and unique local histories. This tapestry is itself draped over the complex biophysical patterns found in the region. To understand the biota of the region, we need to understand the importance of the interactions between the biophysical and historical components that shaped it.

Such an analysis has not been attempted before in any comprehensive way, largely, I would guess, because of the size and complexity of the task. It is thus with great admiration that I conclude this book, which attempts the seemingly impossible task of taking a broad overview of biological diversity in the Mediterranean basin. Such an overview necessitates a discussion of the geological and biological development of the region and how this has shaped the present day biota. However, a clear picture of the nature of the biota cannot be achieved simply by lists of species, but needs to be considered in terms of a variety of spatial scales and levels of biological organization. The various chapters in this book do just that, including discussions of different biotic groups, regional scale patterns, and detailed discussions of the predominant ecosystem types. Dynamics and function of the ecosystems found in the Mediterranean basin are also covered, and we also get a glimpse into the myriad numbers of complex interactions between different species that maintain ecosystem function and provide endless fascination for biologists.

The 10 000-year 'love story' between humans and nature in the Mediterranean has resulted in vast changes to the ecosystems of the region, ranging from the extinction of species through to the development of ecosystems that now depend on certain types of management for their persistence. The story continues today, and the problems of conserving biotic diversity in Mediterranean areas are many. The clear distinction between the relatively affluent northern Mediterranean and the less affluent southern area renders any generic solution difficult. Changes in land use threaten to alter systems, often in unpredictable ways. It is clear, however, that systems with such a long history of modification by humans will change if traditional management practices cease or are modified. We thus are faced with making choices about what our ecosystems should look like and what we want of them. Not making conscious choices but simply 'letting nature take its course' is in itself a choice, the consequences of which may not be advantageous. For instance, cessation of grazing or harvesting may lead to the local loss of species dependent on regular disturbance and to changes in the appearance and function of the landscape. Making choices about how to manage ecosystems and landscapes demands that we have a clear set of goals relating to the aspects of these systems which we value and wish to retain or improve. Setting such goals and making effective decisions is made easier when suitable information is available. There will always be gaps in our knowledge, but this book goes a long way towards providing a useful set of background information upon which future choices concerning the biological diversity of the Mediterranean region can be based.

While this region is in many ways unique, it holds numerous lessons for other parts of the world, especially where the links between humans and natural ecosystems are less apparent. Virtually nowhere in the world is now free from human impact—we face the same need to make informed choices on the future of all the world's ecosystems. Ultimately, these choices will determine the course of humanity's future too.

References

Abdul Riga, A. M. M. and Bouché, M. (1995). The eradication of an earthworm genus by heavy metals in southern France. *Applied Soil Ecology*, **2**, 45–52.

Achérar, M. and Rambal, S. (1992). Comparative water relations of four Mediterranean oak species. *Vegetatio*, **99–100**, 177–84.

Achérar, M., Lepart, J., and Debussche, M. (1984). La colonisation des friches par le Pin d'Alep (*Pinus halepensis*) Mill.) en Languedoc méditerranéen. *Oecologia Plantarum*, **19**, 179–89.

Aerts, R. (1995). The advantage of being evergreen. *Trends in Ecology and Evolution*, **10**, 402–7.

Affre, L., Thompson, J. D., and Debussche, M. (1997). Genetic structure of continental and island populations of the Mediterranean endemic *Cyclamen balearicum* (Primulacae). *American Journal of Botany*, **84**, 437–51.

Aidoud, A. and Nedjraoui, D. (1992). The steppes of alfa (*Stipa tenacissima*) and their utilisation by sheep. In *Plant-animal interactions in Mediterranean-type ecosystems* (ed. C. A. Thanos), pp. 62–7, University of Athens.

Al Hallani, F., Andreoni, R., Bassil, M., Combe, C., Magaud P., and Seytre, L. (1995). La cédraie de Barouk (Liban). *Forêt Méditerranéenne*, **16**, 171–83.

Alcover, J. A., Moya-Sola, S., and Pons-Moya, J. (1981). *Les quimeres del passat. Els vertebras fossils del Plio-Quaternari de les Baleares i Pitiuses*. Palma de Mallorca.

Amigues, S. (1980). Quelques aspects de la forêt dans la littérature grecque antique. *Revue Forestière Française*, **32**, 211–23.

Amigues, S. (1991). Le témoignage de l'antiquité classique sur des espèces en régression. *Revue Forestière Française*, **33**, 47–57.

Anstett, M. C., Michaloud, G., and Kjellberg, F. (1995). Critical population size for fig/wasps mutualism in a seasonal environment: Effect and evolution of the duration of female receptivity. *Oecologia*, **103**, 453–61.

Aronson J. and Shmida, A. (1992). Diversity along a Mediterranean-desert gradient in response to interannual fluctuations in rainfall. *Journal of Arid Environments*, **23**, 235–47.

Aronson J., Kigel, J., Shmida, A., and Klein, J. (1992). Adaptive phenology of desert and Mediterranean populations of annual plants grown with and without water stress. *Oecologia*, **89**, 17–28.

Aronson, J., Floret, C., Le Floc'h, E., Ovalle, C., and Pontanier, R. (1993a). Restoration and rehabilitation of degraded ecosystems. I. A view from the south. *Restoration Ecology*, **1**, 8–17.

Aronson J., Kigel, J., and Shmida, A. (1993b). Reproductive allocation strategies of desert and Mediterranean populations of annual plants grown with and without water stress. *Oecologia*, **93**, 336–42.

Aronson, J., Floret, C., Le Floc'h, E., Ovalle, C., and Pontanier, R. (1993c). Restoration and rehabilitation of degraded ecosystems. II. Case studies in Chile, Tunisia and

Cameroon. *Restoration Ecology*, **1**, 168–87.

Aronson, J., Dhillion, S, Le Floc'h, E., and Abrams, M. (1998). Rehabilitation of degraded ecosystems in mediterranean France and their reintegration into living landscapes. *Landscape and Rural Planning*, **41**, 273–83.

Arroyo, M. T. K., Cavieres, L., Marticorena, C., and Muñoz-Schick, M. (1994). Convergence in the mediterranean floras in central Chile and California: insights from comparative biogeography. In *Ecology and biogeography of Mediterranean ecosystems in Chile, California and Australia* (ed. M. T. K. Arroyo, M. Fox, and P. Zedler), pp. 43–88. Springer-Verlag, Heidelberg.

Attenborough, D. (1987). *The first eden. The Mediterranean world and man*. Fontana/ Collins, London.

Auerbach, M. and Shmida, A. (1985). Harmony among endemic littoral plants and adjacent floras in Israel. *Journal of Biogeography*, **12**, 175–87.

Auffray, J. C., Tchernov, E., and Nevo, E. (1988). Origin of the commensalism of the house mouse (*Mus musculus domesticus*) in relation to man. *Comptes Rendus de l'Académie des Sciences, Paris*, **307**, 517–22.

Auffray, J. C., Vanlerberghe, F., and Britton-Davidian, J. (1990). The house mouse progression in Eurasia: a palaeontological and archaeozoological approach. *Biological Journal of the Linnean Society*, **41**, 13–25.

Avise, J. C. and Walker, D. (1998). Pleistocene phylogeographic effects on avian populations and the speciation process. *Proceedings of the Royal Society of London B*, **265**, 457–63.

Avishai, M. (1985). The role of botanic gardens. In *Plant conservation in the Mediterranean area* (ed. C. Gomez-Campo), pp. 221–36. Dr. W. Junk, Dordrecht.

Axelrod, D. I. (1975). Evolution and biogeography of Madrean-Tethyan sclerophyll vegetation. *Annals of the Missouri Botanical Garden*, **62**, 280–334.

Bairlein, F. (1991). Nutritional adaptations to fat deposition in the long-distance migratory garden warbler *Sylvia borin*. In *Proceedings of the XXth International Congress of Ornithology*, pp. 2149–58. Christchurch, New Zealand.

Bairlein, F. (1992). Recent prospects on trans-Saharan migration of songbirds. *Ibis*, **134** (suppl. 1), 41–6.

Baker, H. G. and Stebbins, G. L. (1965). *The genetics of colonizing species*. Academic Press, New York.

Baletto, E. and Casale, A. (1991). Mediterranean insect conservation. In *The conservation of insects and their habitats* (ed. N. M. Collins and J. A. Thomas), pp. 121–42. Academic Press, London.

Barbéro, M., Loisel, R., and Quézel, P. (1991). Sclerophyllous *Quercus* forests in the eastern Mediterranean area: ethological significance. *Flora et Vegetatio Mundi*, **9**, 189–98.

Barbéro, M., Loisel, R., and Quézel, P. (1992). Biogeography, ecology and history of Mediterranean *Quercus ilex* L. ecosystems. *Vegetatio*, **99–100**, 19–34.

Barbéro, M., Loisel, R., Quézel, P., Richardson, D. M., and Romane, F. (1998). Pines of the Mediterranean Basin. In *Ecology and Biogeography of Pinus* (ed. D. M. Richardson), pp. 153–70. Cambridge University Press.

Baumann, H. and Kuenkele, S. (1982). *Die wildwachsenden Orchideen Europas*. Kosmos Verlag, Stuttgart.

Beadle, N. C. W. (1966). Soil phosphate and its role in molding segments of the Australian flora and vegetation, with special reference to xeromorphy and sclerophylly. *Ecology*,

47, 992–1007.

Benchekroun, F. and Buttoud, G. (1989). L'arganeraie dans l'économie rurale du sud-ouest marocain. *Forêt méditerranéene*, **11**, 127–36.

Ben-Tuvia, A. (1981). Man-induced changes in the freshwater fish fauna of Israel. *Fisheries Management*, **12**, 139–48.

Ben-Tuvia, A. (1985). The impact of the Lessepsian (Suez Canal) fish migration on the eastern Mediterranean ecosystem. In *Mediterranean marine ecosystems* (ed. M. Moraitou-Apostolopoulo and V. Kiortsis), pp. 367–75. Plenum Press, New York.

Bernhardt, P. and Burns-Balogh, P. (1986). Floral mimesis in *Thelymitra* (Orchidaceae). *Plant Systematics and Evolution*, **151**, 187–202.

Berthold, P. (1993). *Bird migration. A general survey*. Oxford University Press.

Berthold, P., Fiedler, W., Schlenker, U., and Querner, U. (1998). 25-year study of the population development of central european songbirds: a general decline, most evident in long-distance migrants. *Naturwissenschaften*, **85**, 350–3.

Bianchi, C. N. (1996). The state of marine biodiversity. *International Conference on Mediterranean Biodiversity*, pp. 51–61. ENEA, Roma.

Bibby, C. J. (1992). Conservation of migrants on their breeding grounds. *Ibis*, **134** (suppl. 1), 29–34.

Bino, R. J., Dafni, A., and Meeuse, A. D. J. (1982). The pollination ecology of *Orchis galilea* (Bornm. et Schulze) Schltr. (Orchidaceae). *New Phytologist*, **90**, 315–19.

Blanco, J. C. and Gonzalez, J. L. (ed.) (1992). *Libro rojo de los vertebrados de Espana*. Coleccion Tecnica, ICONA, Madrid.

Blondel, J. (1969). *Synécologie des passereaux résidents et migrateurs dans un échantillon de la région méditerranéenne française*. Centre Régional de Documentation Pédagogique, Marseille.

Blondel, J. (1970). Biogéographie des oiseaux nicheurs en Provence occidentale, du Mont-Ventoux à la mer Méditerranée. *L'Oiseau*, **40**, 1–47.

Blondel, J. (1985a). Historical and ecological evidence on the development of Mediterranean avifaunas. *Acta XVIII Congressus Internationalis Ornithlogici*, Vol. 1, pp. 373–86, Moscow.

Blondel, J. (1985b). Habitat selection in island versus mainland birds. In *Habitat selection in birds* (ed. M. L. Cody), pp. 477–516. Academic Press, New York.

Blondel, J. (1986). *Biogéographie évolutive*. Masson, Paris.

Blondel, J. (1987). From biogeography to life history theory: a multithematic approach. *Journal of Biogeography*, **14**, 405–22.

Blondel, J. (1991). Birds in biological isolates. In *Bird population studies, their relevance to conservation and management* (ed. C. M. Perrins, J. D. Lebreton, and G. Hirons), pp. 45–72. Oxford University Press.

Blondel, J. (1995). *Biogéographie. Approche écologique et évolutive*. Masson, Paris.

Blondel, J. and Aronson, J. (1995). Biodiversity and ecosystem function in the Mediterranean basin: human and non-human determinants. In *Mediterranean-Type Ecosystems. The Function of Biodiversity* (ed. G. W. Davis and D. M. Richardson), pp. 43–119. Springer-Verlag, Berlin, Heidelberg.

Blondel, J. and Farré, H. (1988). The convergent trajectories of bird communities in European forests. *Oecologia*, **75**, 83–93.

Blondel, J. and Isenmann, P. (1981). *Guides des Oiseaux de Camargue*. Delachaux et Niestlé, Neuchâtel.

Blondel, J. and Vigne, J. D. (1993). Space, time, and man as determinants of diversity of birds and mammals in the Mediterranean region. In *Species diversity in ecological communities* (ed. R. E. Ricklefs and D. Schluter), pp. 135–46. Chicago University Press.

Blondel, J., Vuilleumier, F., Marcus, L. F., and Terouanne, E. (1984). Is there ecomorphological convergence among Mediterranean bird communities of Chile, California and France? *Evolutionary Biology*, 18, 141–213.

Blondel, J., Chessel, D., and Frochot, B. (1988). Bird species impoverishment, niche expansion and density inflation in Mediterranean island habitats. *Ecology*, 69, 1899–917.

Blondel, J., Dias, P., Maistre, M., and Perret, P. (1993). Habitat heterogeneity and life history variation of Mediterranean blue tits. *The Auk*, 110, 511–20.

Blondel, J., Catzeflis, F., and Perret, P. (1996). Molecular phylogeny and the historical biogeography of the warblers of the genus *Sylvia* (Aves). *Journal of Evolutionay Biology*, 9, 871–91.

Bocquet, G., Widler, B., and Kiefer, H. (1978). The Messinian model—a new outlook for the floristics and systematics of the Mediterranean area. *Candollea*, 33, 269–87.

Böhme, W. and Wiedl, H. (1994). Status and zoogeography of the herpetofauna of Cyprus, with taxonomic and natural history notes on selected species (genera *Rana*, *Coluber*, *Natrix*, *Vipera*). *Amphibia and Reptilia*, 10, 31–52.

Bonhomme, F. (1986). Evolutionary relationships in the Genus *Mus*. *Current topics in microbiology and immunology*, 127, 119–24.

Bonhôte, J. and Vernet, J.-L. (1988). La mémoire des charbonnières. Essai de reconsitution des milieux forestiers dans une vallée marquée par la métallurgie (Aston, Haute-Ariège). *Revue Forestière Française*, 40, 197–212.

Borg-Karlson, A. K. and Tengoe, J. (1986). Odour mimetism? Key substances in the *Oprhys lutea-Andrena* pollination relationship (Orchidaceae-Andrenidae). *Journal of Chemical Ecology*, 12, 1927–41.

Bouché, M. (1972). Lombriciens de France, écologie et systématique. *Annales de Zoologie et d'Ecologie animale*, 72 (suppl. 2). INRA, Paris.

Boursot, P., Auffray, J. C., Britton-Davidian, J., and Bonhomme, F. (1993). The evolution of house mice. *Annual Review of Ecology and Systematics*, 24, 119–52.

Braudel, F. (1949). *La Méditerranée et le monde méditerranéen à l'époque de Philippe II*. 2 vols. Armand Colin, Paris.

Braudel, F. (1985). *La Méditerranée. L'espace et l'histoire*. Flammarion, Paris.

Bretagnolle, F. (1993). Etude de quelques aspects des mécanismes de la polyploidisation et de ses conséquences évolutives dans le complexe polyploide du dactyle (*Dactylis glomerata* L.). Unpublished D. Phil. thesis. University of Paris XI, Orsay.

Briggs, J. C. (1974). *Marine zoogeography*. McGraw-Hill, New York.

Brisebarre, A. M. (1978). *Bergers des Cévennes*. Berger-Levrault, Paris.

Britton, R. H. and Crivelli, A. J. (1993). Wetlands of southern Europe and North Africa: Mediterranean wetlands. In *Wetlands of the world*, I (ed. D. F. Whigam), pp. 129–94. Kluwer Academic Publishers, Dordrecht.

Britton, R. H. and Johnson, A. R. (1987). An ecological account of a Mediterranean salina: the Salin de Giraud, Camargue (France). *Biological Conservation*, 42, 185–230.

Britton-Davidian, J. (1990). Genetic differentiation in *Mus musculus domesticus* populations from Europe, the Middle East and North Africa: Geographic patterns of colonization events. *Biological Journal of the Linnean Society*, 41, 27–45.

Brosset, A. (1990). L'évolution récente de l'avifaune du nord-est Marocain: pertes et gains depuis 35 ans. *Revue d'Ecologie (Terre et Vie)*, 45, 237–44.

Brown, J. H. and Gibson, A. C. (1983). *Biogeography*. C. V. Mosby Company, Saint-Louis.

Callot, G. and Jaillard, B. (1996). Incidence des caractéristiques structurales du sous-sol sur l'entrée en production de *Tuber melanosporum* et d'autres champignons mycorhiziens. *Agronomie*, 16, 405–19.

Casevitz-Weulersse, J. (1992). Analyse biogéographique de la myrmécofaune Corse et comparaison avec celle des régions voisines. *Compte Rendus des Séances de la Société de Biogéographie*, 68, 105–29.

Chaline, J. (1974). Palingenèse et phylogenèse chez les campagnols (Arvicolidae, Rodentia). *Comptes-rendus de l'Académie des Sciences de Paris*, D278, 437–40.

Charrier, A. (1995). France maintains strong tradition of support for biodiversity activities worldwide. *Diversity*, 11, 89–90.

Chauvet, J. M., Brunel Deschamps, E., and Hillaire, C. (1995). *La Grotte Chauvet*. Le Seuil, Paris.

Cheylan, G. (1984). Les mammifères des îles de Provence et de Méditerranéenne occidentale: un exemple de peuplement insulaire déséquilibré? *Revue d'Ecologie (Terre et Vie)*, 39, 37–54.

Cheylan, G. (1991). Patterns of Pleistocene turnover, current distribution and speciation among Mediterranean mammals. In *Biogeography of Mediterranean invasions* (ed. R. H. Groves and F. di Castri), pp. 227–62. Cambridge University Press.

Cheylan, M. and Poitevin, F. (1998). Conservazione di rettili e anfibi. In *La gestione degli ambienti costieri einsulari del Mediterraneo* (ed. X. Monbailliu and A. Torre), pp. 275–336. Edizione del Sole, Alghero.

Chouquer, G., Clavel-Lévêque, M., and Favory, F. (1987). Le paysage révélé: l'empreinte du passé dans les paysages contemporains. *Mappemonde*, 4, 16–21.

Cody, M. L. (1975). Towards a theory of continental species diversity: bird distribution over Mediterranean habitat gradients. In *Ecology and evolution of communities* (ed. M. L. Cody and J. M. Diamond), pp. 214–57. Harvard University Press, Cambridge, MA.

Cody, M. L. and Diamond, J. M. (1975). *Ecology and evolution of communities*. Harvard University Press, Cambridge, MA.

Cody, M. L. and Mooney, H. A. (1978). Convergence versus nonconvergence in Mediterranean-climate ecosystems. *Annual Review of Ecology and Systematics*, 9, 265–321.

Cohen, C. R. (1980). Plate tectonic model for the Oligo-Miocene evolution of the western Mediterranean. *Tectonophysics*, 68, 283–311.

Cohen, D. and Shmida, A. (1993). The evolution of flower display and reward. *Evolutionary Biology*, 68, 81–120.

Collins, N. M. and Thomas, J. A. (ed) (1991). *The conservation of insects and their habitats*. Academic Press, London.

Couvet, D., Atlan, A., Belhassen, E., Gliddon, C., Gouyon, P. H., and Kjellberg, F. (1990). Co-evolution between two symbionts: the case of cytoplasmic male-sterility in higher plants. *Oxford Surveys in Evolutionary Biology*, 7, 225–48.

Covas, R. and Blondel, J. (1998). Biogeography and history of the Mediterranean bird fauna. *Ibis*, 140, 395–407.

Cowling, R. M., Rundel, P. W., Lamont, B. B., Arroyo, M. K., and Arianoutsou, M. (1996). Plant diversity in mediterranean-climate regions. *Trends in Ecology and Evolution*, 11, 362–6.

Crick, H. Q. P. and Jones, P. J. (1992). The ecology and conservation of Palaearctic-African migrants. *Ibis*, **134** (supp. 1).

Crivelli, A. J. (1996). *The freshwater fish endemic to the northern Mediterranean region.* Tour du Valat, Arles.

Crivelli, A. J. and Maitland, P. S. (1995). Endemic freshwater fishes of the northern Mediterranean region. *Biological Conservation*, **72**, 337 pages.

Crocq, C. (1990). *Le Casse-noix moucheté.* Lechevallier and Chabaud, Paris.

Cronquist, A. (1981). *An integrated system of classification of flowering plants.* Columbia University Press, New York.

Cuzin, F. (1996). Répartition actuelle et statut des grands mammifères sauvages du Maroc (Primates, Carnivores, Artiodactyles). *Mammalia*, **60**, 101–24.

Dafni, A. and Bernhardt, P. (1990). Pollination of terrestrial orchids of southern Australia and the Mediterranean region. Systematic, ecological and evolutionary implications. *Evolutionary Biology*, **24**, 193–252.

Dafni, A. and O'Toole, C. (1994). Pollination syndromes in the Mediterranean: generalisations and peculiarities. In *Plant-animal interactions in Mediterranean-type ecosystems* (ed. M. Arianoutsou and R. H. Groves), pp. 125–35. Kluwer Academic Publishers, Dordrecht.

Dafni, A. and Shmida, A. (1996). The possible implications of the invasion of *Bombus terrestris* (L.) (Apidae) at Mt. Carmel, Israel. In *The conservation of bees* (ed. International Bee Research Association), pp. 183–200. Linnean Society of London.

Dafni, A., Ivri, Y., and Brantjes, N. B. M. (1981). Pollination of *Serapias vomeracea* Briq. (Orchidaceae) by imitation of holes for sleeping male bees (Hymenoptera). *Acta Botanica Neerlandica*, **30**, 69–73.

Daget, P. (1977). Le bioclimat méditerranéen: caractères généraux, modes de caractérisation. *Vegetatio*, **34**, 1–20.

Damesin, C., Rambal, S., and Joffre, R. (1998). Co-occurrence of trees with different leaf habit: a functional approach on Mediterranean oaks. *Acta Oecologica*, **19**, 195–204.

Darwin, C. (1859). *The origin of species.* John Murray, London.

Davis, M. B. (1976). Pleistocene biogeography of temperate deciduous forests. *Geoscience Canada*, **13**, 13–26.

Davis, P. H. (1965). *Flora of Turkey and the east Aegean islands.* Vol. 1. Edinburgh University Press.

Dawson, T. and Fry, R. (1998). Agriculture in Nature's image. *Trends in Ecology and Evolution*, **13**, 50–1.

Dayan, T. (1996). Weasels from the iron age of Israel: a biogeographic note. *Israel Journal of Zoology*, **42**, 295–8.

De Bonneval, L. (1990). *D'un taillis à l'autre. La déshérance d'un patrimoine forestier communal (Valliguières, Gard). 1820–1990.* INRA, Unité d'Ecodéveloppement, Montfavet, France.

Debussche, M. and Isenmann, P. (1992). A Mediterranean bird disperser assemblage: composition and phenology in relation to fruit availability. *Revue d'Ecologie (Terre et Vie)*, **47**, 411–32.

Debussche, M., Debussche, G., and Affre, L. (1995). La distribution fragmentée de *Cyclamen balearicum* Willk: analyse historique et conséquence des activités humaines. *Acta Botanica Gallica*, **142**, 439–50.

Defleur, A., Bez, J. F., Crégut-Bonnoure, E., Desclaux, E., Onoratini, G., Radulescu, C., *et al.* (1994). Le niveau moustérien de la grotte de l'Adaouste (Jouques, Bouches-du-Rhône). Approche culturelle et paléoenvironnements. *Bulletin du Muséum Anthro-*

pologique et de Préhistoire de Monaco, 37, 29–35.

De Lattin, G. (1967). Grundriss der Zoogeographie. F. Fischer, Stuttgart.

Delaugerre, M. (1988). Statut des Tortues marines de la Corse et de la Méditerranée. Vie et Milieu, 37, 243–64.

Delaugerre, M. and Cheylan, M. (1992). Atlas de répartition des Batraciens et reptiles de Corse. Parc Naturel Régional de la Corse, E. P.H. E., Ajaccio.

De Lillis, M. (1991). An ecomorphological study of the evergreen leaf. Braun-Blanquetia, 7, 1–127.

De Lope, F., Guerrero, J., and de la Cruz, C. (1984). Une nouvelle espèce à classer parmi les oiseaux de la Péninsule Ibérique: Estrilda (Amandava) amandava (Ploceidae, Passeriformes). Alauda, 52, 312.

Dennis, R. L. H., Williams, W. R., and Shreeve, T. G. (1991). A multivariate approach to the determination of faunal structures among European butterfly species (Lepidoptera: Rhopalocera). Zoological Journal of the Linnean Society, 101, 1–49.

Dennis, R. L. H., Shreeve, T. G., and Williams, W. R. (1995). Taxonomic differentiation in species richness gradients among European butterflies (Papilionoidea, Hesperioidea): contribution of macroevolutionary dynamics. Journal of Biogeography, 18, 27–40.

Dias, P. C. (1996). Sources and sinks in population biology. Trends in Ecology and Evolution, 11, 326–30.

Dias, P. C., Verheyen, G. R., and Raymond, M. (1996). Source-sink populations in Mediterranean blue tits: evidence using single-locus minisatellite probes. Journal of Evolutionary Biology, 9, 965–78.

Dias, P. and Blondel, J. (1996). Local specialization and maladaptation in Mediterranean blue tits, Parus caeruleus. Oecologia, 107, 79–86.

Di Castri, F. (1981). Mediterranean-type shrublands of the world. In Mediterranean-type shrublands (ed. F. di Castri, D. W. Goodall, and R. L. Specht), pp. 1–52. Collection Ecosystems of the World, Vol. 11. Elsevier, Amsterdam.

Di Castri, F. (1996). Mediterranean diversity in a global economy. International symposium on Mediterranean biodiversity, 21–30, Roma, Italy, ENEA.

Di Castri, F. (1998). Politics and environment in mediterranean-climate regions. In Landscape degradation and biodiversity in mediterranean-type ecosystems (ed. P. W. Rundel, G. Montenegro, and F. Jaksic), pp. 407–32. Ecological series 136, Springer-Verlag, Berlin.

Di Castri, F. and di Castri, V. (1981). Soil fauna of mediterranean-climate regions. In Mediterranean-type shrublands (ed. F. di Castri, D. W. Goodall, and R. L. Specht), pp. 445–78. Collection Ecosystems of the World, Vol. 11. Elsevier, Amsterdam.

Di Castri, F. and Mooney, H. A. (1973). Mediterranean type ecosystems. Origin and structure. Springer-Verlag, Heidelberg.

Di Castri, F., Goodall, W., and Specht, R. L. (ed.) (1981). Mediterranean-type shrublands. Collection Ecosystems of the World, Vol. 11. Elsevier, Amsterdam.

Di Castri, F., Hansen, A. J., and Debussche, M. (ed.) (1990). Biological invasions in Europe and the Mediterranean Basin. Kluwer Academic Publishers, Dordrecht.

Di Martino, A. and Raimondo, F. M. (1979). Biological and chorological survey of the Sicilian flora. Webbia, 34, 309–35.

Drake, J. A., Mooney, H. A., di Castri, F., Groves, R. H., Kruger, F. J., Rejmanek, M., and Williamson, M. (1989). Biological invasions. A global perspective. John Wiley, Chichester.

Dufaure, J.-J. (1984). La mobilité des paysages méditerranéens. Revue Géographique des

Pyrénées et du Sud-Ouest, **Supp. 1**, 1–387.

Du Merle, P. (1978). Le massif du Ventoux, Vaucluse. Eléments d'une synthèse écologique. *La Terre et la Vie*, **Suppl. 1**, 1–314.

Du Merle, P., Jourdheuil, P., Marro, J. P., and Mazet, R. (1978). Evolution saisonnière de la myrmécofaune et de son activité prédatrice dans un milieu forestier: les interactions clairière-lisière-forêt. *Annales de la Société Entomologique de France*, **14**, 141–57.

Du Puy, B. and Jackson, P. W. (1995). Botanic gardens offer key component to biodiversity conservation in the Mediterranean. *Diversity*, **11**, 47–50.

Economidis, P. S. (1991). *Check list of freshwater fishes of Greece*. Hellenic Society for the Protection of Nature, Athens.

Eisikowitch, D., Gat, Z., Karni, O., Chechik, F., and Raz, D. (1992). Almond blooming under adverse conditions. A compromise between various forces. In *Plant-animal interactions in Mediterranean-type ecosystems* (ed. C. A. Thanos), pp. 234–40, University of Athens.

Emberger, L. (1930a). La végétation de la région Méditerranéenne. Essai d'une classification des groupements végétaux. *Revue Génerale de Botanique*, **42**, 641–62.

Emberger, L. (1930b). Sur une formule climatique applicable en géographie botanique. *Comptes Rendus de l'Académie des Sciences, Paris*, **191**, 389–90.

Escarré, J., Houssard, C., Debussche, M., and Lepart, J. (1983). Evolution de la végétation et du sol après abandon cultural en région méditerranéenne: étude de la succession dans les garrigues du Montpéllierais (France). *Acta Oecologica—Ecologia Plantarum*, **4**, 221–39.

Espadaler, X. and Lopez-Soria, L. (1991). Rareness of certain Mediterranean ant species: fact or artifact? *Insectes Sociaux*, **38**, 365–77.

Etienne, M., Napoleone, M., Jullian, P., and Lachaux, J. (1989). Elevage ovin et protection de la forêt méditerranéenne contre les incendies. *Etudes et Recherches*, **15**, 1–46.

Evenari, M., Shanan, L., and Tadmor, N. (1982). *The Negev. The challenge of a desert*. Harvard University Press, Cambridge, MA.

Flahaut, C. (1937). *La distribution géographique des végétaux dans la région Méditerranéenne Française*. Lechevalier, Paris.

Floret, C. and Pontanier, R. (1982). L'aridité en Tunisie présaharienne. Climat, sol, végétation et aménagement. *Travaux et Documents de l'ORSTOM*, 150, Paris.

Fons, R. (1975). Premières données sur l'écologie de la Pachyure étrusque *Suncus etruscus* (Savi, 1822) et comparaison avec deux autres crocidurinae: *Crocidura russula* (Hermann, 1780) et *Crocidura suaveolens* (Pallas, 1811) (Insectivora Soricidae). *Vie Milieu*, **25**, 315–60.

Fox, B. J. and Fox, M. D. (1986). Resilience of animal and plant communities to human disturbance. In *Resilience in Mediterranean-type ecosystems* (ed. B. Dell, A. J. M. Hopkins, and B. B. Lamont), pp. 39–64. Dr. W. Junk, Dordrecht.

Fragman, O. and Shmida, A. (1998). *Bulbous plants of the holy land*. Sutlands, London.

Francour, P., Boudouresque, C. F., Harmelin, J. G., Harmelin-Vivien, M. L., and Quignard, J. P. (1994). Are the Mediterranean waters becoming warmer? Information from biological indicators. *Marine Pollution Bulletin*, **28**, 523–6.

Gamisans, J. and Marzocchi, J.-F. (1996). *La Flore endémique de la Corse*. Edisud, Aix-en-Provence.

Gass, I. G. (1968). Is the Troodos Massif of Cyprus a fragment of Mesozoic ocean floor? *Nature*, **220**, 39–42.

Gaston, K. (1996). Biodiversity congruence. *Progress in Physical Geography*, **20**, 105–12.

Gaussen, H. (1954). Théorie et classification des climats et microclimats. *VIIème Congrès International de Botanique*, pp. 125–30.

Georgoudis, A. (1995). Animal genetic diversity plays important role in Mediterranean agriculture. *Diversity*, 11, 16–19.

Gilpin, M. E. and Soulé, M. E. (1986). Minimum viable populations: the process of species extinctions. In *Conservation biology: the science of scarcity and diversity* (ed. M. E. Soulé), pp. 13–34. Sinauer, Sunderland, MA.

Ginocchio, R. and Montenegro, G. (1992). Effects of insect herbivory on plant architecture. In *Plant-animal interactions in Mediterranean-type ecosystems* (ed. C. A. Thanos), pp. 7–21. University of Athens.

Gintzburger, G., Rochon, J. J., and Conesa, A. P. (1990). The French mediterranean zones: sheep rearing systems and the present and potential role of pasture legumes. In *The role of legumes in the farming systems of the Mediterranean area* (ed. A. E. Osman, M. H. Ibrahim, and M. A. Jones), pp. 179–94. ICARDA, Aleppo, Syria.

Gomez, J. M., Zamora, R., Hodar, J. A., and Garcia, D. (1996). Experimental study of pollination by ants in Mediterranean high mountains and arid habitats. *Oecologia*, 105, 236–42.

Gomez-Campo, C. (1985). *Plant conservation in the Mediterranean area*. Dr. W. Junk, Dordrecht, Boston, and Lancaster.

Gomez-Campo, C. and Herranz-Sanz, J. M. (1993). Conservation of Iberian endemic plants: the botanical reserve of La Encantada (Villarrobledo, Albacete, Spain). *Biological Conservation*, 64, 155–60.

Gonzales, J. J., Garzon, P., and Merino, M. (1990). Censo de la poblacion espanola de cernicalo primilla. *Quercus*, 49, 6–12.

Gould, S. J. and Lewontin, R. C. (1979). The sprandels of San Marco and the Panglossian paradigm: a critique of the adaptationist programme. *Proceedings of the Royal Society of London B*, 205, 581–98.

Gouyon, P. H. and Couvet, D. (1987). A conflict between two sexes, females and hermaphrodites. In *The evolution of sex and its consequences* (ed. S. C. Stearns), pp. 245–61. Birkhauser Verlag, Berlin.

Gouyon, P. H., Vernet, P., Guillerm, J.-L., and Valdeyron, G. (1986). Polymorphisms and environments: the adaptive value of the oil polymorphisms in *Thymus vulgaris* L. *Heredity*, 57, 59–66.

Granjon, L. and Cheylan, G. (1989). Le sort de rats noirs (*Rattus rattus*) introduits sur une île, révélé par radio-tracking. *Comptes Rendus de l'Académie des Sciences, Paris*, 309 (Series III), 571–5.

Granval, P. (1988). Approche écologique de la gestion de l'espace rural: des besoins de la Bécasse à la qualité des milieux. Unpublished D. Phil. thesis. University of Rennes.

Granval, P. and Muys, B. (1992). Management of forest soils and earthworms to improve woodcock (*Scolopax* sp.) habitats: a literature survey. *Gibier Faune Sauvage*, 9, 243–55.

Grenon, M. and Batisse, M. (1989). *Futures for the Mediterranean Basin: the Blue Plan*. Oxford University Press.

Greuter, W. (1991). Botanical diversity, endemism, rarity, and extinction in the Mediterranean area: an analysis based on the published volumes of Med-Checklist. *Botanica Chronica*, 10, 63–79.

Greuter, W. (1994). Extinction in Mediterranean areas. *Philosophical Transactions of the Royal Society, London. Series B*, 344, 41–6.

Greuter, W., Burdet, H. M. and Long, G. (eds) (1984). *Med-Checklist*. Vol. 1.

Conservatoire et Jardin Botanique de Genève, Genève.

Grillas, P. and Roché, J. (1997). *Végétation des marais temporaires. Ecologie et gestion.* Tour du Valat, Arles.

Grime, J. P. (1979). *Plant strategies and vegetation processes.* John Wiley, Chicester.

Groves, R. H. (1986). Invasion of mediterranean ecosystems by weeds. In *Resilience in mediterranean-type ecosystems* (ed. B. Dell, A. J. M. Hopkins, and B. B. Lamont), pp. 129–45. Dr. W. Junk, Dordrecht.

Grubb, P. J. (1977). The maintenance of species-richness in plant communities: the importance of the regeneration niche. *Biological Review,* 52, 107–45.

Haffer, J. (1977). Secondary contact zones of birds in Northern Iran. *Bonner Zoologische Monographien,* 10, 1–64.

Hammond, P. M. (1995). The current magnitude of biodiversity. In *Global Biodiversity Assessment* (ed. V. H. Heywood), pp. 113–128. Cambridge University Press, Cambridge.

Harlan, J. R. and Zohary, D. (1966). Distribution of wild wheats and barley. *Science,* 153, 1074–80.

Harper, J. L. (1982). After description. In *The plant community as a working mechanism* (ed. E. I. Newman), pp. 11–25. Blackwell Scientific, Oxford.

Hawkes, J. G. (1995). Centers of origin for agricultural diversity in the Mediterranean: from Vavilov to the present day. *Diversity,* 11, 109–11.

Hays, J. D., Imbrie, J., and Shackleton, N. J. (1976). Variations in the Earth's orbit: pacemaker of the ice ages. *Science,* 194, 1121–32.

Henry, P. M. (1977). The Mediterranean: a threatened microcosm. *Ambio,* 6, 300–7.

Hepper, N. (1981). *Bible plants at Kew.* HMSO, London.

Herrera, C. M. (1984). A study of avian frugivores, bird-dispersed plants, and their interactions in Mediterranean scrublands. *Ecological Monographs,* 54, 1–23.

Herrera, C. M. (1992). Historical effects and sorting processes as explanations for contemporary ecological patterns: character syndromes in Mediterranean woody plants. *American Naturalist,* 140, 421–46.

Herrera, C. M. (1995). Plant-vertebrate seed dispersal systems in the Mediterranean: ecological, evolutionary, and historical determinants. *Annual Review of Ecology and Systematics,* 26, 705–27.

Higgins, L. G. and Riley, N. D. (1988). *A field guide to the butterflies of Britain and Europe.* Collins, London.

Hildebrand, E. E. (1987). Die struktur von waldböden—ein gefährdetes fliessgleichgewicht. *Allgemeine Forst Zeitschrift,* 16–17, 424–6.

Hobbs, R. and Saunders, D. A. (ed.) (1992). *Reintegrating fragmented landscapes: towards sustainable production and nature conservation.* Springer-Verlag, New York.

Hobbs, R., Richardson, D. M., and Davis, G. W. (1995). Mediterranean-type ecosystems: opportunities and constraints for studying the function of biodiversity. In *Mediterranean-type ecosystems. The function of biodiversity* (ed. G. W. Davis and D. M. Richardson), pp. 1–42. Springer-Verlag, Berlin.

Hockin, D. C. (1980). The biogeography of the butterflies of the Mediterranean islands. *Nota Lepidoptera,* 3, 119–25.

Hopper, S. D. (1992). Patterns of plant diversity at the population and species level in south-west Australian mediterranean ecosystems. In *Biodiversity of Mediterranean ecosystems in Australia* (ed. R. J. Hobbs), pp. 27–46. Surrey Beatty, Perth, Australia.

Houston, J. M. (1964). *The western Mediterranean world. An introduction to its regional landscapes*. Frederick A. Praeger, New York.

Hsü, K. J. (1971). Origin of the Alps and western Mediterranean. *Nature*, **233**, 44–8.

Hsü, K. J. (1972). When the Mediterranean dried up. *Scientific American*, **227**, 27–36.

Huntley, B. (1988). European post-glacial vegetation history: a new perspective. In *Acta XIX Congressus Internationalis Ornithologici*, Vol. 1 (ed. H. Ouellet), pp. 1061–77. National Museum of Natural Sciences, Ottawa.

Huntley, B. (1993). Species-richness in north-temperate zone forests. *Journal of Biogeography*, **20**, 163–80.

Huntley, B. and Birks, H. J. B. (1983). *An atlas of past and present pollen maps for Europe: 0–13 000 years ago*. Cambridge University Press.

IUCN (1985). *United Nations list of national parks and protected areas*. IUCN, Gland.

Izhaki, I., Walton, P. B., and Safriel, U. N. (1991). Seed shadow generated by frugivorous birds in an eastern Mediterranean scrub. *Journal of Ecology*, **79**, 575–90.

Jacquard, A. (1991). *Voici le temps du monde fini*. Seuil, Paris.

Jaenkie, J. (1991). Mass extinction of European fungi. *Trends in Ecology and Evolution*, **6**, 174–5.

Janzen, D. H. (1980). When is it coevolution? *Evolution*, **34**, 611–12.

Joffre, R. and Rambal, S. (1993). How tree cover influences the water balance of Mediterranean rangelands. *Ecology*, **74**, 570–82.

Joffre, R., Vacher, J., de los Llanos, C., and Long, G. (1988). The dehesa: an agrosilvopastoral system of the Mediterranean region with special reference to the Sierra Morena area of Spain. *Agroforestry Systems*, **6**, 71–96.

Jürgens, K. D., Fons, R., Peters, T., and Sender, S. (1996). Heart and respiratory rates and their significance for connective oxygen transport rates in the smallest mammal, the Etruscan shrew *Suncus etruscus*. *Journal of Experimental Biology*, **199**, 2579–84.

Kanyamibwa, S., Schierer, A., Pradel, R., and Lebreton, J.-D. (1990). Changes in adult survival rates in a western European population of the white stork *Ciconia ciconia*. *Ibis*, **132**, 27–35.

Kaplan, D. Y. (1992). Responses of mediterranean grassland plants to gazelle grazing. In *Plant-animal interactions in Mediterranean-type ecosystems* (ed. C. A. Thanos), pp. 75–9. University of Athens.

Kaplan, D. Y. and Gutman, M. (1989). Food composition of the mountain gazelle and cattle in the southern Golan. *Journal of Zoology*, **36**, 154.

Keeley, J. E. (1991). Seed germination and life history syndromes in the California chaparral. *Botanical Review*, **57**, 81–116.

Keeley, J. E. and Swift (1995). Biodiversity and ecosystem functioning in mediterranean-climate California. In *Mediterranean-type ecosystems. The function of biodiversity* (ed. G. W. Davis and D. M. Richardson), pp. 121–83. Springer-Verlag, Heidelberg.

Kjellberg, F., Gouyon, P. H., Ibrahim, M., Raymond, M., and Valdeyron, G. (1987). The stability of the symbiosis between dioecious figs and their pollinators: a study of *Ficus carica* L. and *Blastophaga psenes*. *Evolution*, **91**, 117–22.

Kolars, J. (1982). Earthquake-vulnerable populations in modern Turkey. *Geographical Review*, **72**, 20–35.

Kowalski, K. and Rzebik-Kowaska, B. (1991). *Mammals of Algeria*. Polish Academy of Sciences. Institute of Systematics and Evolution of animals. Wroclaw, Warszawa, and Krakow.

Kraiem, M. M. (1983). Les poissons d'eau douce de Tunisie: inventaire commenté et répartition géographique. *Bulletin de l'Institut National des Sciences et Techniques*

Océanographiques, 10, 107–24.

Küchler, A. W. (1964). *Potential natural vegetation of the conterminous United States*. American Geographic Society, New York.

Kuhnholtz-Lordat, G. (1938). *La Terre Incendiée. Essai d'agronomie comparée*. Maison Carrée, Nîmes, France.

Kuhnholtz-Lordat, G. (1958). *L'écran vert*. Mémoires du Muséum National d'Histoire Naturelle, 9, 1–276. Paris.

Kullenberg, B. and Bergström, G. (1976). The pollination of *Ophrys* orchids. *Botaniska Notiser*, 129, 11–19.

Kurtz, C. and Luquet, P. (1996). The traffic in Mediterranean birds of prey. In *Biologia y conservacion de las rapaces Mediterraneas, 1994*. Monografias, N° 4. SEO, Madrid.

Kusler, J. A. and Kentula, M. E. (1990). *Wetland creation and restoration: the status of the science*. Island Press, Washington, DC.

Lambrechts, M. M., Blondel, J., Hurtrez-Boussès, S., Maistre, M., and Perret, P. (1997). Adaptive inter-population differences in blue tit life-history traits on Corsica. *Evolutionary Ecology*, 11, 599–612.

Lamont, B. B. (1982). Mechanisms for enhancing nutrient uptake in plants, with particular reference to mediterranean South Africa and Australia. *Botanical Review*, 48, 597–689.

Lanza, B. and Poggesi, M. (1986). *Storia naturale delle isole satelliti della Corsica*. Istituto Geografico Militare, Firenze.

Larsen, T. B. (1986). Tropical butterflies of the Mediterranean. *Nota Lepidoptera*, 9, 63–77.

Latham, R. E. and Ricklefs, R. E. (1993). Continental comparisons of temperate-zone tree species diversity. In *Species diversity in ecological communities: historical and geographical perspectives* (ed. R. E. Ricklefs and D. Schluter), pp. 294–314. Chicago University Press.

Lavelle, P. (1988). Earthworm activities and the soil system. *Biology and Fertility of Soils*, 6, 237–51.

Laville, L. and Reiss, F. (1992). The Chironomid fauna of the Mediterranean region reviewed. *Netherlands Journal of Aquatic Ecology*, 26, 239–45.

Lavorel, S., Lepart, J., Debussche, M., Lebreton, J.-D., and Beffy, J.-L. (1994). Small scale disturbances and the maintenance of species diversity in Mediterranean old fields. *Oikos*, 70, 455–73.

Le Floc'h, E., Neffati, M., Chaieb, M., Floret, C., and Pontanier, R. (1995). Un essai de réhabilitation en zone aride. Le cas de Menzel Habib (Tunisie). In *L'homme peut-il refaire ce qu'il a défait?* (ed. R. Pontanier, A. M'Hiri, J. Aronson, N. Akrimi, and E. Le Floc'h), pp. 139–60. John Libbey Eurotext, Paris.

Le Floc'h, E., Aronson, J., Dhillion, S., Guillerm, J. L., Grossmann, A., and Cunge, E. (1998). Biodiversity and ecosystem trajectories: first results from a new LTER in southern France. *Acta Oecologica*, 19, 285–93.

Le Houérou, H. N. (1981). Impact of man and his animals on Mediterranean vegetation. In *Mediterranean-type shrublands* (ed. F. di Castri, D. W. Goodall, and R. L. Specht), pp. 479–517. Collection Ecosystems of the World, Vol. 11. Elsevier, Amsterdam.

Le Houérou, H. N. (1985). Pastoralism. In *Climate impact assessment* (ed. R. W. Kates, J. H. Ausubel, and M. Berberian), pp. 155–85. John Wiley, New York.

Le Houérou, H. N. (1990). Global change: vegetation, ecosystems and land use in the southern Mediterranean Basin by the mid twenty-first century. *Israel Journal of Botany*, 39, 481–508.

Le Houérou, H. N. (1991). La Méditerranée en l'an 2050: Impacts respectifs d'une éventuelle évolution climatique et de la démographie sur la végétation, les écosystèmes et l'utilisation des terres. *La Météorologie*, **36**, 4–37.

Lebreton, J.-D. and Isenmann, P. (1976). Dynamique de la population camarguaise de Mouette rieuse (*Larus ridibundus*): un modèle mathématique. *Terre et Vie*, **30**, 529–49.

Lemée, G. (1967). *Précis de Biogéographie*. Masson, Paris.

Leon, C., Lucas, G., and Synge, H. (1985). The value of information in saving threatened Mediterranean plants. In *Plant conservation in the Mediterranean area* (ed. C. Gomez-Campo), pp. 177–96. Dr. W. Junk, Dordrecht.

Lepart, J. and Debussche, M. (1992). Human impact on landscape patterning: Mediterranean examples. In *Landscape boundaries. Consequences for biotic diversity and ecological flows* (ed. A. J. Hansen and F. di Castri), pp. 76–106. Springer-Verlag, New York, Berlin.

Lescure, J. (1992). Les amphibiens du pourtour méditerranéen. *Bulletin de la Société Herpétologique de France*, **64**, 1–14.

Lesins, K. A. and Lesins, I. (1979). *The genus Medicago (Leguminosae): a taxogenetic study*. Dr. W. Junk, Dordrecht.

Linhart, Y. B. and Thompson, J. D. (1995). Terpene-based selective herbivory by *Helix aspersa* (Mollusca) on *Thymus vulgaris* (Labiatae). *Oecologia*, **102**, 126–32.

Little, R. J. (1983). A review of floral food deception mimicries with comments on floral mutualism. In *Handbook of experimental pollination biology* (ed. C. E. Jones and R. J. Little), pp. 294–309. Scientific and Academic Editions, New York.

Llédo, M. J., Sanchez, J. R., Bellot, J., Boronat, J., Ibañez, J. J., and Escarré, A. (1992). Structure, biomass and production of a resprouted holm-oak (*Quercus ilex* L.) forest in NE Spain. *Vegetatio*, **99–100**, 51–9.

Lumaret, J. P. (1995). Desiccation rate of excrement: a selective pressure on dung beetles (Coleoptera, Scarabaeoidea). In *Time scales of biological responses to water constraints* (ed. J. Roy, J. Aronson, and F. di Castri), pp. 105–18. SPB Academic Publishing bv, Amsterdam.

Lumaret, J. P. and Kirk, A. (1987). Ecology of dung beetles in the French mediterranean region (Coleoptera: Scarabaeidae). *Acta Zoologica Mexicana*, **24**, 1–55.

Lumaret, R. (1988). Cytology, genetics, and evolution in the genus *Dactylis*. *Critical Reviews in Plant Sciences*, **7**, 55–91.

Lumaret, R., Yacine, A., Berrod, A., Romane, F., and Li, T. X. (1991). Mating system and genetic diversity in holm oak (*Quercus ilex*) L. Fagaceae). In *Biochemical markers in the population genetics of forest trees* (ed. S. Fineschi, M. E. Malvoti, F. Cannata, and H. H. Hattemer), pp. 145–53. S.P.B. Academic Publishing, The Hague.

Lynes, H. (1909–1910). Observations on the migration of birds in the Mediterranean. *British Birds* **3**, 36–51, 69–77, 99–104, 133–150.

MacArthur, R. H., Diamond, J. M., and Karr, J. R. (1972). Density compensation in island faunas. *Ecology*, **53**, 330–42.

Macgregor, H. C., Sessions, S. K., and Arntzen, J. W. (1990). An integrative analysis of phylogenetic relationships among newts of the genus *Triturus* (family Salamandridae), using comparative biochemistry, cytogenetics and reproductive interactions. *Journal of Evolutionary Biology*, **3**, 329–73.

Magnin, G. (1987). *An acount of the illegal catching and shooting of birds in Cyprus during 1986*. International Council for Bird Preservation, Cambridge.

Magnin, G. (1991). Hunting and persecution of migratory birds in the Mediterranean region. In *Conserving migratory birds* (ed. T. Salathé), pp. 59–71. International Council for Bird Preservation, Cambridge.

Maillard, C. and Raibaut, A. (1990). Human activities and modifications of ichtyofauna of the Mediterranean sea: effects on parasitosis. In *Biological invasions in Europe and the Mediterranean Basin* (ed. F. di Castri, A. J. Hansen, and M. Debussche), pp. 297–305. Kluwer Academic, Dordrecht.

Manicacci, D., Couvet, D., Belhassen, E., Gouyon, P. H., and Atlan, A. (1996). Founder effects and sex ratio in the gynodioecious *Thymus vulgaris*. *Molecular Ecology*, 5, 63–72.

Marchand, H. (1990). *Les forêts Méditerranéennes. Enjeux et perspectives.* Les fascicules du Plan Bleu, 2. Economica, Paris.

Margaris, N. S. (1981). Adaptive strategies in plants dominating mediterranean-type ecosystems. In *Mediterranean-type shrublands* (ed. F. di Castri, D. W. Goodall, and R. L. Specht), pp. 309–16. Collection Ecosystems of the World, Vol. 11. Elsevier, Amsterdam.

Margaris, N. S. and Vokou, D. (1982). Structural and physiological features of woody plants in phryganic ecosystems related to adaptive mechanisms. *Ecologia Mediterranea*, 8, 449–59.

Martin, J. and Gurrea, P. (1990). The peninsular effect in Iberian butterflies (Lepidoptera: Papilionoidea and Hesperioidea). *Journal of Biogeography*, 17, 85–96.

Martin, P. S. (1984). Prehistoric overkill: the global model. In *Quaternary extinctions* (ed. P. S. Martin and R. G. Klein), pp. 354–403. University of Arizona Press, Tucson.

Mathez, J., Quézel, P., and Raynaud, C. (1985). The Maghreb countries. In *Plant conservation in the Mediterranean area* (ed. C. Gomez-Campo), pp. 141–57. Dr. W. Junk, Dordrecht, Boston, and Lancaster.

McCulloch, M. N., Tucker, G. M., and Baillie, S. R. (1992). The hunting of migratory birds in Europe: a ringing recovery analysis. *Ibis*, 34 (supp. 1), 55–65.

McGlone, M. S. (1996). When history matters: scale, time, climate and tree diversity. *Global Ecology Biogeography Letters*, 5, 309–14.

McNeely, J. A., Gadgil, M., Leveque, C., Padoch, C., and Redford, K. (1995). Human influences on biodiversity. In *Global biodiversity assessment* (ed. V. H. Heywood and R. T. Watson), pp. 711–821. Cambridge University Press.

McNeil, J. R. (1992). *The mountains of the Mediterranean world, an environmental history.* Cambridge University Press.

Médail, F. and Quézel, P. (1997). Hot-spots analysis for conservation of plant biodiversity in the Mediterranean Basin. *Annals of the Missouri Botanical Garden*, 84, 112–27.

Médail, F. and Verlaque, V. (1997). Ecological characteristics and rarity of endemic plants from S. E. France and Corsica. Implications for biodiversity conservation. *Biological Conservation*, 80, 269–81.

Medus, J. and Pons, A. (1980). Les prédécesseurs des végétaux méditerranéens actuels jusqu'au début du Miocène. *Naturalia Monspeliensia*, 23, 11–20.

Meliadou, A. and Troumbis, A. (1997). Aspects of heterogeneity in the distribution of diversity of the European herpetofauna. *Acta Oecologica*, 18, 393–412.

Menzel, R. and Shmida, A. (1993). The ecology of flower colours and the natural colour vision of insect pollinators: the Israeli flora as a case study. *Biological Review*, 68, 81–120.

Mesléard, F., Lepart, J., and Tan Ham, L. (1995). Impact of grazing on vegetation dynamics in former ricefields. *Journal of Vegetation Science*, 6, 683–90.

Michaud, H., Toumi, L., Lumaret, R., Li, T. X., Romane, F., and di Giusto, F. (1995). Effect of geographical discontinuity on genetic variation in *Quercus ilex* L. (Holm oak). Evidence from enzyme polymorphism. *Heredity*, 74, 590–606.

Mills, J. (1983). Herbivory and seedling establishment in post-fire southern California chaparral. *Oecologia*, 60, 267–70.

Molero, J. and Rovira, A. M. (1998). A note on the taxonomy of the Macaronesian *Euphorbia obtusifolia* complex (Euphorbiaceae). *Taxon*, 47, 321–32.

Monk, C. D. (1966). An ecological significance of evergreenness. *Ecology*, 47, 504–5.

Mönkkönen, M. (1994). Diversity patterns in Palaearctic and Nearctic forest bird assemblages. *Journal of Biogeography*, 21, 183–95.

Mooney, H. A. and Dunn, E. L. (1970). Convergent evolution of Mediterranean climate evergreen sclerophyllous shrubs. *Evolution*, 24, 292–303.

Moreau, R. E. (1972). *The Palaearctic-African bird migration system*. Academic Press, London.

Mourer-Chauviré, C., Salotti, M., Pereira, E., Quinif, Y. T., Courtis, J. Y., Dubois, J. N. and La Miza, J. C. (1997). *Athene angelis* n. sp. (Aves Strigiformes), nouvelle espèce endémique insulaire étiente du Pléistocène moyen et supérieur de Corse (France). *Comptes Rendus de l'Académie des Sciences, Paris*, 324, 677–684.

Muntaner, J. and Mayol, J. (1996). *Biologia y conservacion de las rapaces Mediterraneas, 1994*. Monografias N° 4. SEO, Madrid.

Myers, N. (1988). Threatened biotas: 'hot spots' in tropical forests. *The Environmentalist*, 8, 187–208.

Myers, N. (1990). The biodiversity challenge: expanded hot-spots analysis. *The Environmentalist*, 10, 243–56.

Naveh, Z. (1974). Effects of fire in the Mediterranean region. In *Fire and ecosystems* (ed. T. T. Kozlowski and C. E. Ahlgren), pp. 401–34. Academic Press, New York.

Naveh, Z. and Dan, J. (1973). The human degradation of Mediterranean landscapes in Israel. In *Mediterranean-type ecosystems: origins and structure* (ed. F. di Castri and H. A. Mooney). pp. 373–90. Ecological Studies, Vol. 7. Springer-Verlag, Berlin.

Naveh, Z. and Whittaker, R. H. (1979). Structural and floristic diversity of shrublands and woodlands in northern Israel and other Mediterranean areas. *Vegetatio*, 41, 171–90.

Nègre, M. (1931). Les reboisements du massif de l'Aigoual. *Bulletin de la Société d'Etude des Sciences Naturelles de Nîmes*, pp. 1–135.

Nevo, E., Zohary, D., Brown, A. H. D., and Haber, M. (1979). Genetic diversity and environmental associations of wild barley, *Hordeum spontaneum*, in Israel. *Evolution*, 33, 815–33.

Nilsson, L. A. (1978). Pollination ecology and adaptation in *Platanthera chlorantha* (Orchidaceae). *Botaniska Notiser*, 131, 35–51.

Noy-Meir, I. (1988). Dominant grasses replaced by ruderal forbs in a vole year in undergrazed Mediterranean grasslands in Israel. *Journal of Biogeography*, 15, 579–87.

Noy-Meir, I., Gutman, M., and Kaplan, Y. (1989). Response of Mediterranean grassland plants to grazing and protection. *Journal of Applied Ecology*, 77, 290–310.

Olivieri, I., Couvet, D., and Gouyon, P. H. (1990). The genetics of transient populations: research at the metapopulation level. *Trends in Ecology and Evolution*, 5, 207–10.

Oosterbroek, P. and Arntzen, J. W. (1992). Area-cladograms of Circum-Mediterranean taxa in relation to Mediterranean palaeogeography. *Journal of Biogeography*, 19, 3–20.

Orshan, G. (1972). Morphological and physiological plasticity in relation to drought. In *Proceedings of the international symposium on wildland shrub biology and utilization*, pp. 245–54. Utah State University, Logan.

Orshan, G. (ed.) (1989). *Plant pheno-morphological studies in mediterranean-type ecosystems*. Kluwer Academic, Dordrecht.

O'Toole, C. and Raw, A. (1991). *Bees of the world*. Blandford, London.

Ozenda, P. (1975). Sur les étages de végétation dans les montagnes du Bassin Méditerranéen. *Documents de Cartographie Ecologique*, **16**, 1–32.

Palamarev, E. (1989). Paleobotanical evidence of the Tertiary history and origin of the Mediterranean sclerophyll dendroflora. *Plant Systematics and Evolution*, **162**, 93–107.

Panou, A., Jacobs, J., and Panos, D. (1993). The endangered Mediterranean monk seal *Monachus monachus* in the Ionian Sea, Greece. *Biological Conservation*, **64**, 129–40.

Papageorgiou, N. (1979). *Population energy relationships of the agrimi (Capra aegagrus cretica) on Theodorou Island*. Paul Parey Verlag, Hamburg and Berlin.

Pasquet, E. (1998). Plylogeny of the nuthatches of the *Sitta canadensis* group and its evolutionary and biogeographic implications. *Ibis*, **140**, 150–6.

Pasquet, E. and Thibault, J. C. (1997). Genetic differences among mainland and insular forms of the citril finch (*Serinus citrinella*). *Ibis*, **139**, 679–84.

Pearce, F. and Crivelli, A. (1994). *Caractéristiques générales des zones humides méditerranéennes*. MedWet, Arles.

Petit, C. and Thompson, J. D. (1997). Variation in phenotypic response to light availability between diploid and tetraploid populations of the perennial grass *Arrhenatherum elatius* from open and woodland sites. *Journal of Ecology*, **85**, 657–67.

Petit, C., Thompson, J. D., and Bretagnolle, F. (1996). Phenotypic plasticity in relation to ploidy levels and corm production in the perennial grass *Arrhenaterum elatius*. *Canadian Journal of Botany*, **74**, 1964–73.

Pfeffer, F. (1973). *Les animaux domestiques et leurs ancêtres*. Bordas, Paris.

Pickett, S. T. A. and White, P. S. (ed.) (1985). *The ecology of natural disturbance and patch dynamics*. Academic Press, New York.

Pignatti, S. (1978). Evolutionary trends in mediterranean flora and vegetation. *Vegetatio*, **37**, 175–85.

Plitmann, U. (1981). Evolutionary history of the old world lupines. *Taxon*, **30**, 430–7.

Pons, A. (1981). The history of the Mediterranean shrublands. In *Mediterranean-type shrublands* (ed. F. di Castri, D. W. Goodall, and R. L. Specht), pp. 131–8. Collection Ecosystems of the World, Vol. 11. Elsevier, Amsterdam.

Pons, A. and Quézel, P. (1985). The history of the flora and vegetation and past and present human disturbance in the Mediterranean region. In *Plant conservation in the Mediterranean area* (ed. C. Gomez-Campo), pp. 25–43. Dr. W. Junk, Dordrecht.

Population et Sociétés (1997). Tous les Pays du Monde. *Bulletin Mensuel d'Information de l'Institut National d'Etudes démographiques*, **326**, 1–5.

Povz, M. D., Jesensek, D., Berrebi, P., and Crivelli, A. (1996). The marble trout, *Salmo trutta marmoratus*, Cuvier 1817, in the Soca river basin, Slovenia. Tour du Valat publication, 65 pages.

Prendergast, J. R., Quinn, R. M., Lawton, J. H., Eversham, B. C., and Gibbons, D. W. (1993). Rare species, the coincidence of diversity hotspots and conservation strategies. *Nature*, **365**, 335–7.

Primack, R. B. (1993). *Essentials of conservation biology*. Sinauer, Sunderland, MA.

Prodon, R. and Lebreton, J.-D. (1981). Breeding avifauna of a Mediterranean succession: the holm oak and cork oak series in the eastern Pyrénées. I. Analysis and modelling of the structure gradient. *Oikos*, **37**, 21–38.

Prodon, R., Fons, R., and Athias-Binche, F. (1987). The impact of fire on animal communities in the Mediterranean area. In *The role of fire in ecological systems* (ed. L. Trabaud), pp. 121–57. SPB Academic, The Hague.

Prosperi, J. M., Guy, P., and Balfourier, F. (ed.) (1995). *Ressources génétiques des plantes fourragères et à gazon*. Bureau des Ressources Génétiques and Institut National de la Recherche Agronomique, Paris.

Pullin, A. S. (ed.) (1995). *Ecology and conservation of butterflies*. Chapman and Hall, London.

Quézel, P. (1976a). Le dynamisme de la végétation en région méditerranéenne. *Collana Verde*, **39**, 375–91.

Quézel, P. (1976b). Les forêts du pourtour méditerranéen. *Note Technique MAB*, **2**, 9–34.

Quézel, P. (1978). Analysis of the flora of Mediterranean and Saharan Africa. *Annals of the Missouri Botanical Garden*, **65**, 479–534.

Quézel, P. (1985). Definition of the Mediterranean region and origin of its flora. In *Plant conservation in the Mediterranean area* (ed. C. Gomez-Campo), pp. 9–24. Dr. W. Junk, Dordrecht.

Quézel, P. (1995). La flore du bassin méditerranéen: origine, mise en place, endémisme. *Ecologia Mediterranea*, **21**, 19–39.

Quézel, P. and Médail, F. (1995). La région circum-méditerranéenne, centre mondial majeur de biodiversité végétale. In *Actes des 6èmes rencontres de l'Agence Régionale pour l'Environnement, Provence-Alpes-Côte d'Azur*, pp. 152–60. Gap (France).

Quézel, P., Barbéro, M., Bonin, G., and Loisel, R. (1990). Recent plant invasion in the Circum-Mediterranean region. In *Biological invasions in Europe and the Mediterranean Basin* (ed. F. di Castri, A. J. Hansen, and M. Debussche), pp. 51–60. Kluwer Academic, Dordrecht.

Ramade, F. (1997). *Conservation des écosystèmes méditerranéens*. Economica, Paris.

Rambal, S. (1987). Evolution de l'occupation des terres et ressource en eau en région méditerranéenne karstique. *Journal of Hydrology (Amsterdam)*, **93**, 339–57.

Randi, E. (1996). A mitochondrial cytochrome B phylogeny of the *Alectoris* partridges. *Molecular Phylogenetics and Evolution*, **6**, 214–27.

Raunkiaer, C. (1934). *The life form of plants and statistical plant geography*. Oxford University Press.

Raven, P. H. (1964). Catastrophic selection and edaphic endemism. *American Naturalist*, **98**, 336–8.

Raven, P. H. (1973). The evolution of Mediterranean floras. In *Mediterranean-type ecosystems: origin and structure* (ed. F. di Castri and H. A. Mooney), pp. 213–24. Springer-Verlag, Heidelberg.

Raven, P. H. and Axelrod, D. I. (1974). Angiosperm biogeography and past continental Movements. *Annals of the Missouri Botanical Garden*, **61**, 539–673.

Raven, P. H. and Axelrod, D. I. (1978). Origin and relationships of the California flora. *University of California Publications in Botany*, **72**, 1–115.

Reifenberg, A. (1955). *The struggle between the desert and the sown*. The Publishing Department of the Jewish Agency, Jerusalem.

Reijnders, P. J. H. (1997). A mass mortality hits the already critically endangered Mediterranean monk seal. *Species*, **29**, 49–50.

Reille, M. and Pons, A. (1992). The ecological significance of sclerophyllous oak forests in the western part of the Mediterranean basin: a note on pollen analytical data. *Vegetatio*, **99–100**, 13–17.

Reille, M., Triat, H., and Vernet, J. L. (1980). Les témoignages des structures actuelles de végétation méditerranéenne durant le passé contemporain de l'action de l'homme. *Naturalia Monspeliensia*, **23**, 79–87.

Rosenzweig, M. L. (1995). *Species diversity in space and time*. Cambridge University Press.

Ross, J. D. and Sombrero, C. (1991). Environmental control of essential oil production in Mediterranean plants. In *Ecological chemistry and biochemistry of plant terpenoids* (ed. J. B. Harborne and F. A. Tomas-Barberan), pp. 83–94. Clarendon, Oxford.

Roy, J. (1990). In search of the characteristics of plant invaders. In *Biological invasions in Europe and the Mediterraean basin* (ed. F. di Castri, A. J. Hansen, and M. Debussche), pp. 333–52. Kluwer Academic, Dordrecht.

Roy, J. and Sonié, L. (1992). Germination and population dynamics of *Cistus* species in relation to fire. *Journal of Applied Ecology*, **29**, 647–55.

Sadoul, N. (1996). Dynamique spatiale et temporelle des colonies de Charadriiformes dans les salins de Camargue: implications pour la conservation. Unpublished D. Phil. thesis. University of Montpellier II.

Sari, D. (1977). *L'homme et l'érosion dans l'Ouarsenis*. University of Alger, Alger.

Sarrazin, F. and Barbault, R. (1996). Reintroduction: challenge and lessons for basic ecology. *Trends in Ecology and Evolution*, **11**, 474–8.

Schemske, D. W. (1981). Floral convergence and pollinator sharing in two bee pollinated tropical herbs. *Ecology*, **62**, 946–54.

Schluter, D. and Rickelfs, R. E. (1993). Convergence and the regional component of species diversity. In *Species diversity in ecological communities* (ed. R. E. Ricklefs and D. Schluter), pp. 230–40. The University of Chicago Press.

Schulze, E. D. (1982). Plant life forms and their carbon, water and nutrient relations. In *Physiological plant ecology II. Water relations and carbon assimilation, Encyclopedia of plant physiology, New Series, Vol. 12B* (ed. O. L. Lange, P. S. Nobel, C. B. Osmond, and H. Ziegler), pp. 615–77. Springer-Verlag, Heidelberg.

Scott, J. M., Csuti, B., and Davis, F. (1991). Gap analysis: an application of Geographic Information Systems for wildlife species. In *Challenges in the conservation of biological resources: a practitioner's guide* (ed. D. J. Decker, M. E. Krasny, R. Goff, C. R. Smith, and D. W. Gross), pp. 167–79. Westview Press, Boulder.

Sealy, J. R. (1949). *Arbutus unedo. Journal of Ecology*, **37**, 365–88.

Seligman, N. G. and Perevolotsky, A. (1994). Has intensive grazing by domestic livestock degraded Mediterranean Basin rangelands? In *Plant-animal interactions in Mediterra-nean-type ecosystems* (ed. M. Arianoutsou and R. H. Groves), pp. 93–104. Kluwer Academic, Dordrecht.

Sezik, E. (1989). Turkish orchids and saleps. In *Orchidées botaniques du Monde entier*, pp. 181–9. Société Française d'Orchidophilie, Paris.

Shay, C. T., Shay, J. M., and Zwiazek, J. (1992). Paleobotanical investigations at Kommos, Crete. In *Plant-animal interactions in Mediterranean-type ecosystems* (ed. C. A. Thanos), pp. 382–9. University of Athens.

Shmida, A. (1981). Mediterranean vegetation in California and Israel: similarities and differences. *Israel Journal of Botany*, **30**, 105–23.

Shmida, A. (1984). Endemism in the flora of Israel. *Botanische Jahrbücher für Systematik*, **104**, 537–67.

Shmida, A. and Aronson, J. (1986). Sudanian elements in the flora of Israel. *Annals of the Missouri Botanical Garden*, **73**, 1–28.

Shmida, A. and Dafni, A. (1990). Blooming strategies, flower size and advertising in the 'lily-group' geophytes in Israel. *Herbertia*, **45**, 111–23.

Shmida, A. and Ellner, S. (1983). Seed dispersal on pastoral grazers in open Mediterranean chaparral, Israel. *Israel Journal of Botany*, **32**, 147–59.

Shmida, A. and Werger, M. J. A. (1992). Growth form diversity on the Canary Islands. *Vegetatio*, **102**, 183–99.

Shmida, A. and Whittaker, R. (1984). Convergence and non-convergence of mediterranean type communities in the old and the new world. *Tasks for Vegetation Science*, **13**, 5–11.

Simmons, A. H. (1988). Extinct pygmy hippopotamus and early man in Cyprus. *Nature*, **333**, 554–7.

Simpson, B. B. and Neff, J. L. (1981). Floral rewards: alternatives to pollen and nectar. *Annals of the Missouri Botanical Garden*, **68**, 301–22.

Sket, B. (1997). Distribution of *Proteus* (Amphibia: Urodela: Proteidae) and its possible explanation. *Journal of Biogeography*, **24**, 263–80.

Skoula, M., Griffee, P., and Heywood, V. H. (1997). Identification, conservation and use of wild plants of the Mediterranean region—the 'Medusa' network. *Cahiers Options Méditerranéennes*, **23**, 1–4.

Smith, A. G. and Woodcock, N. H. (1982). Tectonic syntheses of the Alpine-Mediterranean region: a review. In *Alpine-Mediterranean Geodynamics* (ed. H. Berckhemer and K. J. Hsü), pp. 15–38. American Geophysical Union, Washington, DC.

Snogerup, S. (1971). Evolutionary and plant geographical aspects of chasmophytic communities. In *Plant life of southwest Asia* (ed. P. H. Davis, P. C. Harper, and I. C. Hedge), pp. 157–70. The Botanical Society of Edinburgh.

Socias y Company, R. (1990). Breeding self-compatible almonds. *Plant Breeding Review*, **8**, 313–38.

Sondaar, P. Y. (1977). Insularity and its effects on mammal evolution. In *Major patterns of vertebrate evolution* (ed. P. C. Goody and B. M. Heckt), pp. 671–707. Plenum Publishing Corporation, New York.

Sondaar, P., Elburg, E., Klein Hofmeijer, G., Martini, F., Sanges, M., Spaan, A., and de Visser, H. (1995). The human colonization of Sardinia: a Late-Pleistocene human fossil from Corbeddu cave. *Comptes-Rendus de l'Académie des Sciences, Paris*, **320** (Series II a), 145–50.

Soulier, A. (1993). *Le Languedoc pour héritage*. Presses du Languedoc, Montpellier, France.

Specht, R. L. and Rundel, P. W. (1990). Sclerophylly and foliar nutrient status of mediterranean-climate plant communities in southern Australia. *Australian Journal of Botany*, **38**, 459–74.

Stamps, J. A. and Buechner, M. (1985). The territorial defense hypothesis and the ecology of insular vertebrates. *Quarterly Review of Biology*, **60**, 155–81.

Stebbins, G. L. (1942). The genetic approach to problems of rare and endemic species. *Madrono*, **6**, 24–58.

Stebbins, G. L. (1971). *Chromosomal evolution in higher plants*. E. Arnold, London.

Stebbins, G. L. and Zohary, D. (1959). Cytogenetic and evolutionary studies in the genus *Dactylis*. I. Morphology, distribution and interrelationships of the diploid subspecies. *University of California at Berkeley Publication in Botany*, **31**, 1.

Suc, J. P. (1980). Contribution à la connaissance du Pliocène et du Pléistocène inférieur des régions méditerranéennes d'Europe occidentale par l'analyse palynologique des dépôts du Languedoc-Roussillon (sud de la France) et de la Catalogne (nord-est de l'Espagne). Unpublished D. Phil. thesis, University of Montpellier II.

Suc, J. P. (1984). Origin and evolution of the Mediterranean vegetation and climate in Europe. *Nature*, 307, 429–32.

Taberlet, P. and Bouvet, J. (1994). Mitochondrial DNA polymorphism, plylogeography and conservation genetics of the brown bear *Ursus arctos* in Europe. *Proceedings of the Royal Society of London B*, 225, 195–200.

Taberlet, P., Fumagali, L., Wust-Saucy, A. G., and Cosson, J. F. (1998). Comparative phylogeography and postglacial colonization routes in Europe. *Molecular Ecology*, 6, 289–301.

Tamisier, A. (1971). Le régime alimentaire des sarcelles d'hiver *Anas crecca* L. en Camargue. *Alauda*, 19, 1–31.

Tamisier, A. and Grillas, P. (1994). A review of habitat changes in the Camargue: an assessment of the effects of the loss of biological diversity on the wintering waterfowl community. *Biological Conservation*, 70, 39–47.

Tarayre, M., Thompson, J. D., Escarré, J., and Linhart, Y. B. (1995). Intra-specific variation in the inhibitory effects of *Thymus vulgaris* (Labiatae) monoterpenes on seed germination. *Oecologia*, 101, 110–18.

Tarayre, M., Saumitou-Laprade, P., Cuguen, J., Couvet, D., and Thompson, J. D. (1997). A comparison of the spatial genetic structure of cytoplasmic (cpDNA) and nuclear (allozymes) markers within and among populations of the gynodioecious *Thymus vulgaris* L. (Labiatae) in southern France. *American Journal of Botany*, 84, 1675–84.

Tchernov, E. (1984). Faunal turnover and extinction rate in the Levant. In *Quaternary Extinctions* (ed. P. S. Martin and R. G. Klein), pp. 528–52. University of Arizona Press, Tucson.

Tchernov, E. (1992). Eurasian-African biotic exchanges through the Levantine corridor during the Neogene and Quaternary. *Courier Forschungsinstitut Senckenberg*, 153, 103–23.

Terrasse, M. (1996). Réintroduction de rapaces dans l'aire Méditerranenne de répartition. In *Biologia y conservacion de las rapaces Mediterraneas, 1994* (ed. J. Muntaner and J. Mayol), pp. 251–9. Monografias, N° 4. SEO, Madrid.

Thanos, C. A. (1992). Theophrastus on plant-animal interactions. In *Plant-animal interactions in Mediterranean-type ecosystems* (ed. C. A. Thanos), pp. 1–5. University of Athens.

Thébaud, C., Finzi, A. C., Affre, L., Debussche, M., and Escarré, J. (1996). Assessing why two introduced *Conyza* differ in their ability to invade Mediterranean old fields. *Ecology*, 77, 791–804.

Thellung, A. (1908–10). La flore adventice de Montpellier. *Bulletin de la Société des Sciences Naturelles de Cherbourg*, 38, 57–72.

Thirgood, J. V. (1981). *Man and the Mediterranean forest*. Academic Press, New York.

Thomas, J. A., Elmes, G. W., Wardlaw, J. C., and Woyciechowski, M. (1989). Host specificity among *Maculinea* butterflies in *Myrmica* ant nests. *Oecologia*, 79, 452–7.

Thompson, J. D., Manicacci, D., and Tarayre, M. (1998). Thirty five years of thyme: a tale of two polymorphisms. *BioScience*, 48, 805–15.

Tomaselli, R. (1981). Main physiognomic types and geographic distribution of shrub systems related to mediterranean climates. In *Mediterranean-type shrublands* (ed. F. di Castri, D. W. Goodall, and R. L. Specht), pp. 95–106. Collection Ecosystems of the World, Vol. 11. Elsevier, Amsterdam.

Tortonese, E. (1985). Distribution and ecology of endemic elements in the Mediterranean faunas (fishes and echinoderms). In *Mediterranean marine ecosystems* (ed. M.

Moraitou-Apostolopoulo and V. Kiortsis), pp. 57–83. Plenum Press, New York.

Toumi, L. and Lumaret, R. (1998). Allozyme variation in cork oak (*Quercus suber* L.): the role of phylogeography, genetic introgression by other Mediterranean oak species and human activities. *Theoretical Applied Genetics*, 97, 647–56.

Toutain, F., Diagne, A., and Le Tacon, F. (1988). Possiblités de modification du type d'humus et d'amélioration de la fertilité des sols à moyen terme en hêtraie par apport d'éléments minéraux. *Revue Forestière Française*, 40, 99–107.

Trabaud, L. (1981). Man and fire: impacts on mediterranean vegetion. In *Mediterranean-type shrublands* (ed. F. di Castri, D. W. Goodall, and R. L. Specht), pp. 523–37. Collection Ecosystems of the World, Vol. 11. Elsevier, Amsterdam.

Trabaud, L. and Oustric, J. (1989). Heat requirements for seed germination of three *Cistus* species in the garrigue of southern France. *Flora*, 183, 321–5.

Trabaud, L. V., Christensen, N. L., and Gill, A. M. (1993). Historical biogeography of fire in temperate and mediterranean ecosystems. In *Fire in the environment: its ecological, and atmospheric importance* (ed. P. J. Crutzen and J. G. Goldammer), pp. 277–95. John Wiley, New York.

Traveset, A. (1992). Production of galls in *Phillyrea angustifolia* induced by cecidomyiid flies. In *Plant-animal interactions in Mediterranean-type ecosystems* (ed. C. A. Thanos), pp. 198–204. University of Athens.

Triat-Laval, H. (1979). Histoire de la forêt provençale depuis 15 000 ans d'après l'analyse pollinique. *Forêt Méditerranéenne*, 1, 19–24.

Tsigenopoulos, C. (1996). Phylogéographie mitochondriale du genre *Barbus* au nord de la Méditerranée. Une approche de la spéciation des poissons d'eau douce d'Europe occidentale. Unpublished leaflet. University of Montpellier II.

Tucker, G. M. and Heath, M. F. (1994). *Birds in Europe: their conservation status*. BirdLife Conservation Series No. 3, BirdLife International, Cambridge.

UNESCO (1963). Bioclimatical map of the Mediterranean zone. *Arid Zone Research*, 21, 1–60.

Vallejo, R. (ed.) (1996). *La restauración de la cubierta vegetal en la comunidad Valenciana*. CEAM, Valencia.

Vallejo, R. and Alloza, J. A. (1998). The restoration of burned lands: the case of eastern Spain. In *Large forest fires* (ed. J. M. Moreno), pp. 91–108. Backhuys, Leiden.

Van Valen, L. (1973). A new evolutionary law. *Evolutionary Theory*, 1, 1–30.

Vaudour, J. (1979). La région de Madrid. Altérations, sols et paléosols. Unpublished D. Phil. thesis. University of Aix-Marseille.

Vavilov, N. I. (1935). Botanical-geographical basis of breeding. In *Origin and geography of cultivated plants* (ed. N. I. Vavilov), pp. 288–333. Nauka, Leningrad.

Vavilov, N. I. (1951). The origin, variety, immunity and breeding of cultivated plants. *Chronica Botanica*, 13, 1–366.

Vernet, J. L. (1973). Etude sur l'histoire de la végétation du sud-est de la France au Quaternaire d'après les charbons de bois principalement. *Paléobiologie continentale*, 4, 1–73.

Vernet, P., Gouyon, P. H., and Valdeyron, G. (1986). Genetic control of the oil content in *Thymus vulgaris* L.: a case of polymorphism in a biosynthetic chain. *Genetica*, 69, 227–31.

Vigne, J. D. (1983). Le remplacement des faunes de petits mammifères en Corse, lors de l'arrivée de l'homme. *Comptes Rendus de la Société de Biogéographie*, 59, 41–51.

Vigne, J. D. (1987). L'extinction holocène du fond (sic) de peuplement mammalien indigène des îles de Méditerranée occidentale. *Mémoire de la Société Géologique de France (nouvelle série)*, **150**, 167–177.

Vigne, J. D. (1988). Apports de la biogéographie insulaire à la connaissance de la place des mammifères sauvages dans les sociétés néolithiques méditerranéennes. *Anthropozoologica*, **8**, 31–52.

Vigne, J. D. (1990). Biogeographical history of the mammals on Corsica (and Sardinia) since the final Pleistocene. In *Biological aspects of insularity* (ed. A. Azzaroli), pp. 370–92. Academia Nazionale dei Lincei, Roma.

Vigne, J. D. (1992). Zooarchaeological and biogeographical history of the mammals of Corsica and Sardinia since the last ice age. *Mammal Review* **22**, 87–96.

Vuilleumier, F. (1977). Suggestions pour des recherches sur la spéciation des oiseaux en Iran. *Terre et Vie*, **31**, 459–88.

Wall, R. and Strong, L. (1987). Environmental consequences of treating cattle with the anti-parasitic dung ivermectin. *Nature*, **327**, 418–21.

Wallace, A. R. (1880). *Island life*. Macmillan, London.

Walter, H. (1968). *Die Vegetation der Erde, Bd. 2: Die gemässigten und arktischen Zonen*. G. Fischer, Jena, Stuttgart.

Walter, H. and Lieth, H. (1960). *Klimadiagramm Weltatlas*. Fisher Verlag, Jena.

Webb, T. and Bartlein, P. J. (1992). Global changes during the last 3 million years: climatic controls and biotic responses. *Annual Review of Ecology and Systematics*, **23**, 141–73.

Westman, W. E. (1988). Species richness. In *Mediterranean type ecosystems, a data source book* (ed. R. L. Specht), pp. 81–92. Kluwer Academic, Dordrecht.

Williams, P. H. and Gaston, K. J. (1994). Measuring more of biodiversity: can higher taxon richness predict wholesale species richness? *Biological Conservation*, **67**, 211–17.

Wilson, E. O. (1992). *The diversity of life.* Allen Lane, The Penguin Press, London.

Woldhek, S. (1979). *Bird killing in the Mediterranean*. European Committee for the Prevention of Mass Destruction of Migratory Birds. Zeist, The Netherlands.

Woodward, F. I. (1993). How many species are required for a functional ecosystem? In *Biodiversity and ecosystem function* (ed. E. D. Schulze and H. A. Mooney), pp. 271–91. Ecological Series 99. Springer-Verlag, Berlin.

Yom-Tov, Y. (1988). The zoogeography of the birds and mammals of Israel. In *The zoogeography of Israel* (ed. Y. Yom-Tov and E. Tchernov), pp. 389–410. Dr. W. Junk, Dordrecht.

Yom-Tov, Y. and Mendelssohn, H. (1988). Changes in the disribution and abundance of vertebrates in Israel during the 20th century. In *The zoogeography of Israel* (ed. Y. Yom-Tov and E. Tchernov), pp. 515–47. Dr. W. Junk, Dordrecht.

Zohary, D. (1969). The progenitors of wheat and barley in relation to domestication and agricultural dispersal in the Old World. In *The domestication of plants and animals* (ed. P. J. Ucko and G. W. Dimpley), pp. 47–66. Aldine, Chicago.

Zohary, D. (1983). Wild genetic resources of crops in Israel. *Israel Journal of Botany*, **32**, 97–100.

Zohary, D. and Hopf, M. (1993). *Domestication of plants in the Old World*. Clarendon, Oxford.

Zohary, D. and Plitmann, U. (1979). Chromosome polymorphism, hybridization and colonization in the *Vicia sativa* group (Fabaceae). *Plant Systematics and Evolution*, **131**, 143–56.

Zohary, D. and Spiegel-Roy, P. (1975). Beginnings of fruit-growing in the Old World.

Science, 187, 319–27.

Zohary, M. (1962). *Plant life in Palestine*. Ronald Press, New York.

Zohary, M. (1973). *Geobotanical foundations of the Middle East*. Gustav Fisher Verlag, Stuttgart.

Zohary, M. and Feinbrun, N. (1966–86). *Flora Palaestina. Parts 1–4*. The Israel Academy of Sciences and Humanities. Jerusalem.

Index

CL

508.
318
22
BLO